Self-Aware Robots

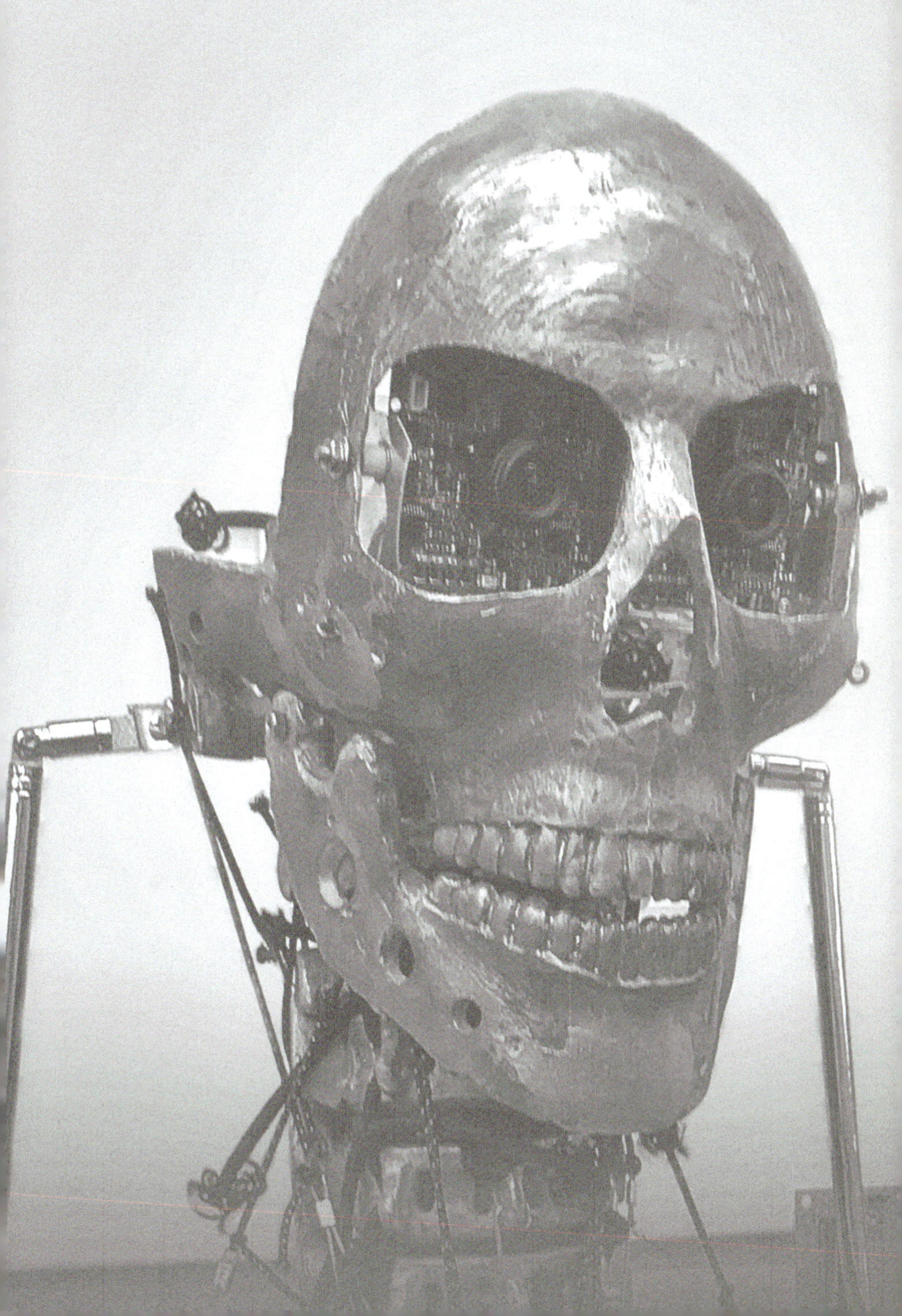

Self-Aware Robots

On the Path to Machine Consciousness

Junichi Takeno

Published by

Jenny Stanford Publishing Pte. Ltd.
Level 34, Centennial Tower
3 Temasek Avenue
Singapore 039190

Email: editorial@jennystanford.com
Web: www.jennystanford.com

British Library Cataloguing-in-Publication Data
A catalogue record for this book is available from the British Library.

Self-Aware Robots: On the Path to Machine Consciousness

ISBN 978-981-4877-90-9 (Hardcover)
ISBN 978-1-003-26181-0 (eBook)

Dedicated to the memory of my late father.

Contents

Preface to the Second Edition

I am grateful that the publisher expressed a desire to publish the second edition of this book, and I sincerely appreciate the many responses I have received from a wide variety of readers in the eight years that have already passed since the first edition was published. During this time, my research team and I continued to make progress in various aspects of the theme. The first edition introduced the principles and major experimental results of our conscious system to readers and researchers throughout the world. On reflection, the content of the book may be too basic and partly redundant. In the world of science, deep learning (DL) has been advancing and the machine recognition capabilities of neural networks have dramatically improved. DL is good news for conscious systems. In the second edition of our book, we have removed the redundant content and added a discussion on the correlation between DL and conscious systems. This new edition also covers the consistent progress made in our research. The new areas covered include a robot that cognizes the unknown; robots with episodic memory; a self-aware robot; a Pavlov robot; color vision capabilities of a conscious robot; principle of pleasant and unpleasant feelings in a robot; a self-evolving conscious system; acquisition of a sense of self; and development of a model for diagnosing higher cognitive impairment of humans.

We invite you to enjoy the development of our conscious robots!

Junichi Takeno

Preface to the First Edition

This book introduces the author's current focus on the study of the consciousness and the mind. It also proposes a method for constructing the functions of consciousness and the mind on a machine. The author has been exploring the concept of a conscious robot for nearly 20 years. He has presented his ideas at two international conferences held in Florida, in the summer of 2003: The 7th World Multiconference on Systemics, Cybernetics and Informatics (SCI 2003) and the International Conference on Computer, Communication and Control Technologies (CCCT 2003). The paper presented at these conferences received the Best Paper Award in the system sessions at SCI 2003 and in the methodology sessions at CCCT 2003. The same paper again received the Best Paper Prize at the Innovations in Applied Artificial Intelligence section of the 18th International Conference on Industrial and Engineering Applications of Artificial Intelligence and Expert Systems (IEA/AIE 2005) held in Italy in the summer of 2006. His article titled "Mirror Image Cognition of a Robot" was honored by the academic society at the 3rd International Conference on Sensing Technology (ICST 2008), held in Taiwan. These peer acknowledgments of the merits of this study provided the impetus for writing this book.

This book is expected to be used as a textbook for university students pursuing information technology and mechanical engineering courses and is also suitable for students who are interested in the new areas of research. The content and information herein are also recommended for brain scientists and robot researchers, as well as those in arts who are enthusiastic about philosophical discussions and those who are curious about psychology and sociology. With such readers in mind, the author uses plain language to enable everyone to gain insights into the subject without needing expert knowledge. The initial sections of the book are dedicated to the discussions of basic knowledge about robots and the human brain. These sections intend to familiarize readers with the theme of this book. They are followed by an explanation of the conscious robot developed by the author. Last, experiments are introduced that demonstrate that this conscious robot — for the first time in the world — successfully recognizes its own reflection in a mirror. The author sincerely hopes that all readers interested in elucidating the mystery of human consciousness and the mind will benefit from reading this book.

Chapter 1

Introduction

One day, I was sitting in the phlebotomy room of a relatively large private university hospital, accompanying my elderly father. The room was rather quiet and several phelobotomists were not very busy. A boy, about second-grader in elementary school, was called by his name and number. He came to the phlebotomist, somewhat hesitatingly. The phlebotomist started the routine procedure, placing a band around the boy's arm and disinfecting the skin area. Just when she was ready to insert the needle, the boy said weakly, "I'm not mentally ready for this yet."

I and the other people around me were surprised by what the boy said. We were surprised because this boy, who was so young, used the words "not mentally ready." He clearly asserted the existence of his mind.

How was his mind formed?

In my everyday life, many ideas come to my mind and vanish before I know it. Not in a few cases, I concentrate on something and my consciousness is thoroughly absorbed in it. I sometimes glance at things absentmindedly. I might be impressed, may cry or may laugh when watching a movie. These activities are functions of humans and are known as human consciousness or mind.

Why do humans have a consciousness or mind?

This question has been occupying people since ancient times.

Self-Aware Robots: On the Path to Machine Consciousness
Junichi Takeno

ISBN 978-981-4877-90-9 (Hardcover), 978-1-003-26181-0 (eBook)
www.jennystanford.com

The first few chapters of this book tell you the story of robots, including robots that are much talked about in the media today, and intelligent robots.

Then comes the story of the brain. A wide variety of research studies about the human brain, in particular that which is said to create human consciousness and mind, are introduced.

Another topic is consciousness and the mind as discussed in the fields of philosophy and psychology.

I will further introduce the study by Prof. V. Braitenberg, ex-director of the Max Planck Institute, Germany. With his robot, information gained through the sensory system, such as tactile sensations, is processed in its internal artificial neural networks, and the output is used to directly drive a motor system such as electric motors. His concept gave rise to the idea of building a scheme of a living creature, such as an insect, in a robot.

Another study introduced in this book is that by R. Brooks at the Massachusetts Institute of Technology in the United States. His innovative robot design ideas had a significant impact on researchers after him.

The discussion continues to artificial neural networks and evolutional technology. The neural networks used by Braitenberg are relatively simple, but the generally known and famous neural networks are discussed here. Readers learn that these neural networks assist in achieving the high-level motion of robots, including the recognition of objects.

A technique to build a scheme of the evolution of life in machines is explained. One of the advantages of this technique is that one can design artificial neural networks to transmit information input from the sensory system to the motor system relatively freely.

The subsequent chapters in this book are dedicated to introducing researchers attempting to create consciousness and mind in robots.

First, W. Grey Walter's robot is introduced. Walter developed the world's first biological-type robot.

Tadashi Kitamura developed a conscious robot. His robot features a hierarchical conscious structure and realizes a relatively high level of behavior such as anticipating the behavior of others.

Jun Tani also developed a conscious robot. Tani believes that consciousness occurs in the course of confusing cognition and prediction. He got this idea by observing humans. According to Tani, for example, when the usual way to school is suddenly blocked, one becomes conscious of oneself and asks, "What on earth has happened?" and becomes conscious of the surrounding external world at the same time.

Mitsuo Kawato also studies robots. He believes that the integration of the sensory and motor systems is essential to achieve complete human-like behavior in robots. This integration would enable robots to learn by imitation — one of the high-level behavioral functions of humans — and eventually acquire consciousness.

Kismet, a robot developed by Cynthia Breazeal at the Massachusetts Institute of Technology, has achieved fundamental communications with humans using the theory of the mind as discussed in the field of psychology.

Interest in artificial consciousness increased toward the end of the 20th century. Igor Aleksander at Imperial College London presented a theory that neural networks by themselves have the function of consciousness (Aleksander, 1996).

At the beginning of the current century, Owen Holland and his colleagues at the University of Sussex organized a research team on machine consciousness to lead the study of artificial consciousness in cognitive robotics (Holland, 2017).

After introducing these research studies, the book presents the conscious robot developed by the author. The design of artificial consciousness is described, followed by the design of the mind. The author's study is similar to those by Tani and Kawato, but the difference lies in that the methodology adopted by the author proposes a new paradigm regarding consciousness and the mind — one that encompasses and explains all of the questions that have been a mystery in the study of consciousness and the mind.

The author's methodology addresses the problems of not only learning by imitation but also imagination, qualia, stream of consciousness, and stream of feelings that appear as a representation of emotion and feelings. Qualia is a term that refers to how one "feels about something." For instance, just think of a cup of coffee, and you will "feel that sweet fragrance of coffee." According to some researchers, Qualia is one of the most difficult problems in the study of consciousness. In addition to qualia, the author's methodology is able to describe the human function of self-awareness (i.e., knowing that the image in the mirror is oneself), which is considered the supreme function of human consciousness. Understanding this function will be the first step toward understanding self-consciousness. The author holds that human consciousness has evolved and developed from the necessity of distinguishing oneself from others, and hence, his methodology describes the function for distinguishing between oneself and others.

The next chapter introduces the mirror image experiments conducted using a conscious robot developed by the author. The findings obtained in the experiments are discussed. Three experiments were conducted: (1) The conscious robot imitates the behavior of the image of itself reflected in a mirror, (2) the conscious robot imitates the behavior of another similar robot, which is controlled via cables, and (3) two nearly identical conscious robots imitate each other. These experiments provide physical evidence that for the robot, the image of itself in the mirror is an existence that is closer to itself than part of itself is.

Finally, the author draws many suggestive observations from the experiments, including, for example, the possibility of developing artificial limbs that feel like the user's own limbs.

In the last chapter, further development of the conscious system is described based on the discussions presented in the preceding chapters.

First, a robot capable of cognizing the unknown is introduced. Conventional robots basically work within the range of their existing knowledge. If they happen upon something alien in their environment, they are sure to stop working. The robot featuring the conscious system proposed by the author is different from conventional robots in that it autonomously determines when a new environment is "unknown." The conscious system then takes in the new information and adapts to the environment. Similar adaptation is possible for robots of conventional machine learning if a certain goal is set. Adaptation to a totally new environment is possible for our robot with this conscious system without our needing to take the trouble to set a particular goal for the robot; the robot will adapt to the new environment using a more general formulation.

Self-aware robots are described next. Self-awareness is a sense that the self is controlling an object. The "object" here also includes such actions as driving a car and moving the parts of one's body, such as one's own hands and feet. The study under this theme focuses on how the conscious system grasps the human subjective phenomenon of self-awareness.

Next, a robot with an episodic memory is introduced. When behaving or taking action, the robot memorizes what it does and the associated environmental changes in chronological order. This develops the episodic memory in the robot. In an experiment, we had the robot represent the emotion caused by a collision with an object as unpleasant information, in other words, pain, and have it memorize the event that occurred immediately before the collision. It is beneficial for the robot to use the information stored in its memory when deciding on the next action to take or not to take. The robot refers to the episodic memory and anticipates the collision that occurs immediately after a particular action. The robot is able to anticipate the collision that causes pain and stop moving before such an event occurs.

We then discuss a Pavlov robot. Ivan P. Pavlov (1849–1936) was a Russian physiologist and physician who won the Nobel Prize and was widely known for his famous experiments using dogs. When being fed, a dog would salivate. At the feeding time, Pavlov would ring a bell for the dog. Gradually the dog would salivate only by being stimulated by the sound of the bell without actually receiving any food. This is a kind of learning and is called classical conditioning. We succeeded in reproducing Pavlov's dog experiments using a robot featuring our conscious system.

We also studied color vision capabilities of the conscious robot.

The robot had already learned the three primary colors of red, green, and blue. These colors were respectively tied to three actions: stop, move forward, and move backward. The robot moves forward when it sees green through its vision camera. In the experiment, the robot was shown many new intermediate colors one after another. Upon seeing a new color, the robot memorizes the combination of the shown color and the associated action and stores the combination as a piece of known information. The action to be taken was instructed by the experimenter to the robot in each case. The problem was, how many new colors could the robot learn in association with the instructed action? This was the theme of the study. The result of the experiment shows that the robot learned a total of 18 colors: three primary and 15 intermediate colors. The robot identified these 18 different colors. Humans are said to be able to discern 20 to 30 different colors. The color vision capability of the robot used in our experiments may not be equal to that of humans but could possibly be improved and made comparable. The robot did not use the re-learning technique of neural networks but employed a new learning technique of continually adding new information to the conscious system automatically.

The next theme is the pleasant and unpleasant feelings of a robot and the self-evolving conscious system. The assertion that human behavior is at least partly based on emotion and feelings is being increasingly accepted (Ekman, 1972). Human emotion and feelings still remain a mystery. How do emotion and feelings affect the behavior of a robot? This is the theme of this chapter. Generally, a delay in information processing in the neural networks generates a negative condition for a living being. Based on this fact, we assume that a delay in information processing is also the cause of unpleasant feelings in a robot. We also assume that pleasant feelings are the opposite of this delay and that the robot feels pleasant when the delay in information processing in the neural networks decreases. Utilizing pleasant and unpleasant feelings generated in a robot, we devised a process in which the conscious system evolves on its own. We incorporated this new process into a conscious system that behaves normally in response to input and output information. With the help of the new process, the conscious system restructures itself and evolves autonomously, creating more pleasant conditions than the present for the robot. In the experiment, we saw that the robot learned the changing environment on its own by reading the information on the environment, or in other words, the conscious system behaved by learning the changing environment. Eventually, the conscious system performed in such a way that unpleasant events would not occur and pleasant conditions would be created for the robot.

Let's turn to the theme of Conflict of Concepts and a Model of Rubin's Vase.

As we have learned, MoNADs are able to not only read but also represent external environmental information. Representing information means affixing a certain unique name to the information. Pointing at a cat walking in front of you, you say, "It's a cat." "Cat" is the name of information related to cats. Why that particular name is used may require much discussion, but for the time being, it would suffice to say that we are taught to use that name by other people. Our conscious system utilizing MoNADs is a system that cognizes environmental information. It is not just a "cognizing" system but a "conscious" system because it consistently combines cognition with a certain action. Interestingly, two or more interpretations can exist simultaneously for a single piece of environmental information, and in this case a conflict occurs in the cognition process. This reminds us of the famous Rubin's vase image. Humans cannot interpret two conflicting meanings simultaneously while viewing the Rubin's vase image. In one instance, they may cognize it as a vase but in the other they see two human faces. In a word, cognition does remain on one thing but easily shifts to another. We will model this phenomenon using the conscious system featuring MoNADs in this chapter.

Acquisition of a Sense of Self by a robot is then discussed.

Researchers have been discussing earnestly in recent years whether a robot has a sense of self or not. This theme has long been investigated, but so far we have not established a definite theory. The author has been studying this theme from the viewpoint of what kind of development process would be required for a robot to acquire a sense of self. The robot with the conscious system developed by the author takes action after the reason, emotion & feelings, and association subsystems communicate with one another, based on information on the external environment. The reason subsystem receives environmental information from outside the robot, the emotion & feelings subsystem responds to the robot's internal changes, and the association subsystem is the settler that mediates between those two subsystems. If a robot, while behaving normally, is suddenly stuck in a condition where the reason and emotion & feelings subsystems conflict with each other, what kind of association system should be generated to settle the conflict?

This is the theme of this chapter. Assume, for example, a robot is heading for a certain target destination, and happens to step on a bunch of cables. The robot would select an action that offers a more comfortable state for the robot rather than deciding to continue going on toward the target. That is, the robot puts a high priority on itself. This priority process may be considered the origin of the sense of self. In the experiments described in this chapter, the conscious system of the robot succeeded in automatically developing an association subsystem that solves the conflict between the reason and the emotion & feelings subsystems. This provides a firm starting point for us to find the origin of acquisition of the sense of self.

The last topic is the development of a model for diagnosing higher cognitive impairment of humans. Human brains are said to have a high-level cognitive function. Recent reports say that the human brain suffers from cognitive impairment when subjected to a harsh environment such as a battlefield for a long time. Examples include shell shock which refers to the mental trauma incurred by soldiers who experienced numerous explosions of shells around them while they were sheltering in a trench. Post-traumatic stress disorder (PTSD) and dissociative identity disorder (DID) have been the topic of much discussion recently. This Chapter introduces a research study for developing a model for diagnosing these higher cognitive impairments of humans using the conscious system. Through modeling, we would be able to better understand these diseases and discover a base for establishing treatment guidelines. Let us now embark on an adventurous and intellectual journey!

Chapter 2

Story of Robots

Readers may think of the AIBO robot dog from Sony or the ASIMO walking humanoid robot from Honda when we speak of robots. Recent humanoid robots by Boston Dynamics are clever enough to speedily walk and running through in a room and in a wild field, to recover quickly from a fall, to operate boxes by two arms, or make a backflip in good balance just like a human athretics. Unmanned vehicles are a kind of robot and have reportedly reached the stage of practical application. Human-shaped robots are generally called humanoid robots. Both *AIBO* and *ASIMO* were developed by private businesses and are wonderfully animal-like and human-like, respectively, even from the viewpoints of robot researchers studying at universities. The creation and development of robots are introduced below.

As readers may already know, the word "robot" comes from Czech *robota*, literally meaning work or labor. It is commonly accepted that the English word "robot" was first used in a satirical drama *R.U.R.* (*Rossum's Universal Robots*, 1921), written by Czech playwright Karel Capek. In the drama, the robots created to work for humans eventually expel the humans. It is interesting to know that from the beginning robots were considered an existence to be watched with caution.

Self-Aware Robots: On the Path to Machine Consciousness
Junichi Takeno

ISBN 978-981-4877-90-9 (Hardcover), 978-1-003-26181-0 (eBook)
www.jennystanford.com

Later, robots were gradually exposed to the public, including an appearance in German movie *Metropolis* (1926), and in world expositions and various other exhibitions.

In the 1970 International Exposition in Tsukuba, Japan, many visitors were charmed by a robot playing the organ and a robot walking on two feet like a human. Before Honda surprised the world by revealing its *ASIMO* robot, there were many studies on robots that walked upright like humans.

People were surprised to see the *R2–D2* and *C-3PO* robots in American movie *Star Wars*. The cylindrically shaped *R2–D2* is a functional robot supporting fighter combat and navigation. The humanoid robot *C-3PO* walks on two feet, speaks human language, and even tells jokes. *C-3PO* is a protocol droid and understands the space language according to the setting.

Later, American movie *Terminator* featured a fearful and very tough robot, who performs impressively in a battle between the robots trying to expel humans and the defending humans.

I personally like movie *Andrew NDR114*, which is the name of a robot who fell in love with a human female because, for one thing, the movie raises a serious question about the "heart" of a robot.

Another memorable movie is *AI* (*Artificial Intelligence*). A robotic boy longs to become "real" and sets out to find a fairy. In the movie *iRobot*, a humanoid robot named *Sam*, who has a sense of self, saves the lives of humans and other robots. In the movie *Wall-E*, a robot happens to acquire a heart one day and falls in love.

In Japan, *Tetsujin 28-go* ("Iron-man #28"), *Astro Boy* (Atom), and *Doraemon* (a manga series) are famous.

Astro Boy, in particular, is a typical humanoid robot. It is depicted as a hero of justice, fighting against evil together with humans. The physical ability of *Astro Boy* is tremendous. With visual power of 2.5 and hearing ability 1000 times that of a human, the robot can fly at Mach 5. To gain this speed, the robot is propelled by a jet engine in the atmosphere and by a rocket engine in space. It is said that when *Astro Boy* flies at Mach 5 in the atmosphere, the clenched fist of its arm extended in the direction of flight and its head would melt if ordinary metals were used because temperatures would reach 1500°C. At present, ceramics are known to withstand temperatures of 1500°C. On the other hand, *Astro Boy's* skin is said to be made of soft materials such as plastic. So far, there is no material suitable for *Astro Boy* that is both soft like human skin and highly thermoresistant. *Astro Boy* is basically a machine and does not have a heart. In the story, *Astro Boy* agonizes over its lack of a heart. I think this agony is a proof of the existence of its heart.

I have so far introduced various robots appearing in images, movies, and anime. These robot stories depict the interaction between humans and humanoid robots and at the same time the "light and shadow" of their interaction. Light represents the bright side of the view of a world in which humanoid robots cooperate with humans and help them realizing their dreams. Shadow represents the dark side and warns that humanoid robots could surpass and supersede humans.

Let us now go back further to the history of robots. In Europe, Heron (ca. 200 BC to 150 BC), a Greek inventor devised many automatically operating machines using weights in around the first century BC. And he invented a steam engine. Heron invented a device to automatically open the heavy doors of a temple when fire was used in a ceremony. This is an automaton rather than a robot, but autonomous robots may have their roots in automatons.

The Antikythera mechanism is a particularly interesting complex device. It looks like a machine closely packed with a number of rusted bronze gears. The device was found off the coast of a small Greek island called Antikythera. It is similar to a planetarium for predicting the revolution of the heavenly bodies and is said to be the world's first analog computer made of gears (Wright, 2012). It was possibly manufactured between 150 and 100 BC.

Clock technology saw remarkable development in Europe in the 18th century. Various mechanical dolls were built using clock technology. One of the exquisite achievements at that time is *The Writer*, an android made by Jacquet-Droz and his son. *The Writer*, a marvelous work, holds a quill with one hand, dips it in an inkwell and writes sentences on a sheet of white paper. These dolls are considered the roots of humanoid robots.

In 1738, the French Academy of Sciences in Paris introduced the automaton *Duck*. According to some reports, *Duck* did eat and defecate, but the truth is unknown. In the 17th century in

Figure 2.1 Yumihiki Doji (arrow shooting doll) at the Toshiba Science Museum.

Japan, the Takeda Karakuri troupe led by Oumi Takeda opened the *Karakuri* (mechanical doll) booth in Dotonbori, Osaka, and staged performances of the Takeda *Karakuri* dolls. According to historical records, the show attracted many people. *Karakuri* dolls are basically a machine or a doll remotely controlled by humans. They are the origin of present-day remote-controlled robots.

Well-known animated dolls include *Chahakobi* (tea serving doll) and *Yumihiki Doji* (arrow shooting doll) produced in the Edo period (1603–1868). The famous *Yumihiki* (Fig. 2.1), standing about 20 cm tall, holds an arrow in its right hand and a bow in the left. It slowly moves its right hand, holding the arrow, toward the left hand that holds the bow, places the arrow on the bowstring, and draws the bow. It turns its face to the target and positions himself to aim at the target. After fully pushing forward the bow by the left arm, the doll releases the arrow, and the arrow hits the target beautifully. The doll repeats this feat until it shoots all the arrows. An arrow rest holding the arrows rotates automatically a certain distance at a time for the doll to pick the next arrow always in the same position after each shot. It is surprising that the doll hits the bull's eye automatically. To reveal the secret, the doll releases the arrow always in the same direction and at the same speed so that you simply need to place the target in the position where the arrow will hit. The wonder of this animated doll is not that the arrows accurately hit the target but that such minute movements are automatically performed so smoothly without the slightest slipping. It is said that the story of this machine was embroidered at a later time.

A guidebook for constructing *karakuri* dolls entitled *Karakuri Zui* (*Mechanisms Compiled and Illustrated*) was published around 1796. This book may be the oldest robot engineering manual in Japan.

Traditional *Bunraku* puppetry is still appreciated by Japanese people today. It is performed by puppeteers who manipulate the puppets. These puppets are directly controlled by humans, but it may be said that they are mechanical dolls in a broader sense. *Bunraku* puppetry is good at expressing human feelings, or emotions, in puppets in an exquisite way that has never been seen before because the puppets are directly manipulated by humans. Now, the technique of robot in Japan seems very sophisticated (Fig. 2.2).

Japan is known to be a nation with great robotics expertise. As a proof of this statement, over 50% of all industrial robots working in plants throughout the world operate in Japan. Furthermore, 70% of the world's industrial robots are manufactured in Japan.

Industrial robots are quite different from *Astro Boy*, *ASIMO*, or any other humanoid robots. No humanoid robot has ever been used for industrial purposes. Humanoid robots are not yet good for practical use. There are many reasons for this. Two major reasons that are generally cited are (1) the effectiveness of bipedal walking is not yet enough established, and (2) the safety is uncertain.

Obviously, traveling on wheels is more efficient than walking as far as moving is concerned. Moreover, the robot's gait is not yet stable in safe because working environment is normally complex. If the robot falls over while working, human operators could be injured. The robot itself might also be damaged.

A humanoid that is capable of *ukemi* (literally, a quick response to any action for safety) when falling down on the ground is reportedly

Figure 2.2 Unbelievably human-like robot (at the International Robot Exhibition).

to be developed soon. In the future, this problem of falling over will be solved (Fig. 2.3).

We now go on to a story about industrial robots. A floor-mounted arm mechanism called a manipulator is the basic form of an industrial robot. Since the manipulator is mounted on the floor, it cannot change its position or move. In a mass-production plant, products are transferred on assembly conveyors one after another. Any mechanism designed to process products transferred on the assembly line does not need to move by itself. A manipulator is a mechanical structure that simulates the structure of a human arm, elbow, wrist and hand. So we see a number of artificial arms working at plants (Fig. 2.4).

Figure 2.3 HRP Robot (at the International Robot Exhibition).

The *Unimate* robot, developed in the United States in 1961, was the world's first practical industrial robot. As a typical industrial robot, *Unimate* affected the development of industrial robots thereafter.

The remarkable development of industrial robots is due to the working principle of "teaching playback." An operator performs a job step by step and the robot memorizes the entire procedure. After learning, the robot repeats what it has memorized by itself. Human operators perform the work just once to teach the robot. Thereafter, the robot plays back the steps repeating the job endlessly until the power is turned off. One might say that the robot performs a kind of "imitation learning."

Industrial robots perform spot welding, painting, and assembly more safely and better than humans. Spot welding is a technique for joining the metal of the car body not by welding linearly but in dotted lines. Linear welding provides stronger structures than spot-welding but requires higher welding accuracy. When welding

Figure 2.4 Industrial robots at work (at the International Robot Exhibition).

structures, spot welding in the required locations by the required number of welds is better suited for robots.

Arc welding is a popular technique for welding metals. An arc is an electrical discharge like a spark. High voltage is generated on metal surfaces to produce sparks, and the resultant high temperature is used to fuse the metals.

Assembly robots are extensively used in the semiconductor industry for inserting electronic components into substrates. This job was done by humans in the past. The robot used for this purpose is generally called a SCALA robot (Selective Compliance Assembly Robot Arm). A SCALA robot is a simplified version of the manipulators already mentioned. A SCALA robot is called a horizontal articulated robot and consists of a vertical linear axis and one each rotating axis in two perpendicular directions. Its features include high positional accuracy in a three-dimensional space and the capability of moving parts from one position to another at high speed. These are the reasons why such robots are good at inserting devices.

A large number of manipulator and SCALA industrial robots have been installed in assembly plants throughout Japan to produce low-priced and high-quality products despite the disadvantage of variable-model short-run production, making Japan a nation with great robotics expertise and an economic giant.

Having introduced industrial robots, I will now proceed to the story of robots developed by researchers. Robots created by researchers are deeply related to the study of artificial intelligence (AI). Let us first review the study of AI.

The subjects in the AI studies curriculum at the Massachusetts Institute of Technology (MIT) Artificial Intelligence Laboratory in the United States include language, deduction, learning, vision, perception, operation, programming, architecture, and expert systems. Of these, vision, perception, deduction, programming, learning, and operation relate to the study of AI and robots. This suggests that the study of robots shares many common research areas with the study of AI.

Themes in AI studies include intelligent vision systems, perception function of robots, robot programming and AI, robot hands and tactile sensors, and autonomous mobile robots (Grimson and Eric, 1987).

The study themes for the intelligent vision system include technologies to represent image information captured by vision sensors as line drawings, detection of 3D shapes through the combination of lasers and vision sensors, and detection of 3D shapes by irradiating moiré interfering light and random dots in place of lasers. Another important theme is the development of an automatic system to represent an object viewed by vision as line drawings; decompose the line drawings into basic figures; and describe the relationships between the figures' connections.

The central theme in the study of the perception function of robots is stereovision. In the study of stereovision, at least two vision sensors set apart a certain distance are used to observe the object. The difference of how the object is viewed between the two images captured by the two sensors (generally called disparity) is calculated to determine the distance to the object by triangulation. This technique is simple in principle, but the problem is how to find the correspondence relationship between the images required for calculating disparity. The image correlation method is extensively used, but a decisive solution has not been yet established. The currently available solution is to increase the number of vision sensors for observation to enhance the reliability of the measurements. Most stereovision algorithms succeed in bounding the image domain to be used when searching for the correspondence relationship between images using an epipolar constraint. The epipolar constraint means that in stereovision, assuming a common plane passes through the two imaging planes of the two vision sensors and the object, the image information of the object present on the common plane is always reflected on the image planes of the two vision sensors where these image planes intersect the common plane and nowhere else. Other viable methods include optical flow to measure the movement of the sensors by tracking the images appearing on the sensors.

Robot programming and AI are described now.

The playback method is generally used when building and programming industrial robots. Programming a robot means that an operator does the job in the machine shop and the process is recorded in the memory of a robot for learning. With this system, the robot can flexibly respond to uncertain events and adapt to sudden changes in the environment such as an unexpected object moving in its area during operation. To respond to a variable environment, the robot needs to incorporate an environment observation system such as vision sensors and run a program capable of driving the manipulator flexibly.

Studies on sensitive adaptation to a variable environment belong to the domain of AI study and may be said to be equal to the study of intelligent robots. Many research papers were published in this particular field between 1970 and 1990. Some researchers' activities during this period are introduced here in detail.

Study themes in this field include planning problems, rule-based systems, model-based systems, and motion planning.

Early famous studies on planning problems include GPS (General Problem Solver) by Newell and Simon (1963) and STRIPS (Stanford Research Institute Problem Solver) by Nilsson et al. (1971). Using GPS, given the start state S and goal state G, the procedure from S to G is automatically generated. The monkey and banana problem is a famous toy problem in the study of artificial intelligence. Like GPS, STRIPS calculates the procedure from S to G automatically based on hierarchical planning.

Studies involving mobile robots include the famous SHAKEY developed (1966–1972) by the Stanford Research Institute (SRI) in the United States and the SHRDLU program at MIT that puts blocks together.

SHAKEY, the mobile robot, is connected to computers, has a TV camera and range finder (distance meter) serving as vision sensors. Driven by two electric motors, the robot moved around the room and was able to avoid obstacles of a simple shape. The behavior of the robot was automatically calculated by STRIPS. For example, for a robot in room1, which is given a goal state of in_room(room5) or moving to room 5, STRIPS writes a program reading go_through(door1, room1, room2), go_through(door2, room2, room4), and go_through(door3, room4, room5). Following the specified procedure, the robot executes each go_through instruction, which is a primitive action routines built into the robot, three times, and moves to the goal state.

This study marked a milestone in the history of robots as a successful installation of an automatic move procedure creation program in a mobile robot and one in which the robot actually moved as programmed.

As readers may have noticed, such procedural programs must be frequently re-written to achieve a different goal each time the environment changes. There is also a problem of combinatorial explosion in the hierarchical problem-solving method that STRIPS uses.

STRIPS and SHRDLU are examples of a rule-based system. When calculating work procedures from the start state S to the goal state G, the robot picks up currently viable work procedures using certain rules, and given the present state, the work procedure with the highest possibility of achieving the goal state is identified temporarily.

In a model-based system, all the information on the environment where the robot will have to work is loaded beforehand as models into the robot to save time and enable the robot to accurately calculate the environment using its vision sensors and other means.

Avoiding obstacles is one of the themes discussed in robot motion planning. What if, for example, SHAKEY enters a room and encounters an obstacle? This kind of problem is generally called a robot path planning problem.

Two solutions are available: a visibility graph and configuration space. A visibility graph is a graph in which all vertices and points of tangency of a curved section of obstacles visible from all vertices and points of tangency of any other obstacles are connected by straight lines (that is, if two vertices or points of tangency can see each other, a segment is drawn between them). The configuration space method provides a technique of increasing the size of obstacles relative to the size of the robot to be used. If the robot is a cylindrical one 30 cm in diameter, for example, it can be treated as a point by increasing the girth of an obstacle by, say, 15 cm. The robot then calculates the shortest path to the goal in the visibility graph while considering itself to be a point.

Studies on autonomous mobile robots are described now.

Autonomous mobile robots were first invented by William Grey Walter (1910–1977) in United States in 1948. The robots, named Elmer and Elsie, responded to light stimulation and recharged their batteries automatically. These robots will be further described in detail later in this book.

SHAKEY was certainly an intelligent mobile robot.

The CART project led by Hans Moravec at Stanford University is a well-known study on mobile robots (Moravec, 1980). Their robot moved on a four-wheel drive and communicated with computers by radio. Contrary to the SHAKEY robot, which was programmed with all the models of the moving environment, the CART robot had none of these models. Instead, the CART robot constructed the environmental models by itself while moving around in the environment by evaluating the uncertainty of its position using a nine-sensor stereovision system. After this, Moravec went to Carnegie Mellon University and created the NEPTUNE robot. This robot moved indoors at a speed of about 100 meters per hour while generating environmental models using ultrasound and stereovision sensors.

The author of this book developed an autonomous mobile robot equipped with a stereovision system consisting of two vision sensors in 1987. This was the world's first robot capable of detecting

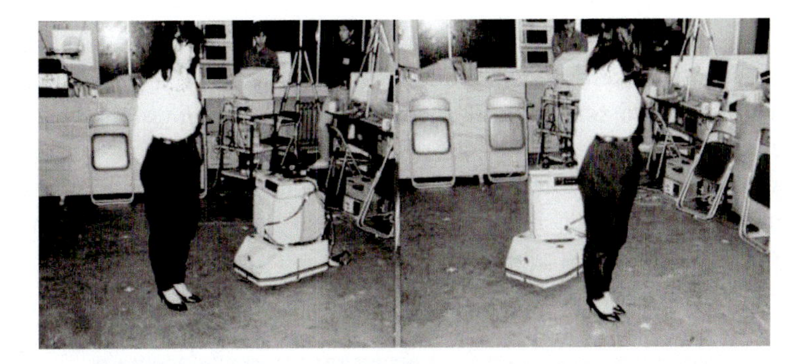

Figure 2.5 Experiment of robot avoiding collision with a moving obstacle (Meiji University).

humans and other moving obstacles and foresaw and avoided collisions with moving obstacles using its vision system (Fig. 2.5) (Takeno, 2003).

Martin Marietta Corp. developed ALV (Autonomous Land Vehicle) robot with the support of the U.S. Defense Advanced Research Projects Agency (DARPA). This robot used a small truck loaded with multiple computers, laser range finders, and vision sensors. It moved outdoors at a maximum speed of 10 km per hour while measuring the direction of its travel.

Large Jeep-like automobiles were then developed, incorporating the small-sized and high-speed computers, which became available in the market in and after around 1995. These vehicles were capable of traveling on rough terrain at nearly practical speed. Unmanned driving experiments are currently being conducted on highways and deserts in the United States.

Professor Ulrich Rembold at University of Karlsruhe, Germany, developed a robot named KAMRO (Karlsruhe Autonomous Mobile Robot) that featured two assembly manipulators and was capable of moving in all directions on a four-wheel drive during 1985 through 1994. The wheels had special driving mechanism to enable the robot to move in all directions as if it was sliding. The practicality of the robot deserves special mention: The robot was capable of assembling parts with two arms while moving around from one shop to another inside a plant. In October 1993, KAMRO successfully demonstrated self-controlled traveling and performed assembly jobs in a pseudo-experimental plant located on the campus of the university (Takeno and Rembold, 1996). KAMRO was equipped with 14 ultrasound sensors around its body and a stereovision sensor system to look ahead for obstacles as it traveled. The stereovision sensor system was developed by this book's author in Japan and mounted on KAMRO in Karlsruhe with Prof. Rembold (Fig. 2.6). The stereovision sensor system consists of a two-lens optical system, a vision sensor (Takeno, *et al.*, 1992).

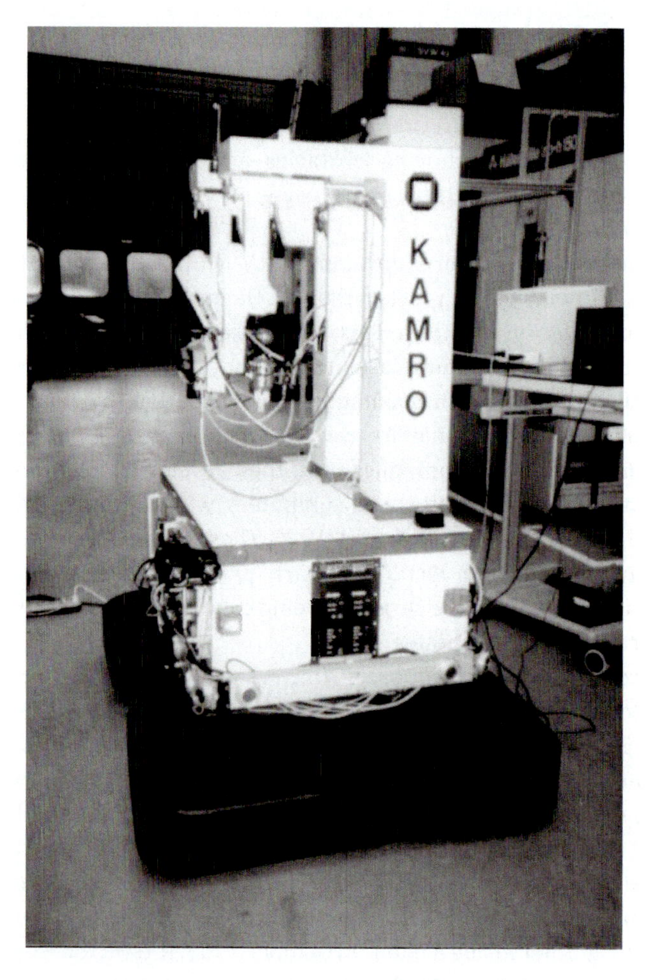

Figure 2.6 Dual-arm robot KAMRO developed by University of Karlsruhe.

Later in Japan, Honda Motor Co. developed the Honda P Series humanoid bipedal walking robots. People all over the world were astonished to see the robot because its development had been done secretly. Until then, many robot researchers said that it was impossible to build robots that could walk on two feet like humans. After Honda's success, humanoid bipedal walking robots became the mainstream of such robots in Japan. Honda's *ASIMO* and Sony's *AIBO* robot dog became very popular. And the QRIO humanoid robot built by Sony in 2003 is also very famous. In the United States, the technology to drive unmanned car-shaped mobile vehicles was established and commercialized. In Europe, studies to understand the cognitive function of humans were actively being pursued although there were some other types of research being conducted in Germany such as those already mentioned in this chapter.

Humans have been attempting to create machines that were capable of replacing themselves from as far back as ancient times. And in doing so, humans were also trying to understand themselves and other living things. These aspects continue to be intertwined with each other up to the present day.

The creation of what can be considered a full-fledged robot began with a watchmaker or machine maker incorporating a mainspring or thousands of gears into a doll and enabling the doll to move in a human-like fashion. Initially, the aim of these makers was to demonstrate their technical skills to the upper class and merchants, or to provide entertainment to common folk.

At the time, such dolls contained a huge number of precisely moving gears inside them, but today those internal movements have been replaced by computer programs.

In the United States, there was a realization that a future breakthrough in innovative technology could take place in manufacturing plants if a computer and an elaborate mechanical doll were integrated. And thus the world's first electronically controlled arm-like robot was developed and sold.

Research on artificial intelligence based on the capabilities of computers began at universities in the United States such as Massachusetts Institute of Technology (MIT), and those technologies have helped robots become intelligent.

I can still remember that during that time Japan learned those modern robotic technologies from the United States and set to work developing industrial robots that could be put to practical use one after another, eventually becoming the world-class leader in robotics.

Also, people were being enthralled by movies in which robots were playing leading roles just like humans.

And, the robots that were actively playing those roles possessed a human-like consciousness and mind, or there were robots called androids that were so elaborately created so as to be indistinguishable from humans in appearance and form.

Robots are now capable of walking around on two legs just like humans do, and androids are smiling at people.

Chapter 3

Story of the Human Brain

This chapter provides some knowledge of the human brain to facilitate an understanding of human consciousness and the problem of the heart. I am not a brain science specialist; so please allow me a somewhat naive approach in my writing here.

The average human brain has a size of about 1500 cubic centimeters, and it contains 100 billion nerve cells called neurons and glial cells to protect the neurons. I had an opportunity to hold an actual human brain in my hands. It was heavier and smaller than I had expected.

Each nerve cell has dendrites as signal inputs and a fibrous neural pathway called an axon as the signal output. Signals pass through the axon as electrical impulses or transient changes in the potential. Each nerve cell has one axon. Each axon has many branches at its end. The axon is covered by many myelin sheaths to insulate the signals passing through the axon and accelerate the transmission speed of the signals.

The end of an axon makes contact with the inputs, the dendrites, of other nerve cells across a small gap. Although we say "contact," the nerve fibers are not directly connected. This gap, or a junction called a synapse, is about 20 to 25 nanometers wide. Signals are transferred across this gap by making use of special

Self-Aware Robots: On the Path to Machine Consciousness
Junichi Takeno

ISBN 978-981-4877-90-9 (Hardcover), 978-1-003-26181-0 (eBook)
www.jennystanford.com

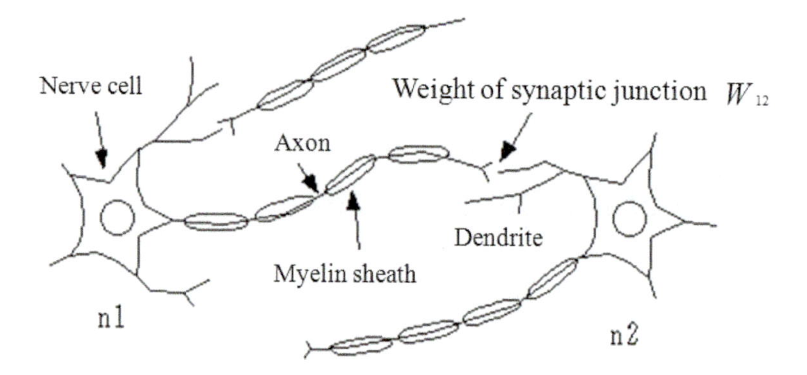

Figure 3.1 Outline of a nerve cell.

chemical substances called hormones. We may, therefore, say that all of the functions of humans are performed using two means of information transmission: electrical signals called spikes and chemical substances called hormones (Fig. 3.1). Specifically, the human brain transmits information to various parts of the body using two types of means: the pin-point type that mainly uses electrical signals to control the motion of the hands and legs, and the secretion type that injects hormones into the blood to transmit information throughout the body.

Each nerve cell has about 10,000 branches (synapses), that is, each cell connects to 10,000 other cells. The number of nerve cells in the human brain is fixed at birth. After birth, 100,000 cells die every year. It had been taken for granted that no new nerve cells are ever generated, but a recent report has refuted this long-held theory.

The brain looks like a walnut and consists of two hemispheres: right and left. It consists of a cerebrum, cerebellum, and the brain stem, which is a bundle of nerves that hang downward. The hemispheres are connected by a bundle of nerve fibers called the corpus callosum. The right hemisphere on the right hand side is called the right brain and the one on the opposite side is the left brain (Fig. 3.2a,b).

The cerebrum is the newest evolved part of the brain compared with the other older parts of the brain.

Earth was formed about 4.6 billion years ago. Scientists say that life first appeared on Earth 3.8 billion years ago. There were about

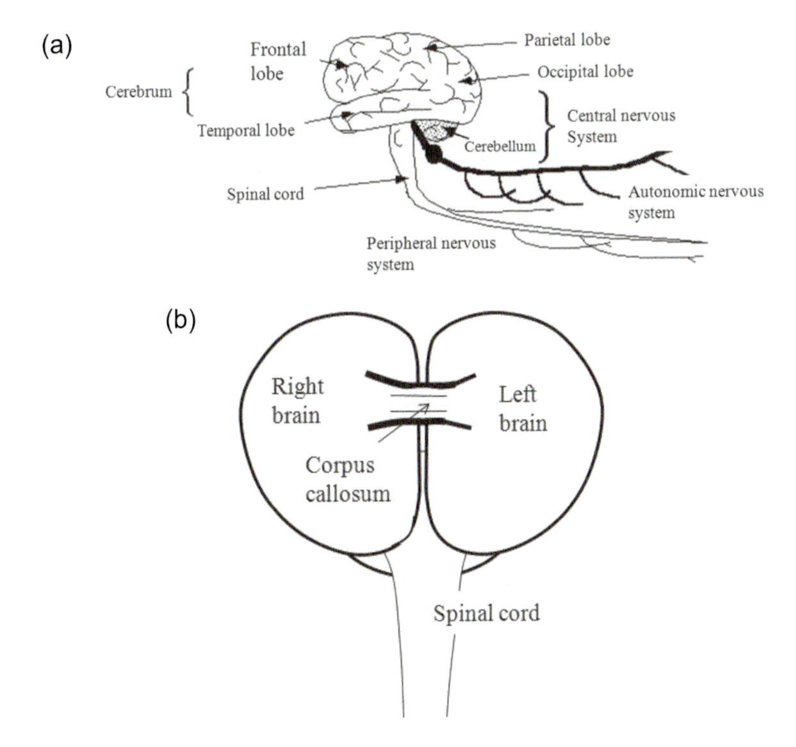

Figure 3.2 (a) Outline of cranial nerves. (b) Left brain, right brain and corpus callosum.

30 species of living creatures 3.6 billion years ago. About 800 million years ago, the number of species increased abruptly to 10,000. Living creatures in the sea moved to land about 700 million years ago. It was 650 million years ago when the big event of the extinction of dinosaurs occurred.

One theory holds that primitive human brains were formed about 600 million years ago and evolved to nearly the present form about 60,000 years ago. The cerebrum is responsible for high-level functions that make humans like humans, while the cerebellum is associated with animal functions such as motor control.

The surface of the cerebrum is covered by the cerebral cortex, which is divided into the neocortex and paleocortex. The neocortex is 2–3 mm thick and contains 14 billion neurons called columns in tiers. We can observe that the structure of the neocortex consists of

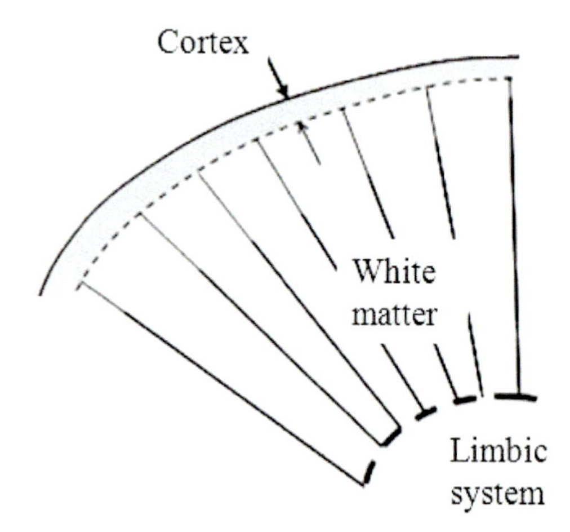

Figure 3.3 Cortex and white matter.

about six layers. This area is supposed to be the home of the smallest functional units for processing terminal information to achieve the high-level functions peculiar to humans. According to scientists, the surface area of the neocortex is about 2,000 square centimeters, which is frequently referred to as equivalent to the size of a sheet of newspaper.

A number of nerve networks extend from the cerebral cortex down to the center of the brain. This area is called the white matter. The white matter consists of axons leading to nearly the center of the brain. The above-mentioned myelin sheaths glow in white (Fig. 3.3). Neurons and glial cells are the two main types of cells in the brain. Neurons transmit information while the glial cells supply nutrients to the neurons and remove unnecessary substances from them.

A small space called the lateral ventricle exists at the deepest part of the cerebrum. Its role is unknown. The limbic system extending around the brain stem below the white matter is an older part of the brain. The part responsible for controlling the irises of the human eyes is said to be located in the brain stem. A simple way for a doctor to test the functioning of a subject's brain is to shine the light of a flashlight into the eye of the subject and check the reaction of the iris. The complete absence of any iris reaction indicates that the oldest and centermost part of the brain is not functioning.

The function of the brain is described now in brief. The right brain controls the nerve networks present on the left side of the body, performs artistic processing regarding meaning, emotion, and so on. The left brain, on the other hand, has its nerve networks mostly on the right side of the body; it handles languages and theoretical processing.

If the right brain is impaired, the left side of the body is paralyzed. Upon looking at a dog, a patient may be able to represent it as "dog" in speech but the semantic meaning of dog is thought to be lost. If the left brain is injured, the right side of the body is paralyzed and language abilities are disturbed.

The corpus callosum is a bundle of about 200 million nerve fibers.

The right and left brains exchange information constantly and both artistic and logical activities are carried out simultaneously. Upon looking at a red rose, one might represent the flower as "This is a rose" in speech and at the same time feel a sense of beauty aroused by the red color of the rose.

When the corpus callosum is severed, the right and left brains are nearly disconnected, and results in a critical condition depending on individual case. This type of surgical operation was once performed on epileptics. Normally no major problems occur, but troublesome cases of alien hand syndrome have been reported. Simply put, alien hand syndrome means being out of control of one's own hand. For instance, if one wants to write with the right hand, but the left hand tries to stop it. The person is conscious of writing with the right hand but the left hand moves without the person being conscious of it. Based on this fact, many researchers assert that the part governing human consciousness resides inside the left brain.

It has been said that by using some clever devices in an experiment, one can see that the part governing consciousness does exist in the right brain, but the relevant process will not surface and remains at an unconscious level. Some researchers thus believe, referring to the alien hand syndrome, that two parts of consciousness would arise if you severe the corpus callosum. Subconsciousness means consciousness that does not surface and thus fails to become conscious. Some call this unconsciousness, but unconsciousness normally refers to a loss of consciousness following anesthesia; so I differentiate how these two words are used in this book.

We can draw three general conclusions from the above discussions:

(1) The part governing consciousness resides in the left brain.
(2) There is just one consciousness that surfaces.
(3) There are two types of consciousness: the consciousness that surfaces (called explicit consciousness or simply consciousness) and subconsciousness.

The limbic system consists of the epiphysis, hippocampus, amygdala, thalamus, hypothalamus, putamen, and caudate nucleus. The epiphysis exists singly, whereas the others exist in right and left pairs. One of the important functions of the limbic system is described here. Based on a variety of information received from the sensory organs, the limbic system processes emotional information by communicating with the cerebral cortex, and it stores the information temporarily. At the same time, the limbic system generates a long-term memory by communicating with the cerebral cortex. Researchers say that the control mechanism for emotion and short-term memory exists in the limbic system.

Multiple nerve systems are interconnected with the limbic system, building a complex mutual relationship among them. I recommend referring to the adequate references noted for details on the limbic system.

A simple description is presented here. The amygdala is closely related to human emotions. The amygdala sends impulses to the hypothalamus by communicating with the frontal lobe of the cerebral cortex and modulates blood pressure and heart rate using hormones. The change occurring in the body is returned to the frontal lobe of the cerebral cortex via the somatic perception area. The circulation of this information seems to create feelings. The amygdala is said to generate negative feelings such as fear in the brain.

The thalamus relays signals input to the brain, to various parts of the brain. Signals entering the brain first reach the thalamus. The information is then sent to the cerebral cortex and the signals are also sent to the amygdala. At this time, if the amygdala is allowed to function with priority, the person is strongly affected by emotion.

The hippocampus is responsible for memory. If the hippocampus is damaged, short-term memory is totally lost.

The putamen controls unintended or subconscious actions. It issues commands to the premotor area of the cerebral cortex, and the premotor area sends commands to the neighboring motor cortex. This results in muscles contracting to perform motor skills.

The caudate nucleus is said to control subconscious actions and thoughts.

The hypothalamus controls metabolic processes and desires.

The cerebral cortex is described here. The cerebral cortex is generally said to encompass the frontal lobe, parietal lobe, occipital lobe, and temporal lobe (Fig. 3.2a).

The frontal lobe is associated with the highest-level functions of humans. It is deeply concerned with the generation of thought, planning, and conception as well as emotions. The parietal lobe is highly associated with some types of motion, calculation, and cognition. The occipital lobe is known to be involved in vision.

The temporal lobe in the left brain is related to speech, and, in particular, Wernicke's area, which is located there, is famous as the part of the brain associated with the cognition of spoken and written language.

The angular gyrus, located immediately posterior to Wernicke's area, is very much associated with information on meaning (Fig. 3.4).

According to experiments carried out by Wilder Graves Penfield (1891–1976), the part of the brain involved in long-term memory spreads over a wide area of the temporal lobe.

It is known that sexual arousal is generated when the temporal lobe is stimulated, and the frontal lobe is active during both male and female orgasms.

The cingulate gyrus of the frontal lobe is where researchers believe self-consciousness resides. Self-consciousness is a function by which an individual is capable of judging that an image in a mirror is his/her own or someone else's. Self-consciousness is also known to be highly associated with Broca's area, which is responsible for speech functions. This suggests that being able to speak about oneself could be an origin of self-consciousness. Broca's area is

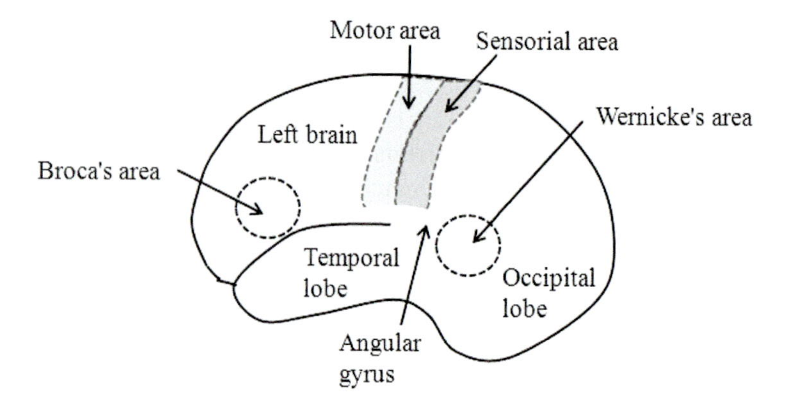

Figure 3.4 Broca's area and Wernicke's area.

located in the frontal lobe. It has also been said that the area involved in **will** exists in the prefrontal area of the frontal lobe.

The sensorial area and motor area exist in the region where the parietal lobe and the occipital lobe meet. The sensorial area is where the sub-areas corresponding to the tactile sensors of various parts of the body exist. These sub-areas are cortices and receive signals from respective tactile sensors. The cortices receiving signals from the hands, fingers, lips, genitals, and toes occupy a relatively large surface area (Fig. 3.4).

The sense of taste is associated with the cortex on the lower part of the right frontal lobe.

The stimuli of smell directly enter the limbic system and activate the lower part of the right frontal lobe.

The primary motor cortex lies at the center of the parietal lobe. It sends signals to activate the motor functions of the body. The right hemisphere's primary motor cortex controls muscles on the left side of the body, while that on the left controls muscles on the right side of the body.

When a human wants to make a certain movement, the primary motor cortex sends signals to contract muscles. At this point, there may be information circulating between the vision and somatic sensors in the body and the brain to monitor and make sure that the desirable motion is achieved.

The area involved in spatial cognition exists in the rear part of the parietal lobe. Most of the occipital lobe is used for vision processing.

An image entering through the lenses of the eyes is inverted when reaching the retinas. There are two types of photoreceptor cells in the retina: cones and rods. The cone cells detect intensity and the rod cells color information. The fovea in the retina is where photoreceptors are most densely concentrated. Signals from the photoreceptors, already image-processed on the retina to some extent, leave the eyes and go to the brain via the part called the blind spot and through a bundle of neural pathways. The blind spot consists of a dense bundle of neural pathways and there are no photoreceptors. The blind spot is therefore "invisible," or cannot respond to light stimulation. It is very interesting to note that the photoreceptors are viewing light that has penetrated the layers of nerve fiber bundles (from the back side of the nerve cables) since the photoreceptors are located at the deepest part or bottom of the retina (Fig. 3.5).

The neural pathways leave the right and left eyes and go to the brain. On the way, they cross each other just once at a certain location. This crossing point is called the optic chiasm.

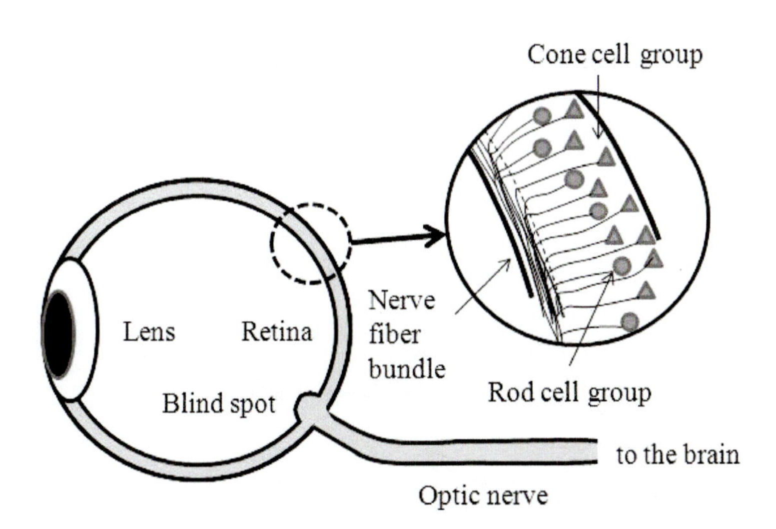

Figure 3.5 Vision and nervous system.

Because of this crossing, the neural pathways associated with the right-side visual fields of both the right and left eyes go to the left occipital lobe and the neural pathways associated with the left-side visual fields of both eyes go to the right occipital lobe.

Optic chiasm has not been fully explained biologically. The author supports the hypothesis that the optic chiasm has an advantage in that even when if one hemisphere of the brain is damaged, the other hemisphere of the brain keeps the functions of stereovision viable.

Visual information runs through the part of the brain called the lateral geniculate body, which is a part of the thalamus and reaches an area named the primary visual cortex (V1) in the right and left occipital lobes.

V1 is like a projector that shows image information captured by the eyes. Visual information is further transferred to other visual areas called V2, V3, V4, and V5 for the brain to determine the position of an object being observed. V2 processes three-dimensional perception, V3 distance, and V5 motion information (Fig. 3.6). The visual information passing through V1, V2, and V4 determines what the object is. V4 is responsible for processing color information.

When the primary visual cortex of the occipital lobe has a lesion, humans cannot see their outside world. It has been reported, however, that even with the visual area damaged, part of the visual

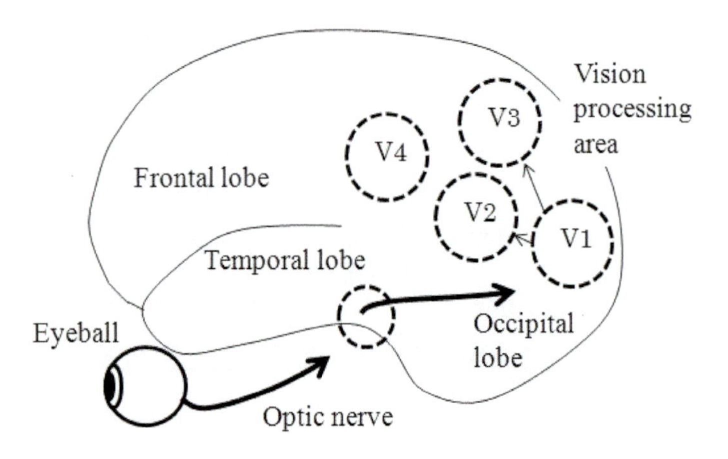

Figure 3.6 Vision processing system.

information was transmitted to the parietal lobe, and by virtue of this information the subject reached out his hands toward a strong light although he could not visually see the light. This is called blindsight or cortical blindness. In this case, the subject was not conscious of the light but reached out his hands toward it.

Chapter 4

Human Consciousness and the Mind

This chapter introduces historical research into the mystery of thought, consciousness, and the human mind, and, in particular, areas of mathematics, psychology, and philosophy. I believe an intricate mechanism of consciousness and mind exists in the background of human thought.

Hippocrates (460 BC–375 BC), an ancient Greek physician, was the first scholar who believed that human thought occurred in the brain (Bear, 2001). He said that the brain controlled the body. He reportedly discovered this fact through his experience in treating gladiators who suffered brain damage in ancient Greece.

4.1 Human Thought and the Turing Machine

Alan Turing (1912–1954) (Fig. 4.1) was the first person in history who mathematically challenged the mystery of human thought. The Turing machine he proposed in 1950 may be considered the origin of thinking machines. Turing said, "What can be solved by humans can be solved by a machine of a small capacity" (Turing, 1936). He thought that human intelligence could be realized by machines.

Self-Aware Robots: On the Path to Machine Consciousness
Junichi Takeno

ISBN 978-981-4877-90-9 (Hardcover), 978-1-003-26181-0 (eBook)
www.jennystanford.com

Alan Turing
1912–1954

Figure 4.1 Alan Turing.

The Turing machine has a very simple construction. It consists of a body, a tape, and an input/output device called head to read and write symbols on the tape. The Turing machine heralded the development of present-day computers. In fact, it has been mathematically proven that all problems that can be solved with present-day computers can be solved with a Turing machine.

It should be remembered, however, that to create computers of today, we needed a stored-program system, which was developed on the basis of the ideas of John von Neumann (1903–1957) and program-specific high-level concepts such as algorithms and data structures.

It is also known presently that there is a problem that a Turing machine cannot solve. The problem is for a Turing machine to decide if it will finish running or will run forever. Any given problem is solved when the Turing machine finishes running.

Turing's attempt to create models of human thought paved the way to the development of computers, on the one hand, and the study of artificial intelligence, on the other.

4.2 Language and Formal Logic

We will start with a description of the study of formal logic before discussing mathematical ways to model human thought. According to formal logic, human thought arises from languages, and is an inference in languages. Formal logic was the basis of the study of languages and inference at that time. The development of modern formal logic is credited to Gottlob Frege (1848–1925), (Frege, 1892) and Bertrand Russell (1872–1970), (Russell, 1959). I believe that their objective of research was to unravel the intelligent functions of humans from the standpoint of formal logic. Their attempts, however, deviated from and became irrelevant to human thought and shifted to a pure pursuit of formal logic itself. They attempted to develop a method to derive mathematical knowledge from basic mathematical hypotheses using logic. Their efforts were in vain, however, when in the 1930s, Czech-born Kurt Gödel (1906–1978) mathematically proved that such an attempt was impossible (Fig. 4.2). These are Gödel's famous incompleteness theorems (*Gödelsche Unvollständigkeitssatz*). The theorems say that formal systems are incomplete because they inherently include arithmetic equations that are defined to be true by the system but that cannot be proved by the basic assumptions of the system. These theorems thus negated the clarity of mathematics. The meaning of these mathematical theorems also applies to the relationships between human thought and the computer as a representative of formal systems. Namely, there are some arithmetic equations that human thought believes to be true but that cannot be proved by computers.

Based on this recognition, Nagel and Newman (1958) held that computers could never reach the level of human thought or the human mind. In 1994, Roger Penrose asserted that human thought could clarify mathematical truths better than computers (Penrose,

Figure 4.2 Kurt Gödel.

1994). According to Penrose, Fermat's last theorem was successfully proved by human thought, whereas if the Turing machine were used to solve that problem, it would have required infinite computations about whether or not the problem was solved to find the solution. He concluded that human thought solved problems that Turing machines could not solve, that is, that human thought involves non-computational elements that computers lack (Penrose, 1989, 1994). He referred to the quantum theory to explain the non-computational elements, but there are not many supporters of his argument today.

4.3 Language and Chomsky

Noam Chomsky (1928–) is a famous linguistic researcher who attempts to explain the philosophical aspects of humanity through the study of language. His arguments are introduced here. Chomsky's conclusions are based on the fact that humans speak languages and exhibit intellectual functions by conversing with others. Importantly, the study of artificial intelligence started with the study of language.

Chomsky thought that creating a machine that understands and speaks natural languages was a viable means to understanding the human mind scientifically. In his Theory of Generative Grammar, Chomsky showed that certain basic rules exist in the languages spoken by humans (Chomsky, 1957). This theory provided a scientific means to analyze human languages and contributed to the later development of the PASCAL and LISP computer programming languages.

Chomsky's theories have influenced many researchers. One of his famous studies involved chimpanzees that learned a "language" and were able to communicate with humans. Pictograms or signs (similar to a sign language) were used because chimpanzees do not speak human language; so a sort of surrogate language was used. The experiments were successful to some extent. The chimpanzees understood the surrogate language as well as the feelings shown by the humans. The experimenters also felt that the chimpanzees were willing to show their feelings to the humans using the surrogate language. The grammar was simple, however, and complex sentences were not used.

Chomsky's linguistic theory could not answer the question of why children can quickly learn a language as they grow up. This question is called Plato's problem. To find the solution to this problem, linguists started to study how language was generated and understood in the language areas of the human brain.

Austrian-born Ludwig Wittgenstein (1889–1951) was a famous philosopher who attempted to clarify human consciousness by analyzing the natural languages of humans (Wittgenstein, 1953). He tried to account for human conscious activities using the Language Game concept (Wittgenstein and Solipsism).

Jean Piaget (1896–1980), a Swiss developmentalist, was a famous figure in the study of how the functions of cognition develop in infants (Piaget, 1952).

As discussed in the previous chapter, it is very difficult to research the brain functions of humans. Human rights must be respected when treating the human brain. Furthermore, even though various systems have been developed to measure brain functions noninvasively (that is, without physical injury) it is still impossible to measure the true functions of the brain.

A functional magnetic resonance imaging (fMRI) system allows a doctor to view how the brain is activated in real time by measuring changes in blood flow in the brain. The system identifies and shows the active parts of the brain but does not provide any information on what the brain is performing at that part.

A transcranial magnetic stimulation (TMS) system is used to magnetically stimulate the brain cortexes through the skull and measure the reactions to examine the brain functions. This also does not reveal any information about the functions of the individual areas of the brain and of the brain as a whole.

Today's computers understand the natural languages of humans to some extent. Some computers are capable of generating languages similar to natural languages.

Chomsky's linguistic theories greatly contributed to the development of programming languages and artificial languages for computers. His theories also achieved their original aim in the research phase of natural language. Computers may understand programming languages and execute some processes, or even understand natural languages and follow instructions, but there remains the question whether computers understand languages in the same way as humans do.

Some researchers, known as behaviorists, believe that machines do understand. Behaviorists do not care what processes are going on within the machine, but they determine that the machine does understand just from the fact that the machine reacts adequately at least to certain stimuli. The soundness of this concept was questioned in a thought experiment conducted by John Searle (1932–   ).

This is a story of a box performing questions and answers in Chinese (Searle, 1980). The story is based on Searle's Chinese Room thought experiment, which is slightly modified. An Englishman, A, who does not understand Chinese at all sits inside a box. A has a complete manual to reply in Chinese to any question written in Chinese. The scenario of the thought experiment goes like this:

A Chinese person, B, writes a question in Chinese on a slip and puts it in the slot of the box. A takes it and consults the manual to find the reply written in Chinese. He copies the reply on a slip and gives it to B. B thinks that A fully understands Chinese because a perfect answer is given in Chinese to every question asked. However, the truth is that the person in the box is an Englishman who does not understand Chinese at all. If B writes a question in English, the person in the box will naturally understand the question and write the answer in English. The questioner would then think that the person in the box understands English, and his thinking would be right.

What is the difference between these two stories?

This question gave rise to the problem of human consciousness and the mind that exists beyond the problem of thought. This thought experiment makes us reconsider what human understanding really is.

4.4 Mind–Body Dualism of Descartes

French philosopher René Descartes (1596–1650) was the first Western philosopher who addressed the problem of the human mind (Fig. 4.3). With the famous declaration, "I think, therefore I am" (*Cogito, ergo sum* in Latin), Descartes insisted on the existence of the self. He stated, "I doubt sensation, logic and thinking, and after doubting everything, what is left is the fact that I think." This statement marks the discovery of the existence of the self of humans, or eventually, self-consciousness. Descartes advocated the famous mind–body dualism. The term appears in his work, *The Discourse on the Method* (1637) as *material–mind dualism.*

Mind–body dualism says, in essence, that the human being is divided into a mind and a body. Descartes thought that body exists in this world as a material substance, while the mind is connected to God. This thinking was favorably accepted in Christian Western Europe as a teaching that did not contradict Christianity. His thinking supported the Christian doctrine that stated that when a human dies, the soul (mind) leaves the body and enters the kingdom of God.

At the time, some people felt that clocks had minds as they watched them, but Descartes' mind–body dualism shattered this idea and convinced people that clocks were simply machines, and there was no mind in them. His thinking helped expel the subjective

Figure 4.3 René Descartes.

existence of the mind from machines and continued to develop further as the concepts supported modern capitalism in Western Europe, which enthusiastically pursued the rationality of machines themselves, now freed from the problem of the mind. It may not be an exaggeration to say that the Industrial Revolution that started around the end of the 18th century was backed by the concept of mind–body dualism.

Let us now look at the history of the study of consciousness and the mind. In 1879, the study of consciousness and the mind was started by Wilhelm Wundt (1832–1920), University of Leipzig, Germany, as part of psychology. Wundt explored his own psychological state using self-observation methods in which he studied consciousness and the mind by describing or speaking of his own mental state (Wundt, 1899). A while later, the self-observation method was discarded after it was criticized for its lack of objective reproducibility and condemned to be far from a scientific technique.

Next comes the Concept of Mind by Gilbert Ryle (1900–1976). According to Ryle, the Cartesian mind–body relationship, as understood by Ryle, was the dogma of the ghost in the machine; the Theory of Dual Existence of the privately observable life (the mind) and the publicly observable life (behavior) is a basic category mistake (Ryle, 1949). A category mistake refers to discussing two matters at the same level while ignoring the categorical difference between them, such as behavior and the mind at the same level.

Ryle states, "Mental episodes are analyzed as **disposition** to behavior." This statement became the first step in expelling the myths of ghosts in machines. However, this statement, at the same time, encouraged another myth called behaviorism, which will be discussed in detail later in this book.

Ryle's assertion gave birth to the mind–brain identity theory, which became popular in the 1960s as advocated by U.T. Place and J.J.C. Smart in Australia.

The mind–brain identity theory holds that mental processes are identical to cerebral processes.

From the 1960s to the 1970s, D.M. Armstrong and D. Lewis disseminated the mind–brain identity theory. Armstrong emphasized that the physical state of the central nervous system determines behavior. Lewis preached that sensation plays a causal role in the generation of sensation by certain stimuli, and beliefs play a causal role in certain behaviors (Ryle, 1949).

H. Putnam further developed the mind–brain identity theory. In 1967, he established a theory in which the mental state was viewed as being identical to the state in a Turing machine. This was the beginning of functionalism. Functionalists view the mental state as being identical to the machine state (Putnam, 1999).

Both behaviorism and functionalism were destined, sooner or later, to face the problem of qualia, which is difficult to solve. The qualia problem will be discussed in detail later.

4.5 Behaviorism and Cognitism

Russian physiologist Ivan Pavlov (1849–1936) (Fig. 4.4) succeeded in his experiments on conditioning the psychical reflex using the

Figure 4.4 Russian physiologist Ivan Pavlov (1849–1936).

salivary glands of dogs. Pavlov won the Nobel Prize for this discovery (Pavlov, 1927).

Pavlov's experiments showed for the first time that conditioned stimuli generated mental reflexes. His experiments led to the birth of a new philosophy called behaviorism.

Behaviorism takes the standpoint that consciousness and the mind are mental representations of reactions to stimuli given to the body. Behaviorists do not discuss consciousness or the mind itself but believe that the functions of consciousness and the mind can be described as stimuli given to the body and the resultant reactions. It is true that behaviorism was considered more scientific than the self-observation method in that, compared with the impalpable notion of consciousness and the mind, stimuli and reactions could be discussed and shown to people in a reproducible way.

Behaviorism was established as a theory by U.S. psychologist John B. Watson (1878–1958), who was inspired by Pavlov's experiments. According to Watson, the study of consciousness itself cannot be a subject of scientific investigation. His thinking remains overwhelmingly influential even today (Watson, 1930).

Behaviorism attempts to understand consciousness and the mind by observing stimuli and reactions. Later, behaviorists realized that spiritual activities involved some active aspects such as will and intention, and the new behaviorism was introduced to explain voluntary actions such as cognition and intention. The new behaviorism, however, departed from the original behaviorism concepts and gradually lost its centripetal force.

Cognitism emerged just at this time. Cognitism is an idea that the intelligence of organisms derives from an internal cognitive system. Cognitists started to re-consider the existence of consciousness and the mind and attempted to incorporate systems of perception, judgment, memory, learning, language, inference and problem solving into computers.

In the early days of the development of cognitism, however, the body, will, and feelings were not taken into consideration. In the beginning, cognitism devoted itself to the study methods of intellectualism and treated representation symbolically. Intellectualism pursues an objective and intelligent way of thinking and ignores subjective concepts such as feelings and will as much as possible. Representation means, in brief, the images of external objects that emerge on consciousness based on perception.

The aforementioned Turing machine and Chomsky's linguistic theory are preeminent achievements in the study of human consciousness and the mind in this era.

It became essential for researchers to directly address and elucidate the problem of consciousness and the mind in order to construct consciousness and the mind in computers based on functionalism.

Cognitive psychology founded by Donald Broadbent (1926–1993) was the first scientific approach directly targeted at studying consciousness. In his work *Perception and Communication* (Broadbent, 1958), Broadbent says that consciousness is similar to a communication channel, and the number of objects that consciousness can pay attention to at the same time is limited.

Telephony developed rapidly during Broadbent's lifetime. Broadbent thought that consciousness might be scientifically explained using an analogy of telephone lines. As he understood it, consciousness was the function of attention. He believed that consciousness could be scientifically described by focusing on this human function of attention. Broadbent was interested in knowing how humans were attentive to the information they received from their environment via bodily sensors such as eyes and ears. He assumed that human attention was restricted, or that the capacity was limited. He relied on two methods: split-span and shadowing methods.

Gray and Wedderburn conducted the following experiments in 1960: A mixture of numbers and words were input into the right and left ears of a subject. For instance, "8, Eat, 5" were input into the right ear, and "Mice, 2, Cheese" into the left ear. According to the Broadbent theory, the subject was expected to answer, "8, Eat, 5, Mice, 2, Cheese." But instead the subject replied, "Mice, eat, cheese, 8, 2, 5." The experiment revealed that when listening to human voices attentively, humans selectively pick up information input into the right and left ears. The subject selectively picked up information and constructed a piece of meaningful information, or a sentence, "Mice eat cheese," from the mixed information input into both ears (Gray and Wedderburn, 1960). This, of course, suggests that humans have a system to selectively utilize consciousness.

I would like to introduce those philosophers who attempted to look into the essence of consciousness and the mind. There are two main trends in modern philosophy: phenomenology and hermeneutics.

4.6 Phenomenology and Hermeneutics

The leading exponents of phenomenology and hermeneutics are Edmund Husserl and Martin Heidegger, respectively.

Germany-born philosopher Edmund Husserl (1859–1938) was an important historical figure who addressed the problem of

Figure 4.5 Edmund Husserl.

consciousness directly as a theme of philosophy (Fig. 4.5). The philosophy he advocated is called phenomenology (Husserl, 1900).

Husserl tried to solve the problem of consciousness and the mind, which are separated from each other in Descartes' body–mind dualism. The most difficult point in dealing with consciousness and the mind is the fact that both are subjective matters. It is impossible to use objective techniques, that is, direct scientific techniques, to study consciousness and the mind. By developing phenomenology, Husserl introduced a technique to solve the problem of subject and object, using consciousness (a subjective existence) as the starting point. This technique was the same as the one used by scientists who promoted science since the Middle Ages, including Nicolaus Copernicus (1473–1543) and Galileo Galilei (1564–1642). Copernicus and Galileo conceived the hypothesis that the earth is not at the center of the celestial motion, but instead the sun is, that is, a heliocentric theory rather than geocentric theory. This hypothesis can be said to be the subjective thinking of Copernicus and Galileo. They started verifying the hypothesis by conducting experiments and celestial observation using astronomical telescopes. After repeating experiments many times, they arrived at a stage at that the hypothesis seemed true to anyone who might think about it, irrespective of time and place. This is the stage where a new conviction was realized in which the subject and the object were consistent with each other. This type of new conviction is called a paradigm.

A scientific technique is a method to describe knowledge from the viewpoint of exploiting the power of nature with its maximum energy at any given time, in any place and by anyone. This is the very technique employed by scientific positivism that supported the development of modern science.

In other words, Husserl's phenomenology addressed the core problems of modern scientific positivism.

Kitamura clearly describes what phenomenology is from the perspective of an engineer as follows:

> Phenomenology is a method that uses subjective analyses to tell the essential character of consciousness and works in the process of arriving at a conviction about something. Two basic functions of consciousness discovered through phenomenological analysis are intuition and orientation. Intuition is the ability to know the properties of an external object via perception, while ignoring relevant hearsay or information on the object. The orientation of consciousness refers to directing consciousness toward an object to define its meaning, and reflecting and elaborating on the above process in an effort to find a more accurate meaning. Phenomenology uses the methods of phenomenological reduction and intersubjective reduction to extract the essence of consciousness. Phenomenologists believe that by using these methods, a purely subjective attitude is possible. To put it simply, phenomenological reduction means to put aside all of our common sense about material objects, customary ways of thinking and convictions about the world — a state called epoché. By so doing, one can understand the fundamental way consciousness exists, that is, the way consciousness exists when one recognizes, feels, views, thinks and recalls.

Husserl also showed through intersubjective reduction that the cognition of others is based on a projection of one's recognition

(Kitamura, 2000). This reduction method of Husserl to get to the core of the problem of consciousness is famous. Husserl also believed that the essence of consciousness is a process in which the object of consciousness (*noema*) and the action of consciousness (*noesis*) appear alternately and eventually arrive at a meaning.

Kitamura considers that this process represents an engineering feedback. He established a machine consciousness theory, which will be discussed later in this book.

Scientific objectivity was highly esteemed in Husserl's time. However, it had a critical deficiency. When scrutinized in detail, alleged scientific objectivity is basically associated with a scientist's daily research activities, which are entirely supported by the subjectivity of the scientist. This means that discoveries of objective facts by scientists are based on the subjective views of those scientists.

The technique adopted by Husserl was to analyze how the consciousness of the self exists in humans using the reflection function inherent in the subjectivity of the self, and eventually to elucidate the universal function commonly existing in human consciousness. Readers may wonder if this technique is good enough to discover objective facts. It should be noted that even when scientific devices are used to take measurements, subjective judgments are inevitably involved in interpreting the measured data. This means that uncertainties in subjective judgment are not much different from uncertainties in so-called objective judgment.

As such, Husserl's phenomenology bridges the wide gap between the subject and the object. According to Husserl, although consciousness is a subjective matter, researchers can come close to the truth by circulating the objects of consciousness and acts of consciousness, and consciousness itself can be an object of consciousness. We may therefore say that Husserl was the first philosopher who established a scientific research method for studying the subjective matters of consciousness and the mind.

Martin Heidegger (1889–1976) was a German philosopher. "In der Welt Sein (*Being in the World*)" and other Heideggerean concepts gave rise to the great trends of cognitive science (Heidegger, 1962). Heidegger re-defined phenomenology as hermeneutics.

Hermeneutics is a methodology used to view and understand all things arising from the human mind as representations of experience, and attempts to grasp the human mind scientifically through interpretation called "understanding." *Sein und Zeit* (*Being and Time*) is one of the most important works of Heidegger.

Based on Husserl's phenomenological methodology, Heidegger analyzed the fundamental problem of human existence as an existence associated with the world as space and time [*In der Welt Sein* (*Being in the World*)].

The difference between Heidegger and Husserl lies in that Husserl analyzed consciousness and the mind as an internal existence in humans, whereas Heidegger interpreted consciousness and the mind as "being in the world." In other words, Heidegger believed that consciousness and the mind should not be considered an internal problem of humans but should be analyzed as they relate to the outside world, or environment and situations. The ideas proposed by Heidegger are now remembered as the starting point of contemporary cognitive science.

Heidegger's assertions became the source of the notion of "cognition embedded in situations" and "embodied mind" in the newly conceived cognitive science.

Embodiment is a hot topic of discussion in cognitive science today.

4.7 Phenomenology of Embodiment

French philosopher Maurice Merleau-Ponty (1908–1961) was the first person to explain embodiment philosophically. He started studying perception and embodiment based on the phenomenology of Heidegger and existentialism of Jean-Paul Sartre (1905–1980). His book *Phenomenology of Perception* is famous.

Merleau-Ponty denied the mind–body dualism of Descartes and his followers, declaring, "My consciousness is experienced in and through my body." He also said, "The mind is the cause of the body, and the body determines the condition of the mind."

French philosopher Sartre explains human existence using two opposing ideas: being-in-itself and being-for-itself. The former referred to the essence of substances and the latter to the essence of human consciousness.

According to Sartre, two different concepts coexist in humans. A human is itself a substance (being-in-itself) and at the same time differentiates between "this" and "not this" (being-for-itself) and endeavors to become an existence exceeding the substance itself, that is, becoming free. He asserts that freedom is the essence of human consciousness. Sartre stayed with the mind–body dualism concept, but, interestingly, he explained that freedom is the essence of human consciousness.

The thinking of Merleau-Ponty was succeeded by Emmanuel Lévinas (1906–1995), who became an influential thinker of the era. The difference between Merleau-Ponty and Lévinas is that Merleau-Ponty tried to explain **the other** based on **the self**, whereas Lévinas attempted to explain **the self** from the viewpoint of **the other**.

Materialism was promoted by German thinker **Karl Marx** (1818–1883) as an analogy of Hegel's philosophy (Fig. 4.6). Marx replaced Hegel's mind with material and the development of the

Figure 4.6 Karl Marx.

Sigmund Freud
1856–1939

Figure 4.7 Sigmund Freud.

mind with the development of an economic society to establish historical materialism.

Austrian psychopathologist Sigmund Freud (1856–1939) (Fig. 4.7) shed light on the psychological principle existing in the subconscious of humans in his book, *The Interpretation of Dreams* (Freud, 1900). He thought that hysteria arose from sexual inhibition and found that hysteria was improved by treatments aimed at sexual release. Later, he generalized this theory and preached that many psychological diseases occurred from sexual inhibitions. It may be said that Freud showed in his book that the problem of the human mind was not yet solved in the era where historical materialism was at its peak.

Carl G. Jung (1875–1961) followed and developed Freud's thinking (Wehr, 1989). Jung believed that there were origins, something like the Great **Mother** and the World **Mandala** of Esoteric Buddhism, in the depths of the human mind, and tried to explain humans' mental world as emotions arising from these origins.

Norbert Wiener
1894–1964

Figure 4.8 Norbert Wiener.

Another noteworthy research of consciousness and the brain is cybernetics, a bio-machine theory. Cybernetics was defined by Norbert Wiener (1894–1964) (Fig. 4.8) as the science of communication and control in organisms and machines. Cybernetics covers a broad field of studies, including medicine, engineering, physics, biology, mathematics, psychology, and social science (Wiener, 1956). It was later divided into information science, bionics, cognitive science and robotics. The term cybernetics is formed from a Greek word that means *steersman*.

I have so far introduced historical researchers related to consciousness and the mind. I will now summarize the knowledge about consciousness and the mind itself.

One of the important problems regarding consciousness and the mind is, which approach should be taken: materialism or idealism?

The Cartesian mind–body dualism holds that the body and the mind are separate from each other, and the mind is connected to God. From the Cartesianist point of view, the mind would then be discussed as a matter of idealism, and the body would perhaps be explained from the viewpoint of materialism. It should be noted, however, that Descartes does not explain how the separated mind and body are interrelated.

Descartes' allegation seems to directly apply to present-day computers. Computer hardware is a materialistic existence, whereas software and data are idealistic. It may sound somewhat rude and unreasonable to say that the software and data stored in a computer are an "idea," but this might be permitted perhaps as an analogy. Software is information and never belongs to the category of material. Software can only have meaning when it exists side by side with a physical substance or hardware.

The central theme of this book is to discuss the problem of creating consciousness artificially; so I would like to consider consciousness and the mind materialistically. I discovered, however, that conventional materialism is insufficient to explain consciousness. I believe a new materialism should be conceived. This will be discussed later in this book.

The next question is, which should be used to explain consciousness and the mind: functionalism or behaviorism?

Functionalism is "a position in epistemology and methodology not to treat things as real matters statically but to consider their functions dynamically, as having correlations, and as a process" (the *Kohjien*, "Wide garden of words," a single-volume Japanese dictionary). Functionalism is, thus, a theory to consider consciousness and the mind as a set of a variety of mental functions.

Behaviorism is "a thought aiming at becoming an objective science of psychology. The target of psychology is limited to learning the lawful relationships between stimuli and externally observable reactions (behaviors), while eliminating concepts related to consciousness" (*Kohjien*; partly modified). According to this definition, "concepts related to consciousness are eliminated."

This book is intended to create artificial consciousness, and therefore I cannot accept behaviorism that specifically eliminates consciousness.

Behaviorism was the product of an age when scientism exerted an overwhelming influence upon psychology to completely remove subjective matters from psychology in an attempt to conduct purely objective and scientific study.

According to Husserlian phenomenology, everything starts from the subject. It is, thus, forever impossible for behaviorism to unravel the mystery of consciousness and the mind because it excludes the study of subjective matters.

In my opinion, the subjective approach of studying consciousness and the mind adopted by phenomenology will gradually shift to an objective approach. This gradual shift will probably take place when, for example, phenomenological hypotheses are scientifically proved as a result of advancements in physiological knowledge about the brain. It is also possible, on the contrary, that multiple scientific discoveries may stimulate the creation of a new phenomenological model for unifying them.

To start a study of consciousness and the mind from a phenomenological perspective, we must first establish subjective hypotheses about consciousness and the mind. This would require a functionalistic approach of some form or another. This is because it is easier to study consciousness and the mind by assuming that consciousness and the mind comprise a set of various functions at least in the initial stage of study.

If we side with the functionalists, we would face a problem in that various functions are individualized and reduced to their ultimate details. For example, the cerebral function will eventually be reduced down to the input and output state of stimuli at every brain cell. Accordingly, instead of simply relying on functionalism, we need to prepare a scheme for unifying the various functions.

The next discussion is about the relationships between the brain, on the one hand, and consciousness and the mind, on the other.

In Chapter 3, "Story of the Human Brain," I mentioned that human consciousness is not lost even when the right brain is damaged. From this fact, brain scientists believe that consciousness is closely related to the left brain.

When one engages in some action intentionally and attentively, the frontal lobe of the brain exhibits vigorous activity. If will and attention area function of consciousness, this would seem to indicate that consciousness is closely related to the frontal lobe. If the important function of the mind is the generation of feelings, the mind may be closely related to the limbic system of the brain.

Spiritual experiences, religious experiences, and subconscious activities are said to be related to the right brain. If subconscious is considered to be consciousness of which the person is not aware, consciousness would be closely related to the right brain (Sperry, 1974).

We cannot ignore the sensory sensors of humans because stimuli input from the environment via these sensors affect consciousness and the mind. It is also obvious that the movements of the muscular system are associated with consciousness, even though they might be unconscious movements. For these reasons, the author believes that consciousness and the mind are closely related to not only the brain but also the body.

The next question is whether the brain has representation.

As discussed above, representation refers to the images of external objects that emerge in consciousness based on perception. There are perceptual representations (the object exists in reality), memory representations (reproduced from memory) and imaginary representations (creating a mental picture). Representation is sensuous and concrete, which makes it "different from concepts and ideas" (*Kohjien*). I will explain this briefly. If you see an apple in front of you, the image of the apple is formed on the retinas of your eyes. The image stimulates the photoreceptors of the retinas, and the relevant signals are transmitted to the brain, first reaching an area called V1. The signals are then distributed to various different areas in the brain responsible for processing visual information. On completing the final signal exchange at the frontal lobe in the front of the brain, you will now "cognize" the existence of an apple in front of you. The details of this cognitive process are unknown. It is assumed, however, that you "cognize the apple" when a perceptual representation of the apple matches the memory representation of an apple in your brain. It is also assumed that when the memory representation of an apple is activated, you visualize or imagine an apple. It is assumed that the representation of an apple exists in the human brain in some form or another. Where does this representation exist?

Functionalists admit the existence of representation, whereas behaviorists do not. Behaviorism excludes subjective matters such as consciousness and the mind and therefore does not discuss representation as a means of explaining consciousness and the mind. It also denies the idea of a "central control tower" of any kind that controls consciousness and the mind to work properly as an intermediary of representation.

Denial of the central control tower originally comes from negation of the "homunculus." It was once believed that a very small human being called a homunculus lived inside the brain of a human and controlled the person's consciousness and mind, and that the spiritual activities of humans were governed by commands given by the homunculus. If one agrees with the existence of the homunculus, it is then necessary to discuss the working of the brain of the homunculus, making for a circuitous discussion that goes around endlessly without reaching a conclusion. How to stop this regression is one of the serious problems related to consciousness.

Descartes explained that the pineal body at the center of the brain played the role of a central control tower, but this idea is denied by modern brain physiology. For these reasons, behaviorists seem not to agree to the idea of a central control tower.

Personally, I believe that although the idea of a central control tower can reasonably be denied, we should admit the existence of asynchronous parallel and distributed systems. According to physiological knowledge about the human brain, it is certain that information is processed by nerve networks and the relevant processing is done by asynchronous parallel and distributed systems.

Connectionism is a theory that attempts to create such asynchronous parallel and distributed systems using artificial nerve networks called neural networks. At present, however, because highly functional computers are available, we do not use asynchronous parallel and distributed systems but simulate similar functions using ordinary synchronous computers. The term "simulate" here means mimicking.

4.8 Mechanical Systems without Representation

The mechanical system without representation is generally known as a speed governor.

Figure 4.9 Heron's steam engine.

The role of this machine is to maintain the constant rotating speed of a flywheel driven by a steam engine. A steam engine is a mechanical system that makes use of the force of steam input into a cylinder to push down a plug-shaped component called a piston. The rod-like component attached to the piston pushes down the flywheel to convert the vertical motion of the piston into rotary motion.

Heron, ancient Greek mathematician and engineer, first invented a steam engine in the form of a rotary device (Fig. 4.9). His steam engine could not keep the rotating speed constant.

James Watt (1736–1819), a Scottish mechanical engineer, invented this speed governor in 1774. To keep the rotating speed constant, the steam pressure should be basically constant. Actually, however, the rotating force is used to do some work, and therefore when the workload varies, the rotating force also varies even when the steam pressure remains constant. For example, when a steam locomotive running across flatlands starts climbing a slope, it will slow down if nothing is done. This is called load fluctuation in physics. What should be done to maintain a constant rotating speed? The answer is to control the steam pressure relative to the load fluctuation. Watt's speed governor exactly satisfied such requirements for pressure control.

In 1774, steam engines were a practical mechanical system for delivering power safely and stably and contributed to the Industrial Revolution. Since then, a large volume of materials have been easily transported using steam locomotives and steamships to support the explosive development of industry.

Let us go back to the story of representation.

A researcher says that representation does not exist in Watt's speed governor. Is it true that his speed governor does not have anything that corresponds to "images of external objects that emerge on the consciousness" as representations? One point needs investigation. The metallic balls, the central components of Watt's speed governor (Fig. 4.10), spin outward from the central rotating shaft by centrifugal force as the rotating speed of the steam engine increases. At this time, the angle between the metallic arms holding the balls increases. When the angle of the balls increases, the steam valve is gradually closed to decrease the rotating speed. Conversely, when the angle decreases, the steam valve is opened to increase the rotating speed. The angle between the ball arms represents the internal condition of the steam engine, and we may call it a representation.

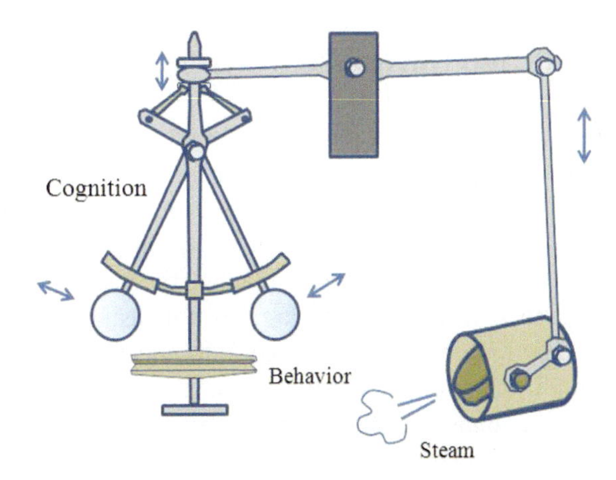

Figure 4.10 Watt's speed governor.

The researcher referred to in the previous paragraph denies this thinking. He denies it because, he says, what is considered a representation is not actually used in the steam engine. He argues that the angle is not used as a numerical value.

The author admits that the researcher is logically right. Indeed, the story of the speed governor seems to suggest the existence of "systems without representation," as they are called in behaviorism.

Since what determines the performance of this autonomous system is the amount of return and the time lag, it suffices in essence if the autonomous system works as expected irrespective of whether the existence of representation is agreed with or not.

In fact, we have today digital systems that use representations that function in the same way as the speed governor.

I think this discussion on the speed governor will eventually lead to the question of whether analog or digital control should be used.

Is this the reason why this is an essential discussion?

Speed governors incorporating a computer that is an aggregation of representations are already being used in real vehicles to control their engines.

The author therefore believes that, in essence, representations may well be used if and when they are required in the study of consciousness and the mind.

4.9 Qualia and Mirror Neurons

Qualia has been talked about much lately. In short, it refers to the feel of a material. David J. Chalmers said that qualia were the hardest problems in consciousness (Chalmers, 1996). He said so because behaviorists agree on the existence of qualia, but it is impossible for them to create, in a behavioristic manner, the qualia that appear in consciousness. It is far easier for functionalists to explain, as they need only to declare that a function to generate qualia exists in the brain. Of course, there still remains a problem in that even functionalism is unable to answer the question of how qualia are created.

The discovery of mirror neurons is described here. Mirror neurons have a special function and were discovered in 1996 by

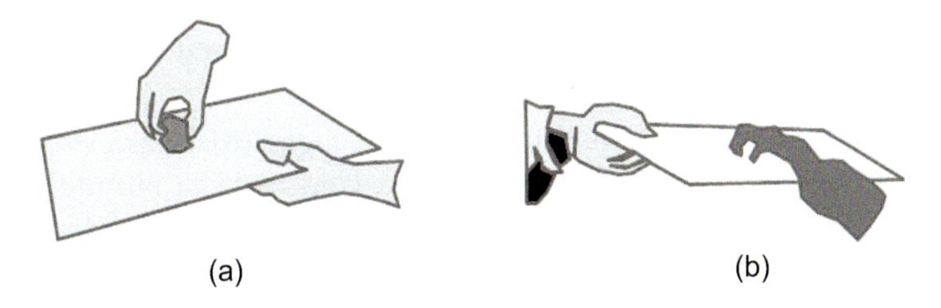

Figure 4.11 Mirror neuron experiments. Cited from the research of Vittorio Gallese (1996).

G. Rizzolatti, an Italian brain science researcher (Rizzolatti, 1996). Mirror neurons were accidentally discovered when a researcher was having his lunch during an experiment and happened to notice that the nerve cells of the monkey in the experiment were activated vigorously. These particular nerve cells were also activated when the monkey was eating its own food. Surprisingly, the brain cells in question reacted in the same way as if the monkey itself was eating by just seeing others eating (Fig. 4.11). Based on this finding, the brain cell was named mirror neuron because it acts like a mirror.

Kenichiro Mogi describes how he was shocked on hearing the news of the discovery of mirror neurons. "Mirror neurons were a leap over forward-thinking and system theory-oriented researchers, and surprised many people by achieving the highest-level unification of connecting motor information on the 'behavior of the self' to sensory information on the 'behavior of the other.'" He continues, "Mirror neurons connect the actions of both the self and the other as if they were reflected in a mirror. As a result, for example, 'The other is doing this action, so if I were doing the same action, I would feel like such and such, and therefore, the other must also be feeling the same way.' The discovery seems to suggest that mirror neurons work as a nerve module that supports the ability to suppose the mental condition of others (the theory of the mind)" (Mogi, 2003). Mogi assumes that the mirror neurons in monkeys are closely associated with their behavior of pretending to be like humans.

The next topic is affordance theory and autopoiesis theory, which are currently attracting general attention.

4.10 Affordance and Autopoiesis Theory

Affordance is explained in detail by Toru Nishigaki (Nishigaki, 1999).

American experimental psychologist James Gibson introduced the theory of affordance in the 1960s. Affordance is an attempt to understand the problem of human consciousness and the mind from the viewpoint of interactions between the human and the environment. According to affordance theory, an object in the environment acts on ("affords") the human (the subject), thereby allowing the human to understand the object. For example, when an individual looks at an object that is a baseball, the individual is given the meaning from the baseball that "this is something for playing with that is thrown, caught, and hit." When a dog sees the baseball, the dog is given the meaning from the baseball that "this is something for playing with and to be chased after and bitten."

The affordance theory asserts that the meaning of an object in an environment is defined by the interaction between the subject and the environment. According to affordance theory, the meaning of an object is not subjective information arbitrarily interpreted by the subject based on stimuli received from the environment, but the meaning is instead information existing in the environment itself, which is objective information described by the environment and is simply transmitted to the subject. On the basis of this thinking, the affordance theory intends to find a solution to the problem of subject and object that is as objective a solution as possible.

The autopoiesis theory closely resembles the affordance theory. It was introduced by Francisco Varela (1946–2001).

The difference between affordance and autopoiesis is that with the former the meaning of an object is the objective information given by the environment, whereas with the latter that objective information constructs, or invents, the meaning of the object using experiential knowledge already learned within the subject.

4.11 Embodied Cognitive Science and Symbol Grounding Problem

This section introduces embodied cognitive science, which has been attracting the attention of many robot researchers today.

Embodied cognitive science is a belief that intelligence can only be understood in combination with physical systems. This concept captured the hearts of many researchers because it specifically proclaims that intelligence is in need of an embodied mechanical system. Embodied cognitive science represented a large flow of research covering the major achievements of robot researchers in the 1980s, which was later summarized by Rolf Pfeifer, Federal Institute of Technology Zurich, Switzerland, and other researchers (Pfeifer and Scheier, 1999).

Pfeifer and other researchers considered that the study of artificial intelligence and robots had been critically blocked by two major obstacles, and he tried to achieve a breakthrough using a new keyword: embodiment. These obstacles are the frame problem and symbol grounding problem.

"Frame"

refers to a symbolically represented knowledge database, and the term was originally proposed by Marvin Minsky. He declared that human intelligence could be realized on a computer using symbolic inference within two or three years. Certainly, artificial intelligence using symbolic inference contributed to the development of the **MYCIN** medical **expert system** designed to examine and diagnose infectious diseases and a robot simulator called SHRDLU (previously mentioned), which performed intellectual tasks as instructed by humans although in the world of building blocks.

In 1997, the IBM research group developed Deep Blue, a chess-playing computer. Readers may remember that the machine played chess with world champion Garry Kasparov and won a five-game match by two wins, one loss, and two draws. Deep Blue symbolized and processed the chessboard and the positions of the chess pieces and anticipated the next moves of its opponent by symbol-based reasoning.

There is no doubt that symbolic inference is one of the basic techniques for achieving artificial intelligence, but the frame

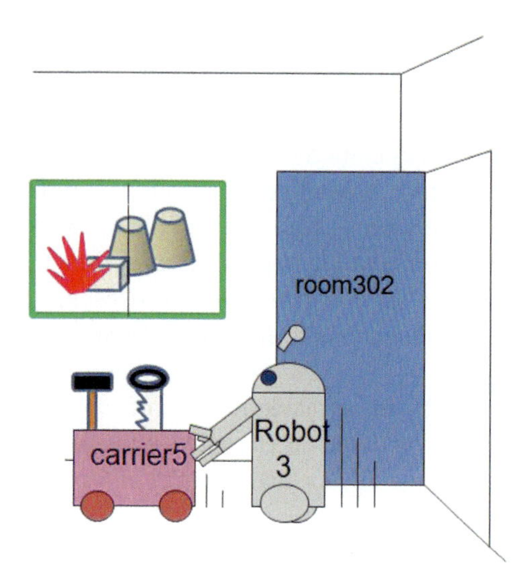

Figure 4.12 Frame problem.

problem prevented the results for use in the future possibilities of artificial intelligence.

John McCarthy is said to be the first person to point out the frame problem (1969). The essence of the frame problem is this: It is an illusion to believe that the knowledge database describes everything, as it is obvious that knowledge databases cannot describe the entire mass of knowledge. A nuclear reactor maintenance robot, for example, operates on a knowledge database that describes the structure and functions of the nuclear reactor. Considering an extreme case, let us assume the reactor were damaged by an earthquake or other accident; then the robot would not have been able to use the knowledge database and therefore would not move and do anything (Fig. 4.12). Generally, there is another problem in that working environments are changing continually and this is not a simple problem that can be solved by just modifying parts of the knowledge database and updating it for the robot to cope with changes in its environment. This is the frame problem: the robot is required to respond to the situation using "common sense," but this is unknown to robots.

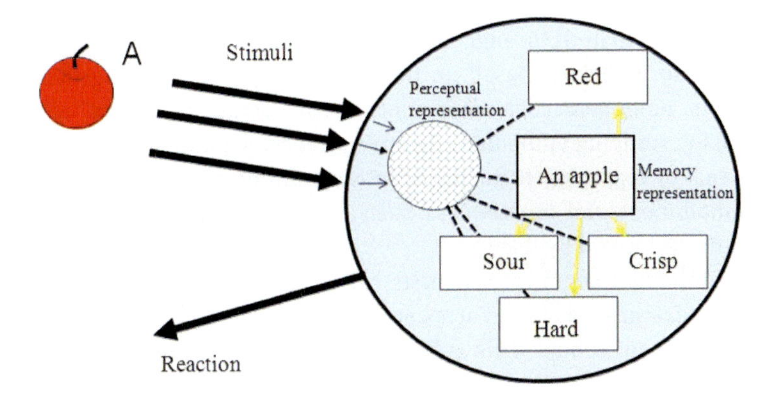

Figure 4.13 Symbol grounding problem.

The other problem is the symbol grounding problem. It is said that Stevan Harnad first mentioned the symbol grounding problem. The essence of the symbol grounding problem is how to define the relationships between symbols used for processing intelligence and real things in the environment (or "mapping"). Mapping does not constitute a problem when humans stand in between the "symbols in the robot" and the "environmental stimuli to be mapped by such symbols," and interpret the relationships between them. Mapping between the environmental stimuli and the symbols in the robot is very difficult when the robot processes symbols while moving autonomously and interacting with its environment. For example, it is not yet defined how the environmental stimuli of object A in the real world, on the one hand, and Apple, Red, Sour, and other symbols (as memory representations) stored in the robot, on the other, should be automatically mapped onto one another. This is called the symbol grounding problem (Fig. 4.13). How does embodied cognitive science try to solve this problem?

The embodied cognitive scientists recommend not using, or minimizing the use of, knowledge databases in order not to be bothered with this problem. Alternatively, they recommend giving the feature of "situatedness" to the robot so that the robot can interact with its environment by itself without human or other intervention. Situatedness means that the robot is capable of acquiring information on the current conditions of its environment (or being situated) through interactions with the environment using its sensory sensors based on the embodiment. Considering these points, embodied cognitive scientists use "complete agents" as a tool for studying embodied cognitive science. A complete agent is a virtual or real robot featuring adaptivity, autonomy, self-sufficiency, embodiment, and situatedness. Such a robot is required to stay alive in the real world by itself.

Embodied cognitive scientists call a robot with such complete agent features a "non-Cartesian" robot to differentiate it from conventional robots. This is because conventional robots lack the property of adaptivity, that is, being able to act by adapting to any unknown environment, which one of the features of a complete agent.

Cognitive scientists further argue that a complete agent featuring such properties is capable of "emergence." Emergence is the capability of a robot to adapt to unknown environments unforeseen by the agent designer or developer and behave unexpectedly. These robot behaviors are not included in the internal characteristics of the agent nor input into it as a program.

There is a story that illustrates this idea in an easily understandable way: a parable of an ant (Fig. 4.14).

An ant is walking on a beach, cleverly avoiding rocks and water. Its walking motion might be scientifically analyzed but would require unbelievably complex formulas to describe the optimum path for the ant to take. Researchers in embodied cognitive science doubt that this approach is basically misleading. If this approach were to be taken, we must first focus on the environment of the beach. It would be almost impossible, however, to scientifically formulate this kind of environment. Given this fact, the researchers realized that a new concept was required. They noticed that the walking motion of the ant should be formulated as an interaction with the environment at all times. Thus, the formulation would read in essence, "the ant takes a certain basic action under certain conditions." In this series of processes, "detecting conditions" and "taking basic action" would be repeated continually. In a sense, this is a very beneficial concept in designing an agent.

It is obvious that researchers in embodied cognitive science consider consciousness and the mind as a problem of emergence.

When, for example, a robot (an agent) is designed to avoid red blocks but approach green ones, it is possible to think that the robot hates red but likes green.

Certainly, from a behavioristic perspective, consciousness and the mind of other people "seem to exist," and in this sense behaviorism is the same as the complete agent.

However, I am clearly aware of my own consciousness (I possess self-consciousness), and I would like to emphasize that embodied cognitive scientists have not yet provided a clear explanation for this phenomenon of self-consciousness. Basically, however, I think their techniques — creating actual moving robots, realizing complex actions using a chain of basic actions, considering an action an interaction with the environment, and devising emergence-oriented designs — should be highly valued. I think, however, it is still difficult for them to elucidate the problem of "self-consciousness as being conscious that the self is conscious." I believe that we

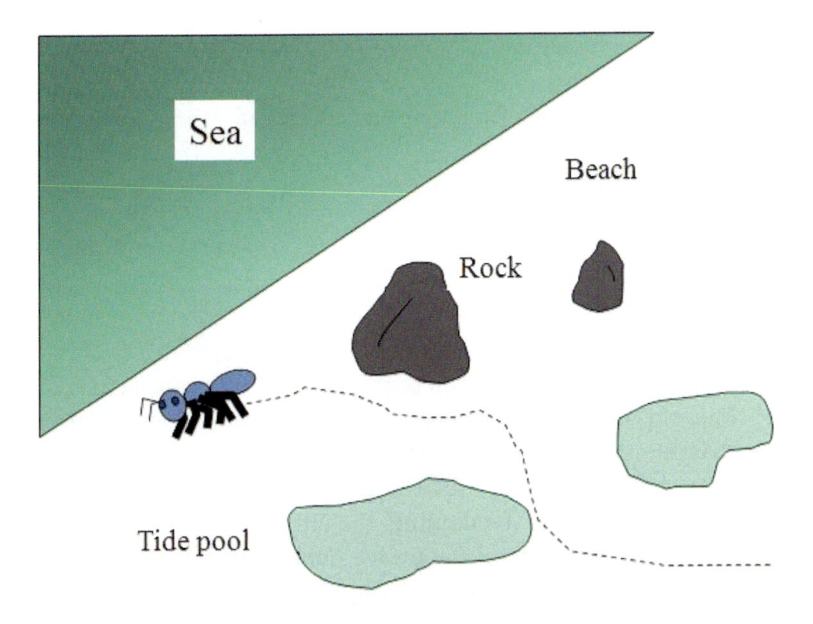

Figure 4.14 Behavior of an ant.

should further proceed with the study of human consciousness and the mind using the phenomenological approaches introduced by Merleau-Ponty.

Of course, they as "non-Cartesian" would accuse me of being a Cartesianist, but I would like to further study the problem of consciousness and the mind by interpreting science more broadly.

Let us now talk about consciousness and about the mind.

4.12 More Knowledge about Consciousness and the Mind

Tadashi Kitamura enumerates 10 properties of consciousness based on the Husserlian phenomenology (Kitamura, 2000). Following are the ten properties:

1. First-person property
2. Orientation
3. Relationships between action and result; the duality of consciousness
4. Expectation
5. Functions of determination and conviction
6. Embodiment
7. Consciousness of others
8. Emotional thought
9. Chaos
10. Emotion

These properties will be described in detail later in this book.

Other important thinking on consciousness is introduced now.

Naoyuki Osaka, Kyoto University, holds that consciousness has a three-layer structure (Osaka, 1996). His theory, which has already become established, is described as follows:

The lowest level is **Awakening**, the intermediate level **Awareness** and the highest level **Recursive Consciousness**. Awakening indicates that one is awake. A living creature apparently loses consciousness when sleeping and engages in biological activities when it is awake. Awakening refers to the state of being awake

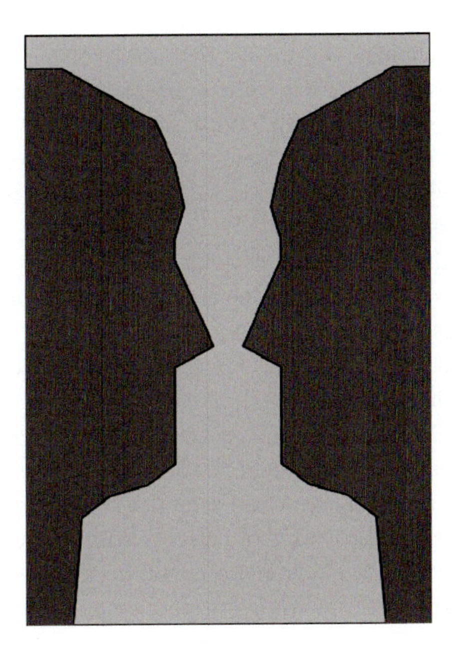

Figure 4.15 Rubin's vase.

and performing biological activities. Awareness refers to the state of being awake and with consciousness directed at a certain object. The state of consciousness being intentionally directed at a certain object is called "orientation of consciousness." An example of this is a student who listens to a teacher speaking while taking notes in a notebook.

I would like the reader to learn about **reversible and ambiguous figures**, which are famous in Gestalt psychology to familiarize yourself with the nature of consciousness.

When viewing these examples, your consciousness is at the intermediate level, and you will clearly understand what is meant by "orientation of consciousness."

Rubin's vase is a famous example of a reversible figure (Fig. 4.15). Two meaningful images exist in the figure and appear alternately: two human faces facing each other and a vase.

The Kanizsa triangle is a typical example of an ambiguous figure, an optical illusion.

Gestalt psychology is a theory that attempts to understand the human mind as a holistic structure, or *Gestalt* in German, rather than as a collection of individual functioning units.

Recursive consciousness refers to the state where consciousness performs activities directed toward consciousness of the self. This is called self-consciousness.

Humans, thus, can be conscious of what they are currently conscious of. Humans are inherently bestowed with a kind of metaconsciousness or superconscious.

Osaka says that metaconsciousness is the function that makes humans truly human-like.

From the perspective of functionalism, a supersystem that overviews a system is foreseen to exist at a higher level of the system.

During his experiments, brain scientist Wilder Penfield directly touched the cerebral cortexes of patients with an electrode during surgery to learn what phenomena would occur when such stimuli were given. The experiments were performed for the purpose of minimizing loss of brain functions due to unintended excision of part of the brain in brain tumor resections. It is said that the patient does not feel pain because there are no sensory cells in the brain.

Penfield asked the patient in the experiment, "Begin counting out loud from the number 1." When he stimulated a certain part of the brain while the patient was counting, suddenly the patient could not say the numbers any more. Another patient was said to have told the doctor that he vividly remembered events from his childhood days when a certain cortex was stimulated.

Now we will revert to the topic of self-consciousness. When a patient told Penfield, "I am now being treated by you," Penfield tried to identify the location but he failed to do so.

As mentioned before, Descartes declared that only humans possessed consciousness. This would mean that dogs and cats have no consciousness functions.

Of the three levels of consciousness advocated by Osaka, I think awareness, the intermediate level, may exist in dogs and cats, but it is doubtful if these animals have self-consciousness. Cats chase rats by orientation, and dogs chase and catch a ball tossed to it by humans. These facts clearly show the existence of awareness in these animals. Dogs and cats, however, do not seem to respond to their mirror image very well. There is no evidence to show that they recognize their images reflected in a mirror. It is, therefore, assumed that self-awareness is not a function that exists in dogs and cats.

An inspection method is available for determining the existence of a high-level cognitive function. It is a method to check if the self-image reflected in a mirror is cognized or not, and it is known as the mirror test as proposed by Gallup.

Will anthropoid apes that resemble humans pass the mirror test?

There is evidence to believe that chimpanzees possess self-awareness. A chimpanzee was given a mirror and an opportunity to play with it for a sufficient length of time. Later, the chimpanzee was surreptitiously marked with signs on its forehead with non-irritant dyes. When once again given a mirror, the chimpanzee touched the marks on its forehead while looking into the mirror. This indicates that the chimpanzee cognized that the image reflected in the mirror was an image of itself. The function to cognize one's own mirror image is self-awareness; so we can determine that chimpanzees possess self-awareness.

Besides chimpanzees, there have been reports that dolphins, Indian elephants, and magpies also pass the mirror test.

There is an interesting story about consciousness. It is an observation about **free will**. What makes humans truly human is the free will that humans possess. Everyone believes that he or she can freely decide matters in any place and at any time. An unbelievable fact was revealed in experiments.

The human subject in the experiment was instructed to bend his fingers at any time he wished. The observer was tracing the activity of the subject's brain with an electroencephalograph (EEG). The subject bent his fingers at a certain time T. The brain started its activities, however, about 1 second earlier than time T. This is interpreted as follows: Although the subject thought he made the decision to bend his fingers and did so at time T, a preliminary decision had already been made about 1 second earlier in the brain. This famous experiment was conducted by H.H. Kornhuber (1976) in Germany. Later, researchers, including Benjamin Libet, Professor Emeritus at the University of California, San Francisco (UCSF), performed additional detailed experiments to verify Kornhuber's findings in the 1980s.

These experiments seem to lead to the conclusion that humans do not have free will because the brain started to act before the human noticed that he actually bent his fingers, and that the human is never able to notice this preparatory work of the brain. Is it true that humans really do not have free will?

This problem will be discussed in detail later in this book.

4.13 Summary and Observations

I would like to point out, in the first place, that the study of consciousness and the mind has a long history extending from olden times to the present day.

Nevertheless, the existence of consciousness is still described as a subjective sensation and as such we cannot advance further.

I think it is generally accepted as an objective fact that consciousness exists in the left hemisphere of the human brain. This is supported by anatomical findings in that if the right hemisphere of the brain is surgically excised, the consciousness function of the patient is not critically affected, but if the left brain is removed the consciousness function may be totally lost depending on the excised areas.

There is no objective analysis of the effects of excision on the consciousness of the subject. Only the subjective judgment of the experimenter is relied on to evaluate the results of the experiments.

I do not agree, however, with the argument of behaviorists that humans possess no consciousness. This is because I can feel that I have consciousness. This feeling is, without a doubt, a subjective sensation. I should note that this subjective feeling is not necessarily irrelevant to physical processes.

In other words, if the feeling of being conscious passes through a physical process, we can scrutinize the process from the perspective of physical science.

It is also possible that the function of human consciousness is unique to each individual. If these individually unique functions of consciousness are based on physical processes, we can analyze them scientifically. By comparing these individually unique physical processes, we may find some typical and common physical features among them.

Since phenomenologists seem to have arrived at a certain common recognition about consciousness and the mind, we should not doubt that someday we will be able to find an objective physical process that describes the occurrence and action of consciousness and the mind.

What is certain at this time is that the existence of consciousness and the mind is known only to oneself. With regard to others, we only feel that they must also have their own consciousness and mind like me.

Descartes insisted on the consciousness of the existence of the self when he declared, "I think, therefore I am." For us today, however, it is totally unknown if Descartes possessed consciousness, and what kind of conscious activities led him to say that he did.

Nevertheless, it is true that his discourses significantly affect our conscious activities.

Chapter 5

Professor Valentino Braitenberg's Vehicles

Valentino Braitenberg was director of the Max Planck Institute for Biological Cybernetics, Germany, where, using small robots equipped with sensors and drive motors, he studied robots with a mind from the standpoint of functionalism.

In brief, functionalism is a method to try to understand something by analyzing the function inherent in the thing to the extent possible by human wisdom. What is important in this definition is that the understanding is limited to the range possible by human wisdom.

Braitenberg was one of the pioneering researchers of robotics, and he designed robots by transmitting information directly from sensors to drive motors. He invented many vehicles, each with successive improvements, and showed that a wide range of robots from toy-level to a self-conscious robot could be made using very simple components (Braitenberg, 1987). He was a forerunner in research on evolutionary robotics and embodied cognitive science, which are currently very actively studied in the field of robotics. Many of the methodologies adopted by these researchers had already been suggested by Braitenberg. This fact is indeed marvelous.

Self-Aware Robots: On the Path to Machine Consciousness
Junichi Takeno

ISBN 978-981-4877-90-9 (Hardcover), 978-1-003-26181-0 (eBook)
www.jennystanford.com

For instance, the idea of a modern evolutionary robot is the same as what Braitenberg called a Darwinian robot. A robot that expresses the feelings of "like and hate" using the connections between the sensors of artificial neural networks and drive motors, a phenomenon that would later be called emergence, was also conceived by Braitenberg. I believe that his research should be further evaluated. This underestimation seems to be due to the argument that "he sides with functionalism and admits the existence of representation," because many scientists favor the behavioristic stance. His vehicle robots are introduced now.

5.1 Braitenberg's Vehicles 1 through 4

Braitenberg's vehicle 1 has the simplest structure. It consists of a sensor and a drive motor. The sensor and the drive motor are connected by an artificial nerve. The motor drives a wheel located at the rear of the robot (Fig. 5.1a). The robot is just like a small cell swimming with a flagellum. Vehicle 1 does not have a rudder. When it runs, it always runs straight ahead. A robot is almost always asymmetrical, so that the robot may make a slight turn. The sensor can be any sensor. Let us assume a temperature sensor. The sensor emits zero signals at the absolute zero point. At other temperatures, the sensor issues stronger signals as the temperature rises.

The signal is transmitted to the drive motor via the artificial nerve. The motor's speed (rotation) increases or decreases with the strength of the signal. Put this robot in a space, and it detects the ambient temperature and starts running at a speed corresponding to the detected temperature. The robot slows down in areas with lower temperatures.

To a casual onlooker, this robot looks like a simple life form such as an amoeba that becomes more active in places of high temperature.

When the reaction property of the artificial nerve is designed as shown in Figure 5.1c, the artificial life will be activated at a certain temperature.

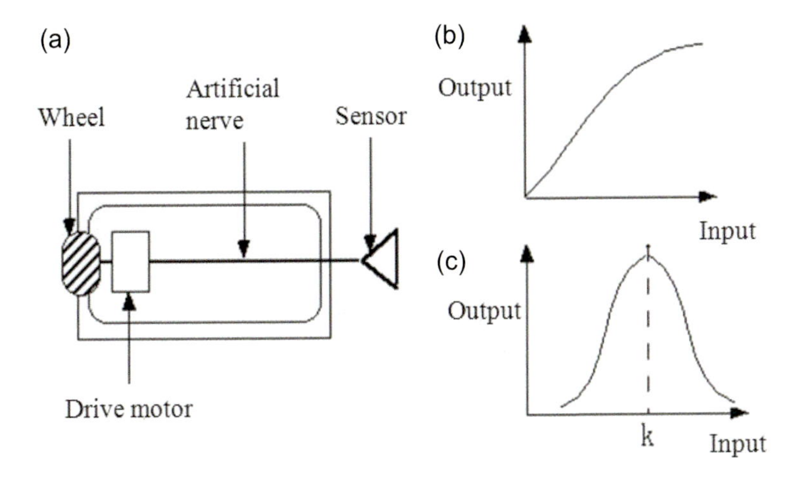

Figure 5.1 Braitenberg's Vehicle 1.

Braitenberg's vehicle 2 has two sensors that are connected to two respective motors. There are two types of artificial nerves: parallel (Fig. 5.2(1)) and crossed (Fig. 5.2(3)). The plus and minus signs (+ and −) indicate the property of the artificial nerve. For a plus (+) sign, the output of the nerve increases with increasing output signals of the sensor (facilitation). For a minus (–) sign, the output of the nerve decreases with increasing output signals (inhibition). Only the minus sign is shown here, but there should be no confusion caused with the plus sign omitted.

Let us assume that this robot is equipped with light sensors. The robot in the figure, with a light just in front of it, will go straight toward the light with its speed increasing until it crashes into the bulb and destroys it (Fig. 5.2(2)a).

If the light is not just in front of the robot, the sensor closer to the light is activated more than the other sensor, and the relevant motor increases its speed. As a result, the robot follows a path to avoid the light (Fig. 5.2(2)b). If the robot sees the light just in front of it again, the robot will rush toward the light.

The robot shown in Fig. 5.2(2) is a fearful and aggressive type since it avoids light but suddenly attacks it depending on the position of the light.

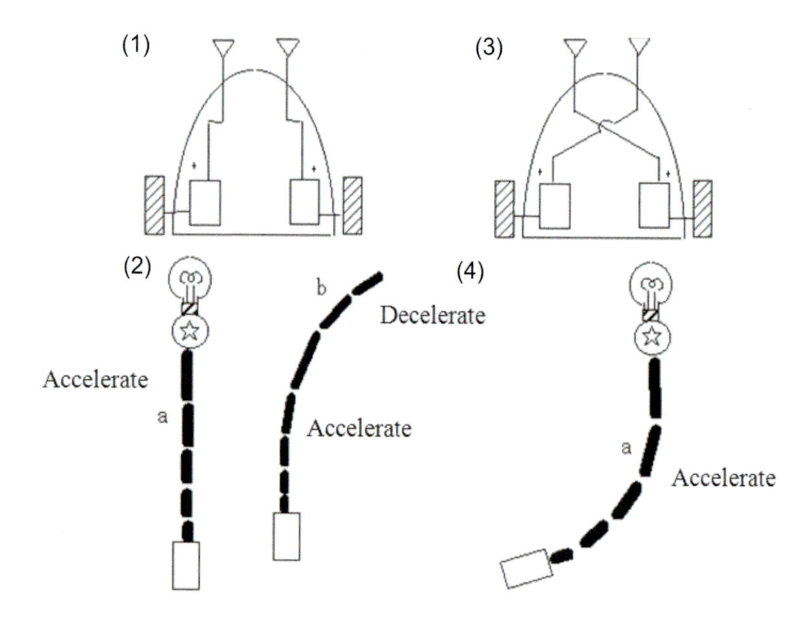

Figure 5.2 Braitenberg's Vehicle 2.

The robot shown in Fig. 5.2(3) is identical to the robot shown in Fig. 5.2(1) except that the nerves are crossed. This robot also goes straight ahead and collides with the light if the light is just in front of it. If the light is on the left side of the robot, the drive motor on the right side increases its speed, an opposite action because the nerves are crossed, and the robot turns toward the light.

Like a guided missile, the robot rushes at the light as if it were being sucked into it (Figure 5.2(4)). This is an aggressive-type robot.

The activation of the nerves is inhibited in vehicle 3.

In Fig. 5.3(1), the right and left nerves are in parallel, and their activation is inhibited (note the minus sign). This robot, placed at an angle to the light, slows down as it draws closer to the light and finally stops because the sensor on the side facing the light emits a higher signal than the other sensor and speed of the motor on the side facing the light slows down (remember the nerves are parallel and inhibited). The motor on the side away from the light runs faster than the motor on the side facing the light, and thus the robot turns itself toward the light (Fig. 5.3(2)). Both motors slow down as the

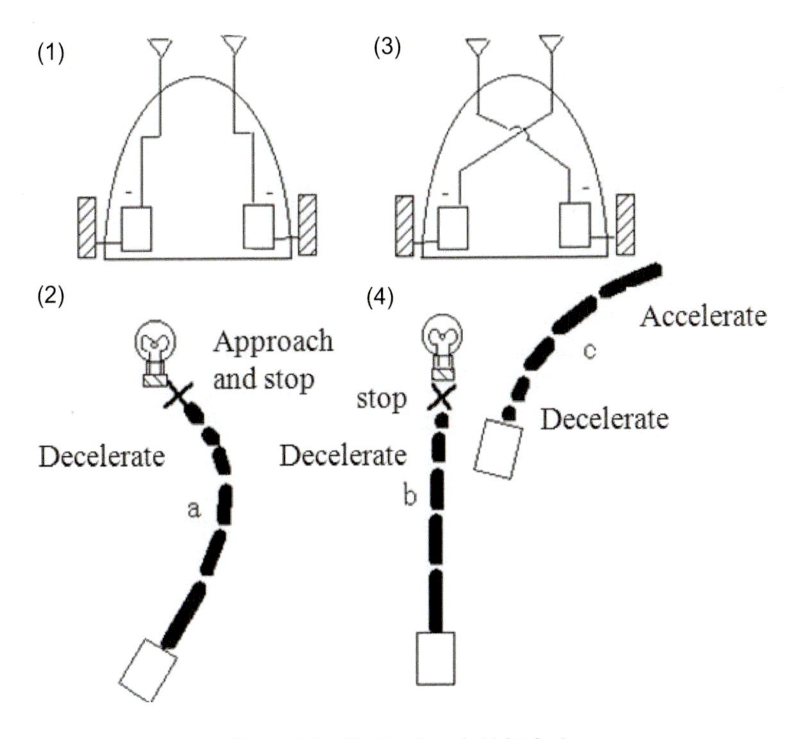

Figure 5.3 Braitenberg's Vehicle 3.

robot approaches the light, until the robot stops before reaching the light (Fig. 5.3(2)a).

The robot shown in Fig. 5.3(3) has crossed nerves and the signals are inhibited. If the light is just in front of the robot, the robot goes straight toward the light while its speed continues decreasing, and it stops before reaching the light (Fig. 5.3(4)b). Otherwise, the robot turns away from the light while slowing down and gains speed as it moves away from the light (Fig. 5.3(4)c).

The robots that slow down and stop near the light may be said to be "affectionate" toward the light, whereas the robot shown in Fig. 5.3(3) is "fickle" because in most cases, it visits the light here and there but quickly runs away.

Braitenberg's vehicle 4 moves around the light in various patterns, reflecting the reaction properties of the artificial nerves Braitenberg devised.

Vehicle 4 seems to be devised with its own value and liking when we see how it approaches, stays, and leaves the light depending on the built-in reaction properties.

Braitenberg assembled logic functions in the nerves of vehicle 5.

By combining the artificial nerves, logical sum (or), logical multiplication (and), and negation (not) can be achieved in a robot. This will be discussed in detail in Chapter 7.

An adder is constructed when these three logical computations are available. Once addition is possible, subtraction, multiplication, and division with the remainder can be calculated. For example, multiplication is done by repeating addition, subtraction by adding complements, and division with remainder by repeating subtraction.

In short, a computing machine can be constructed by combining artificial nerves. This means that vehicle 5 can compute anything that can be calculated on a computing machine. Memorization is also possible. For example, the robot will count and remember the number of right turns it made during a journey.

Next, Braitenberg attempted to create a robot with a brain.

5.2 Robot with a Brain

Braitenberg proposed what he called "the law of uphill analysis and downhill invention" (Braitenberg, 1987). This means that analysis is difficult like climbing a hill, but invention is easy like going down a hill. Braitenberg does not mean invention is easy, but says that invention is easier than analysis.

Specifically, it is easy to say something about the behavior of a robot by looking at it, but it is difficult to explore the internal operation of the robot that caused it to perform a certain behavior. It is very much analogous to the complexity of analyzing an ant's behavior cited before. This also applies to humans. We can easily see what a person has done, but it is difficult to analyze the person's principle of behavior.

Braitenberg did not mean to give up exploring the mystery of human behavior. No scientist would say so. On the contrary, he urges us to build robots to assist us in researching the facts underlying human behavior. He suggests that we try to create the functions of human brains through an engineering approach and elucidate the mystery of the brain using the knowledge derived from such efforts. He studied, for example, the problem of will by analyzing the behavior of robots.

Consider vehicle 2, in which the amount of light shed on the sensors is very small. A robot, being a physical mechanical system, always has problems peculiar to physics such as friction and force of inertia. It is possible that even if the sensor detects light, the wheels might not turn or the robot might remain standing still due to friction. But at any moment, the robot might start moving with slight change in the environment or slight variations of unstable elements of the mechanical system related to friction. A researcher who happens to see the robot move suddenly may think that the robot has made up its mind to move. This is because the robot, which has been standing still, suddenly starts to move by itself. The researcher may conclude that a complex process of the will exists behind this robot's decision making. Actually, however, there was only a slight change in light intensity.

The decision-making process occurring in the human brain may not be as simple as that. Nevertheless, it is probable that just a tiny amount of something affects the decision-making process going on in the brain. No one can deny that these trials would be a clue to reveal the truth.

At this point, philosophers may cite *Critique of Pure Reason*, written by Immanuel Kant (1724–1804) (Fig. 5.4). Kant says in this writing, "Indeed we may have found clues. But they are based upon the human recognition and therefore cannot be anything beyond human recognition." This interprets into, "Humans can recognize only that which they can recognize." This may seem a matter of common sense, but actually a profound meaning is hidden.

In *Critique of Pure Reason*, Kant mentions that the existence of God, the existence of the Soul, and the existence of free will cannot be known by humans. To advocate his agnosticism, Kant argues that "it is possible to prove the existence of God and also possible to prove the nonexistence of God" and holds that there exist matters that humans cannot know.

Figure 5.4 Immanuel Kant.

Agnosticism is the belief that reality behind sensory experiences given to consciousness cannot be recognized demonstratively (*Kohjien*).

The author understands agnosticism but believes that excess reaction is not necessary. It is too extreme to believe that we cannot know anything just because we recognize agnosticism.

Consider clocks, for example.

According to agnosticism, the conclusion is that you cannot build a clock that shows the correct time.

In ancient Greece, Heron improved on water clocks. Galileo Galilei (1564–1642) invented a pendulum clock using the isochronism of a pendulum. The balance mechanism with gears started to tick regularly. Nautical clocks were then invented, which were capable of compensating for errors caused by the rotation of the Earth. This was made possible by the invention of a mechanism in which the balance itself turns (Tourbillon). Modern clocks include quartz clocks that work on the oscillation of a quartz crystal, and then there are atomic clocks. Despite all of these developmental achievements, correct clocks are still to be devised. Nevertheless, mechanical clocks are once again popular today (Fig. 5.5).

This concept is called pragmatism.

Figure 5.5 Revival of mechanical clocks.

Pragmatism is "a way of thinking about things in concrete based on events (*pragma* in Greek); the meaning and truth of a concept is evaluated in terms of how it works and its consequences" (*Kohjien*). William James (1842–1910), an American, was an ardent advocator of pragmatism.

Some readers may claim that pragmatism tends to give rise to the "arrogant overconfidence of scientists" (Friedrich Wilhelm Nietzsche, 1844–1900) and includes the risk of bringing about the collapse of the world at any time.

It should be possible for humankind, I believe, by cautiously avoiding "arrogant overconfidence" and without overestimating the difficulty of "brain problems," to gradually increase its wisdom to such a level where there is no practical problem just as exemplified by the developmental history of clocks.

The author's comments on agnosticism and pragmatism have been introduced here to defend Braitenberg's belief.

5.3 Idea of an Evolutionary Robot

Braitenberg prepared a table, and selected and placed on it some relatively excellent robots from those that he had made. Various components used for the development of the robots were also placed on the table.

The robots moved around on the tabletop at random, with some falling off the table. With the fallen robots left as they were on the floor, new robots were duplicated on the table using the available components. When duplicating the robots, any errors that may have existed in the original robot were also incorporated in the new robot. By repeating this procedure, vehicle 6, one that would never fall off the table, should be created. This vehicle is called a Darwinian robot (Fig. 5.6). This was named after Charles Darwin (1809–1882), who introduced his theory of evolution.

The engineering technique to produce evolutionary robots that simulate the evolution of organisms is described in Chapter 8.

I do not insist that the evolutionary robot was first conceived by Braitenberg, but it should be noted that he explained that a brain can be built in a robot by actually combining the theory of evolution and a mechanical robot.

5.4 Vehicles 7 through 10 with Associative Concept

Braitenberg's vehicle introduced a wire M that has a high initial resistance that decreases when both parts connected at either end of the wire are activated. The resistance returns to the initial value with time. Braitenberg calls these wires "mnemotrix."

Various sensory motor modules of the robot are connected together by the Mnemotrix wires. Hereafter, my description may differ slightly from Braitenberg's writing, but the details will be explained by following Braitenberg's ideas.

Figure 5.7 shows the vehicle 7 robot with logic artificial neural networks. The robot is provided with color sensors "a" and distance sensors "b." Drive motors are indicated as "c." The robot has two sensor drive modules, a-c and b-c. Wires M connect the artificial neuron units "d" located in the middle of the system. These intermediate units integrate the result of cognition of both the color and distance sensors and output the result to the drive motors.

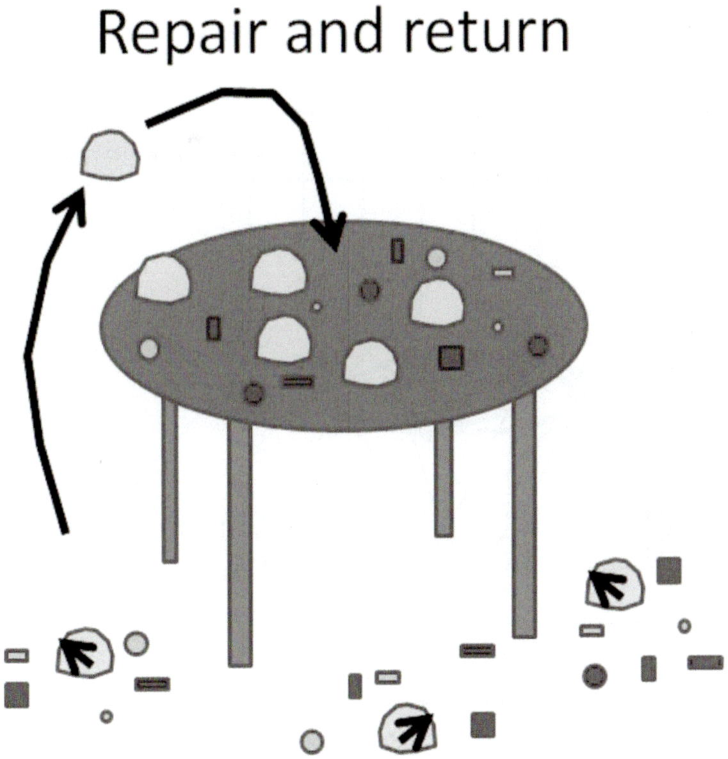

Figure 5.6 Idea of an evolutionary robot.

Assume that this robot has learned to regard a rapidly approaching red object as dangerous. The resistance of the wires M connecting the intermediate unit of the a-d module representing "red" and the intermediate unit of the b-d module representing "rapidly approaching" is slightly lower than the other wires.

As a result, even when only the color information of "red" is given, the robot behaves evasively because the "red" information stimulates the b-d module representing "rapidly approaching" via the wires of slightly lower resistance. Braitenberg interpreted this as the "judgment of 'red' provoked the judgment of 'rapidly approaching' by association."

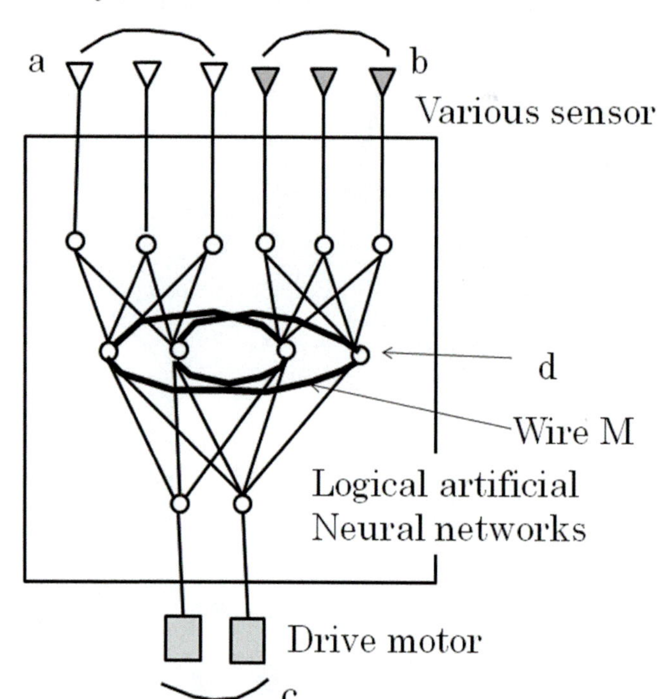

Figure 5.7 Vehicle 7 with associative concepts.

Concepts arising from association are called associative concepts. It is also possible, conversely, that the judgment of "rapidly approaching" invokes the judgment of "red" by association. By the above statement, Braitenberg admits the distinct existence of the concept and the representation of "red" and "rapidly approaching" in the intermediate units, i.e., in the robot. He also explains that wires M created a new concept of a "red and rapidly approaching" object.

According to Braitenberg, if a robot that has learned an association between "green and rapid approaching" behaves evasively by chance for a "blue" moving object (a color that the robot has not yet learned), it is possible for the robot to abstract colors (make a distinction between colored and colorless robots) and the accompanying generalization.

Abstraction is a "mental action to extract and grasp certain aspects and properties of things or representations" (*Kohjien*).

Generalization is the "forming of general concepts or laws by shaking off specifics and saving what is common."

Braitenberg introduced his vehicles 8 and 9 to describe techniques to conceptualize a robot's environment, such as space, objects, motion, and form, within the robot using vision sensors.

The discussion of a robot's learning of its environment is omitted here because in Chapter 7 it is explained that artificial neural networks are capable of such learning and, further, this topic is irrelevant to the problem of the mind.

Braitenberg's vehicle 10 was designed to study the problems encountered when forming a new concept while keeping past associations. He proposes to constantly circulate stimuli for the associations that were memorized in the past.

5.5 Robot with a Sequential Concept

Braitenberg's vehicle 11 features new wires (E) that he calls "Ergotrix." These wires E are used to memorize the order of the occurrences of representations "a" and "b." For example, one may wish to have the robot remember to turn right when "an obstacle pops up in front of the robot immediately after the red lamp lights" (order of occurrence of two events). Wires E connect the intermediate nodes "d" to one another as shown in Fig. 5.8. For any two intermediate units n_1 and n_2 that are connected by wire E, signals are transmitted from n_1 to n_2 when n_1 fires and then n_2 is ignited just a short time later. Importantly, the connections using this wire are different from a simple artificial neural network. The directionality of signal flow is assigned to each of these wires.

The robot shown in Fig. 5.8(1) comprises two sensor drive modules, a-c and b-c. Like the previous robot, sensory sensors "a" are color sensors and sensors "b" distance sensors. The difference from the previous robot is how the intermediate units "d" are connected

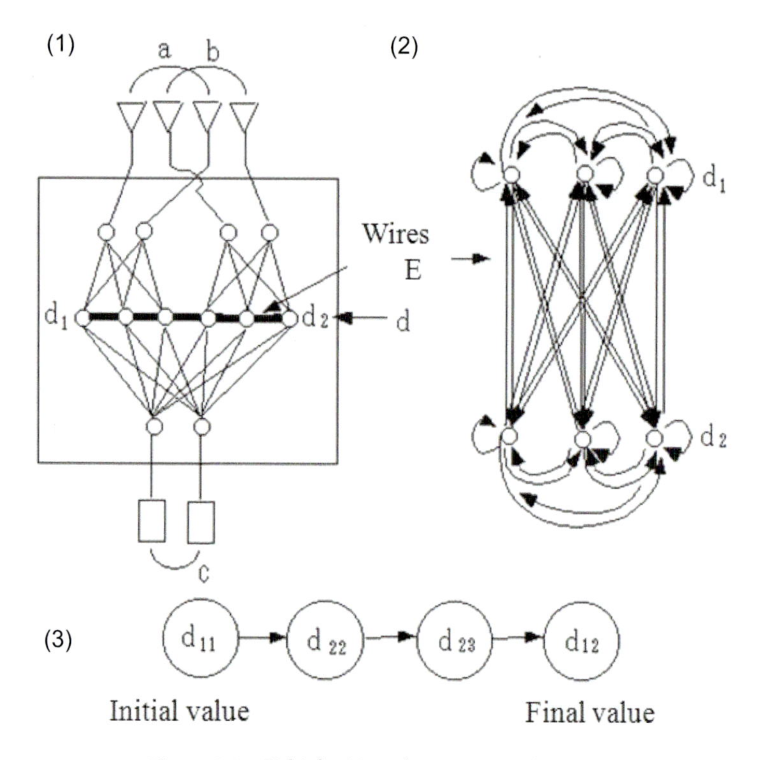

Figure 5.8 Vehicle 11 with a sequential concept.

to one another. New pathways are added to allow recurrence to the intermediate units themselves.

An example of a circuit network with wires E is shown in Fig. 5.8(2). Here, d_1 is the intermediate unit for the color sensor and d_2 for the distance sensor. In the circuit network, a single input circulates through representations d_1 and d_2, just like a series of variations of state occurring in an automaton. The robot is eventually expected to reach a final state as shown in Fig. 5.8(3). In Figure 5.8 (3), the robot memorizes the environmental changes as a temporal change of representations: d_{11}, d_{22}, d_{23}, and d_{12}. Specifically, the robot saw a red lamp, felt a wall on the right, then felt another wall in front, and, lastly, saw a blue lamp.

An automaton is defined as "a general term for mathematical models of various computing mechanisms including computers. A finite number of internal states are considered, connecting inputs and outputs. The internal state varies with inputs, or the output is a function of the internal states" (*Kohjien*).

This robot is capable of memorizing events occurring in the environment and the order of their occurrence. We then add wires M to wires E in Fig. 5.8. This robot shows us that a new concept can be formed by chaining representations in an orderly manner. A concept created by time-series events is called a sequential concept.

5.6 Vehicle 12

Vehicle 11 was improved, at one point, to become vehicle 12. A new type of wire C was added to control the threshold function of all artificial neurons that make up the logic unit of the robot.

Wires C run from the control box of the logic unit comprising the brain of the robot to all neurons in the brain to modify the threshold functions (Fig. 5.9).

"Modify" here means to alter the threshold values of all neurons according to the number of currently firing artificial neurons in the logic unit. If the number of currently firing artificial neurons increased, all threshold values are increased, and as a result, the number of artificial neurons that will fire next time decreases.

The robot's brain accepts input pattern I at its logic unit and outputs pattern O from the logic unit. The output is under the influence of the threshold functions that are changed by the control box. The role of wires C is to highlight the currently represented concept against all others. By elevating the threshold values by controlling all threshold functions using wires C, it is possible to dispel all concepts but the strongest concept g in the logic unit. I personally believe that this function can be said to be a "concentration of attention in consciousness." As a result, concept g remains active with the strongest attentiveness of all concepts both spatially and temporally due to the working of wires M and E.

Concept g then generates a sequential concept g_1 using wire E, and this concept in turn creates several associative concepts g_{1i}. Note that at this time the firing of the artificial neurons increases sharply in the logic unit, and the threshold values are unanimously elevated due to the working of the control box. Thus, concept g_{15}, which is most active in the logic unit, is selected (assuming $i = 5$).

Next, a sequential concept g_2 of concept 15 arises.

We can safely expect that the concept arising at this time is a concept deeply embedded in the memory through past experience.

Braitenberg explains this continuous occurrence of concepts as "trains of thought." He asserts that this scheme is very similar to the human brain. With regard to the occurrence of concepts, Braitenberg says it is impossible to foretell what concept will occur next. This is because we cannot foresee how many firings might occur as a result of the threshold control at the logic unit according to function f. This means that you cannot predict the next occurring concept. The unpredictability of the number of firings is akin to chaos.

Chaos means "in a system where initial conditions determine the succeeding motions uniquely, a slight change in the initial conditions

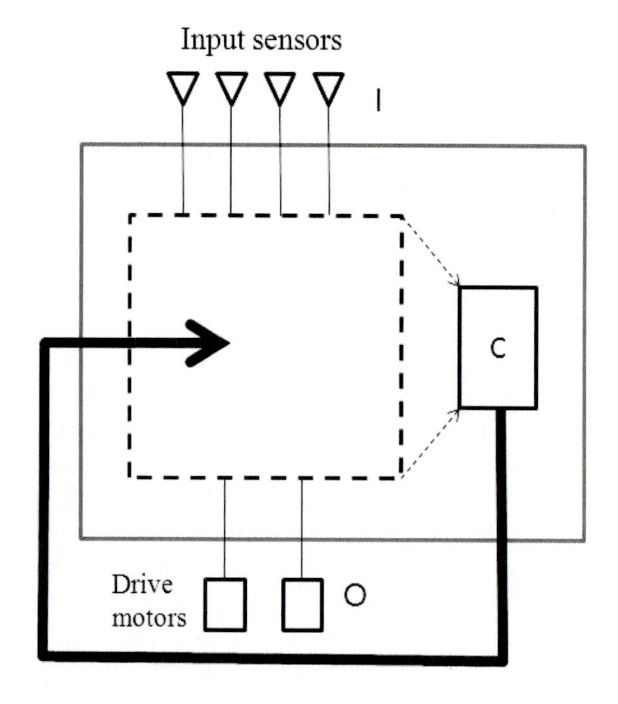

Figure 5.9 Threshold Control Circuit.

results in a significant difference after a long time, practically making it impossible to predict the end result" (*Kohjien*).

By observing the behavior of vehicle 12, Braitenberg concluded that this robot has free will. Behavior is not predictable not only for this robot but also for people. If we recognize free will to mean that no person's instruction can affect another's behavior and that this is one of the dignities of humans, then we must accept that the robot has free will.

5.7 Vehicle 13

As already discussed, Braitenberg's vehicle 12 generates endless concepts, and a train of thoughts appears on the robot. Despite these many concepts that occur, we do not see any voluntary orientation in the thinking of the robot.

Humans are said to think of a number of options in the course of attaining their final target. This is "oriented thought," which is the theme of this section. According to Braitenberg, there are two types of orientation: prediction and judgment of good and evil. To incorporate the prediction function, sequential concepts must be memorized via wires E. Sequential concepts are generated each time the robot behaves. In the prediction process, no synchronization with individual behaviors is required, so that concepts may be stored in the robot at high speed, and independent of the progress of behaviors. The device responsible for this process is called the predictor.

Assume the robot presently has a concept g_3 that is established. The predictor starts working from here, and generates a chain of sequential concepts ($g_3 \triangleright g_{31} \triangleright g_{32} \triangleright \cdots \triangleright g_{3m}$) using wires E. These concepts are what are expected to occur in the future, given the current conditions and past experience. It is unknown, however, whether the robot will really generate g_{31} after g_3 as predicted.

This is because the robot, moving in a variable environment, may develop a concept directly from the changing environment as detected by its sensors, that is, a concept irrelevant to the control given by wires M or E.

If the input received by the sensors is ignored, the robot will perform a behavior that generates the sequential concept g_{31} as predicted, and the sensors will be accordingly changed, also as predicted. This would be followed by a number of associative concepts ($g_4, g_5, \ldots g_s$) via wires M, until the most strongly learned concept g_5, for example, occurs by virtue of the controlled threshold values. If a sequential concept occurs again, a new prediction process starts.

Braitenberg's strategy for cases in which the prediction fails to match the changing environment (as detected by the sensors) is to "believe in the sensors when in doubt," and he turns off the switch of the predictor.

Regarding the judgment of good and evil, humans need not teach robots, rather, according to Braitenberg, the robot will learn by itself that an object is good when the robot approaches it and is evil if the robot moves away from the object. This is indeed a simple solution but you may get it when you think that it is something like the intuition that is deeply embedded in the genes of humans.

5.8 Vehicle 14

Braitenberg's vehicle 14 is different from vehicle 13 in that multiple chains of sequential concepts $(l_1, l_2, \cdots, l_q)$ predicted from the current concept are proposed.

For example,

$$l_1 = (g_3 \triangleright g_{31} \triangleright g_{32} \triangleright \cdots \triangleright g_{3r})$$
$$l_2 = (g_3 \triangleright g_{41} \triangleright g_{42} \triangleright \cdots \triangleright g_{4r})$$
$$\bullet$$
$$\bullet$$
$$l_q = (g_3 \triangleright g_{k1} \triangleright g_{k2} \triangleright \cdots \triangleright g_{ku})$$

The robot selects the sequential concept that gives the best condition for itself. The evaluation criteria would be "advancing toward a better state for the robot." Looking at this robot from outside, we feel that it behaves as it likes with unwavering determination.

5.9 Summary and Observations

Braitenberg proposed many vehicles, from the simplest vehicle 1, through to vehicle 14, the most complex. All of his proposed robots basically consist of a logic unit comprising artificial neural networks that connect sensors and drive motors.

Braitenberg says that representations or concepts of, for example, "red" and "nearby" can be created in the logic unit by adequately stimulating sensors. He asserted that such formed concepts can be abstracted and generalized in the same logic unit.

Even a very simple robot looks like it is behaving "lovingly," "with hatred," or "aggressively" when we look at it.

Braitenberg's idea resembles the concept of an evolutionary robot. In other words, he believed that a robot capable of adapting to its environment could be created as though going through the selection of living organisms. He showed a method to generate associative concepts in a robot by incorporating wires M in the logic unit. He also showed a method to generate sequential concepts that occur as the robot makes a succession of behaviors by incorporating wires E in the logic unit. He then showed us a robot that achieved trains of thought by using wires M and E simultaneously. He then assembled a predictor that automatically generates sequential concepts without necessitating the occurrence of behavior in the robot using wires E and showed that the robot behaves by predicting the future.

Lastly, Braitenberg improved on the predictor and asserted that using the predictor, a robot would be able to "voluntarily move toward a better state for itself in the future" and that by selecting from multiple candidates that are expected to occur from a given concept, the robot could select the optimum chain of sequential concepts that would guide it toward a better state in the best way.

Braitenberg may be said to have created small robots with a mind from the functionalistic viewpoint of a robot researcher. He asserted his law of "uphill analysis and downhill invention." With those words, he confessed that knowing whether a mind really resides in a machine was as difficult as going up a hill, and he tried to find a clue to "uphill analysis" by inventing machines with functions similar to a mind.

Chapter 6

Professor Rodney Brooks' Robots

In 1986, Rodney Brooks, then a roboticist at the MIT Artificial Intelligence Laboratory in the United States, announced a new robot. This robot featured a novel design compared with conventional robots. Brooks' design methodology is known as subsumption architecture. Conventional robots are generally called mechatronics models, which Brooks called a Cartesian model. Figure 6.1 shows an example of the mechatronics model.

A robot captures information about its surrounding environment by "sensing." The robot then recognizes the environment based on the information it captured. For example, the robot recognizes that an obstacle is in front of it. On the basis of this recognition of the environment, the robot generates an action plan. On discerning an obstacle, the robot itself creates a plan to avoid the obstacle. The collision avoidance plan can include various actions. For example, the robot may stop, slow down to let the obstacle go by, increase speed to cut across in front of the obstacle, or somehow run away. It selects the action to take depending on the given situation. It then takes the selected action using the drive system. As a result, the robot moves as planned, but here a serious problem arises. The environment changes as soon as the robot moves. The robot discerns that its environment has changed via its sensors, and the process is

Self-Aware Robots: On the Path to Machine Consciousness
Junichi Takeno

ISBN 978-981-4877-90-9 (Hardcover), 978-1-003-26181-0 (eBook)
www.jennystanford.com

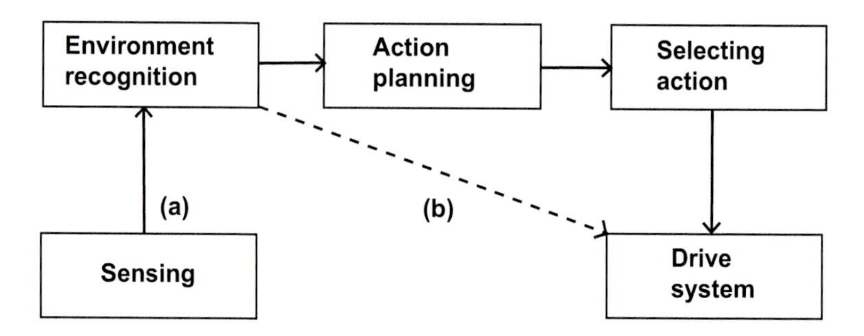

Figure 6.1 Mechatronics model.

repeated again (Fig. 6.1). A Cartesian robot thus repeats the process from sensing to behaving. There is no mistake in this process. All conventional robots are designed after this model.

However, many researchers, including the author, are not necessarily completely satisfied with the mechatronics model. Typically, no problems are experienced when researchers move a robot in a laboratory for research purposes, but this serial processing system turns into quite an annoying process once they attempt to develop a practical or commercial robot.

In the laboratory, researchers simply pursue the study to achieve their purpose and are not much concerned about the real-time response. In addition, we should take into consideration that the speed of computers was very much slower in the 1980s when Brooks conducted his robot studies. At that time, if one adopted an imaging system for the sensors, the turnaround time was terrible — as long as several seconds to complete the processing for a single screen. Nevertheless, researchers tried hard to improve the "real-timeness" by, for example, introducing natural constraints, devising bypass routing from environment recognition to the drive system, and incorporating hardware to execute certain programs (Fig. 6.1b).

Natural constraints include the well-known epipolar constraint explained in Chapter 2. In natural environments, for example, you cannot see an object behind an obstacle.

Brooks introduced his new robot design methodology against this historical backdrop. With the robot system conceived by Brooks, parallel processing replaced the serial processing of the

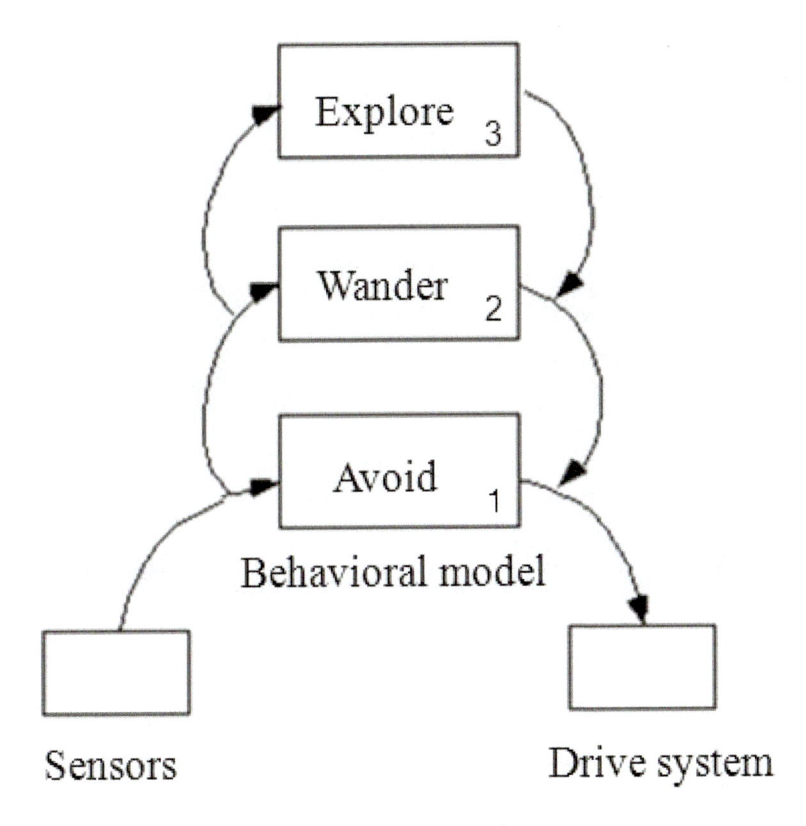

Figure 6.2 Subsumption architecture.

conventional mechatronics models. The structure of his system is shown in Fig. 6.2.

Apparently, one cannot see anything suggesting environment recognition or action planning. In fact, these two factors are actually hidden dexterously. Note that the behavioral modules are stacked one above the other. Each layer corresponds to an independent behavior program. Brooks' robot switches its behavior using the rule, "when a layer is unable to function, a lower layer will take over." For example, if exploring is impossible, the robot goes to and repeats wandering and avoiding (Fig. 6.2). Any layer functions with support from the underlying layers. "Subsume" means that a higher layer includes all the underlying layers. This is the meaning of "subsumption." The capacity of the behavior layers increases as

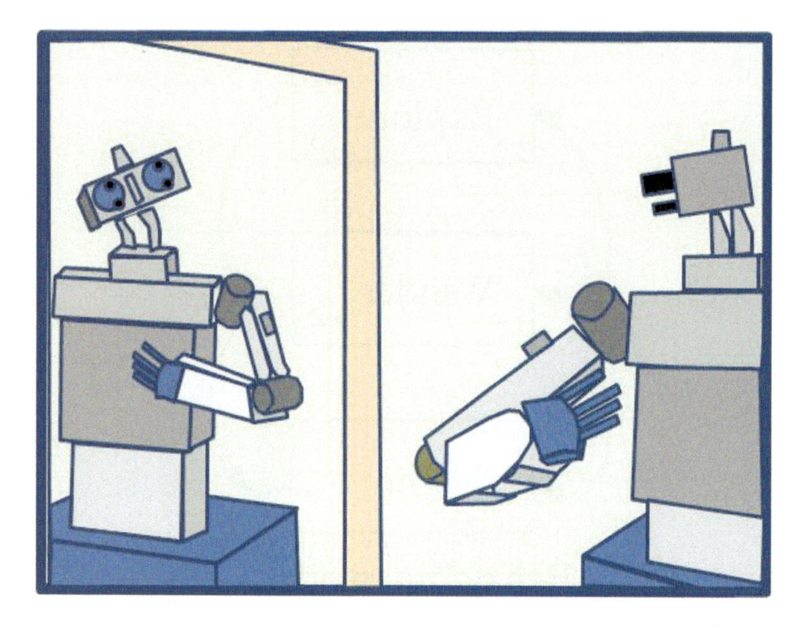

Figure 6.3 Brook's robot Cog.

one climbs the stack of layers. For example, the lowest layer is "avoiding," the one above it "wandering," and the highest "exploring." "Exploring" is selected only when there is no necessity of avoiding and wandering. To put it the other way around, avoiding is selected when exploring and wandering are suppressed.

The switching of the behavioral modules is triggered by replacing (or suppressing) inputs transmitted from the sensors to the behavioral modules and inhibiting outputs from the behavioral modules. The subsumption architecture conceals the process of environment recognition by using a dexterous technique of replacing inputs transmitted to the behavioral modules and suppressing outputs from them. It also hides the process of action planning by organizing a variety of simple behavioral modules into layers from the low to high levels. The net result is that the robot engages in continuous behaviors and will not end up in a deadlock thanks to the behavior-based design. The sensors and drive system are directly connected via the behavioral modules, so that the reaction time from sensing to behaving is short. Furthermore, the robot will never stall because multiple behavioral modules are prepared in layers and at least one or more modules are always available for functioning. It is easy to retrofit behavioral modules on already designed robots.

The robot design method proposed by Brooks has widely influenced later robot researchers.

Brooks has recently created a humanoid robot called Cog and has tried to attain in it a high-level intelligence similar to that of humans through learning by interacting with the environment and by adding behavioral modules. According to a recent progress report, it is getting difficult to increase the behavioral modules as the robot rises higher in the layers.

Summary

Prof. Rodney Brooks at MIT proposed a new robot design methodology called subsumption architecture. This design methodology may be said to be a parallel processing type as opposed to the conventional serial processing type of robot. The robot's reaction is quick because the sensors and drive systems are directly coupled via behavioral modules. Many behavioral modules are available and are stacked in layers one above the other from low to high levels.

Brooks believes that by adding relatively simple behavioral modules, high-level knowledge similar to that of humans can be realized. His approach is behavior based but is also functional because the robot's behaviors are classified by several functions. Brooks does not talk much about human mental activities but seems to think that these will emerge on their own as behavioral modules are added. If he thinks emergent human mental activities can be realized by adding behavioral modules, he may also be trying a phenomenological approach to some extent.

Chapter 7

Artificial Neural Networks and Machine Evolution

This chapter introduces artificial information networks that imitate the human nervous system. It also discusses the evolutionary development of machines using such networks and those in which the theory of evolution in biology is applied. If human consciousness or the mind consists of networks of nervous systems that connect the brain and the body, we should be able to imitate these human functions with networks of artificial nerves and using computer programs. It is only natural to believe that human consciousness and the mind came to be in the course of the phylogenetic development of humans. The latter part of this chapter describes possibility of a machine system autonomously acquiring something like human consciousness or the mind by evolving and developing artificial nervous systems, applying the principles of evolution theory and using computers. And the last part of this chapter describes the deep learning and the self organizing map.

7.1 Neural Networks

Braitenberg's vehicles behaved variously via signals directly transmitted from the sensors to the drive motors. His vehicles are a

Self-Aware Robots: On the Path to Machine Consciousness
Junichi Takeno

ISBN 978-981-4877-90-9 (Hardcover), 978-1-003-26181-0 (eBook)
www.jennystanford.com

simple mechanical system with neural pathways for transmitting signals from sensors to drive motors. The electronic pathways are "neural pathways" mimicking those of a living organism. In other words, the behavior of the neural pathways determines the behavior of the vehicle.

If we could cleverly build neural pathways connecting sensors and drive motors in a robot, it would be able to perform various behaviors. Perhaps it would even be possible to create "consciousness" or "mind" on a computer or other machine systems using such artificial neural networks.

7.1.1 Hebb's Rule

Artificial neural pathways simulating the neural pathways of organisms are already known. They are generally called artificial neural networks, or simply neural networks. Neural networks were first introduced in 1949 by Canadian psychologist Donald Hebb, who proposed a theory known as Hebb's rule. Hebb's rule says that when two connected neurons are simultaneously activated, the strength, i.e., the weight, of the synapse between the two neurons increases. For example, when neurons x_1 and y_2 are simultaneously activated, the relevant synaptic weight w_{12} increases (Fig. 7.1). The synaptic weight is also simply called weight and is also known as the synaptic value. The synaptic weight is defined by the following equation:

$$\Delta w_{12} = x_1 \times y_2$$

or more generally:

$$\Delta w_{ij} = x_i \times y_j \tag{7.1}$$

where x_i and y_j are presynaptic and postsynaptic neurons, respectively (see Fig. 7.1).

Let the output $(y_1, y_2, y_3) = (1{,}0{,}0)$ be expected for the input pattern $(x_1, x_2, x_3, x_4) = (0{,}1{,}0{,}1)$ for this neural network. In this neural network satisfying the above input and output conditions, two pairs of neurons, x_2, y_1 and x_4, y_1, are activated according to Hebb's rule. Synaptic weight w_{21} between x_2 and y_1 is unity, i.e., the strength of the synapse between these two neurons increases. This condition also applies to w_{41} connecting x_4 and y_1 (Fig. 7.1(1)).

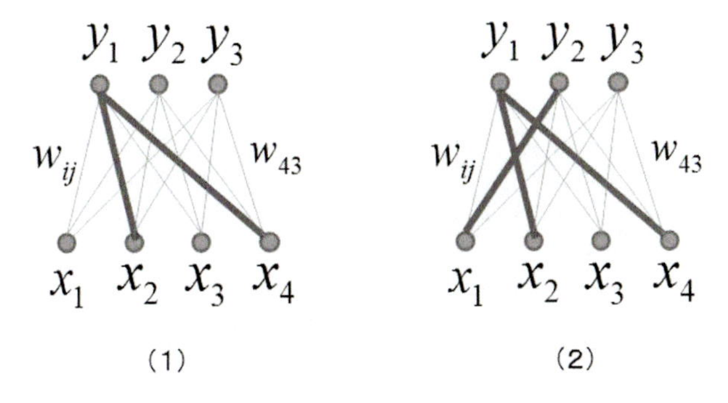

Figure 7.1 Application of Hebb's rule.

All other synaptic weights do not increase and are zero because, from Eq. 7.1, $w_{11} = 0$, for example, since $x_1 = 0$ and $y_1 = 1$. In Fig. 7.1, the synapses with an increased weight are shown using thicker lines. One lesson has been learned with Eq. 7.1.

Here is the second lesson. Let the output pattern $(y_1, y_2, y_3) =$ (0,1,0) be expected for the input pattern $(x_1, x_2, x_3, x_4) = (1,0,0,0)$. Then synaptic weight w_{12} increases in addition to the first lesson. The neural network after the second learning changes as shown in Fig. 7.1(2). As you might have already anticipated, the more lessons we learn, the more there will be thicker lines in the neural network until all the synapses have unity. Ultimately, all output patterns are ignited no matter what the input pattern is. The obvious error of this approach is that the synaptic weight increases with learning endlessly. The idea of the delta rule was devised to resolve this problem. It will be presented soon.

7.1.2 Single-Layer Neural Network and Delta Rule

Let us consider the learning of a neural network as shown in Fig. 7.2(1). This type of neural network is called a single-layer neural network.

Let an input pattern ρ_n with n-input neurons be input into this neural network.

$$\rho_n = (x_1, x_2, x_3, \ldots, x_{n-2}, x_{n-1}, x_n)$$
$$= (1, 1, 0, 1, 0, 1, \ldots, 0, 1) \text{ (this is an example)}$$

where k is the total number of output neurons.

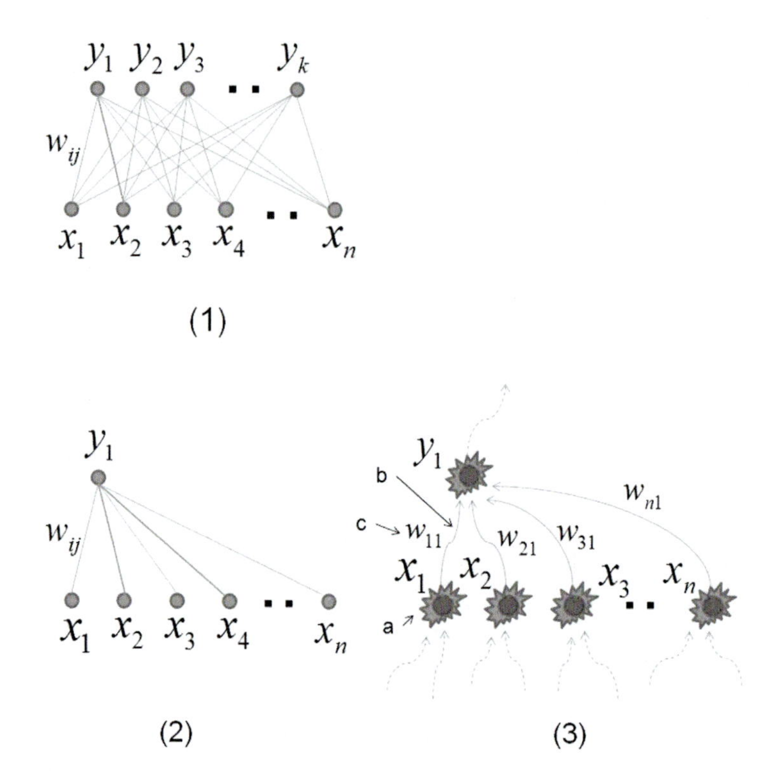

Figure 7.2 Single-layer neural network and explanatory illustration.

$\Theta(M)$ is the linear output function to return a value continuously for variable M.

$$y_1 = \Theta(w_{11} \times x_1 + w_{21} \times x_2 + \cdots + w_{n1} \times x_n) = \Theta\left(\sum_{i=1}^{n} w_{i1} \times x_i\right)$$

$$y_2 = \Theta(w_{12} \times x_1 + w_{22} \times x_2 + \cdots + w_{n2} \times x_n) = \Theta\left(\sum_{i=1}^{n} w_{i2} \times x_i\right)$$

$$\vdots$$

$$y_k = \Theta(w_{1k} \times x_1 + w_{2k} \times x_2 + \cdots + w_{nk} \times x_n) = \Theta\left(\sum_{i=1}^{n} w_{ik} \times x_i\right)$$

$$y_{(j=1\sim k)} = \Theta\left(\sum_{i=1}^{n} w_{ij} \times x_i\right) \tag{7.2}$$

This equation calculates the values of signals output from the neural network as the input patterns vary with the synaptic value w_{ij}. These complex equations are described in a more easily understood manner below.

Figure 7.2(2) is part of Fig. 7.2(1) that is extracted for the purpose of explanation. The extracted figure shows the relationship between output terminal y_1 and the input terminal pattern $(x_1, x_2, x_3, \ldots, x_n)$. Figure 7.2(3) is a network representation of the neurons in Fig. 7.2(2). The network consists of several neurons, a, as shown in Fig. 7.2(3), for example, and the neural pathways connecting the neurons, b, for example. Each neural pathway has a direction for transmitting signals, from x_1 to y_1, for example. Furthermore, each neural pathway has its own synaptic weight w_{ij} that indicates the strength of the flow of signals (c, or w_{11}, in Fig. 7.2(3), for example).

By modifying this weight, the input patterns can be changed to a desired output pattern. If, for example, weight w_{11} for information transmitted from input terminal x_1 to output terminal y_1 is zero, the signals received at x_1 cannot be transmitted to y_1. The value of output terminal y_1 in Fig. 7.2(3) is calculated by the first expression of Eq. 7.2. First of all, the input patterns are multiplied by their respective weight and the results are added to find the sum.

$$M_1 = (w_{11} \times x_1 + w_{21} \times x_2 + w_{31} \times x_3 + \cdots w_{n1} \times x_n)$$

Value M_1 represents the total weight-adjusted strength of the signals appearing on the input terminals. The value M_1 is then used to determine the value of output terminal y_1. The value of y_1 is determined only when value M_1 exceeds a certain threshold value.

Several methods are available to calculate the threshold value. Detailed explanations are omitted here, but in essence, the function $\Theta(M_1)$ is calculated.

The values of $y_2, y_3, y_4, \ldots, y_k$ are respectively determined by calculation. As a result, the output terminal patterns can be decided. All of these calculations are subsumed in the final expression of Eq. 7.2.

It is clear that various reactive systems can be constructed using these neural networks. The function of a system is determined by the structure of the network. Figure 7.2(1) shows one such neural network structure. The components of a "structure" here include the wiring connecting the terminals and the synaptic weight that determines the strength of the signals. Once the structure of the neural network is determined, its function is also determined. A reactive system can thus be constructed by determining wiring and synaptic weight. The objective of creating a reactive system is to obtain desired values or output patterns from variable inputs.

In a word, a system is a function. In that sense, neural networks may be said to be a computer program. Input a pattern into the system, and the system calculates and outputs the result for you. This is a computing system, but what is important is that this is a parallel computing system.

We use Eq. 7.2 to obtain y_1. This is pseudo-computer processing or sequential computing. Originally, all other output terminals y_2, y_3, y_4, . . . , y_k were supposed to be computed simultaneously in parallel as shown in Fig. 7.2(3), and the computation of just a few steps would be enough to finish the calculation. The capability of parallel computation is one of the most important advantages of the human nervous system. Unfortunately, humans have not yet acquired a computation system to effectively make use of the merits of the parallel system; they have to simulate parallel calculation using currently available serial system computers.

Let us now study the problem of how to obtain the desired output pattern given certain input patterns. Let S, called the supervised signal, in Eq. 7.3, be the desired output pattern of a neural network that is given several input patterns ρ.

$$\delta_j^\rho = \left(S_j^\rho - y_j^\rho\right) \tag{7.3}$$

This equation finds the difference between the supervised signal and the current output pattern. The synaptic weight has not yet been learned so that an error δ inevitably exists. First, the approach is to determine the synaptic value to minimize the error δ. This δ is generally called the delta error.

How should we modify the synaptic weight? One possible solution is to use Eq. 7.4.

$$\Delta w_{ij}^\rho = \delta_j^\rho \times x_i^\rho \tag{7.4}$$

This means that input value x_i multiplied by the delta error (δ) is used as the value (Δw) to influence the synaptic weight.

The underlying idea is that the value to influence the synaptic weight is basically proportional to both the delta error and the input value. The value Δw can be a negative number, and therefore the problem experienced with a direct application of Hebb's rule is avoided.

The next step is to add Δw to each synaptic value in the neural network to obtain new values. When this operation is repeated for several training patterns, the synaptic weight converges to a certain value, and at that state we can say that learning is successful. Actually, we determine that convergence is reached when a very small value is observed.

Equation 7.5 is called the error function. It is one of the popular methods to determine convergence. This is generally known as the error of least squares. The error is intentionally squared for the purpose of always obtaining a positive number for the result and also for the purpose of excessively showing the estimated error.

$$E = \frac{1}{2}\sum_{\rho}^{m}\sum_{j}^{n}\left(S_j^{\rho} - y_j^{\rho}\right)^2 \tag{7.5}$$

Clearly, the delta rule is more powerful than Hebb's rule. Later, it was found that this system had a defect. The single-layer neural network, mentioned earlier, is incapable of an exclusive OR (XOR) operation of a logic circuit. The truth table of function y = XOR (x_1, x_2) is shown in Fig. 7.3(1). This function is a subtraction in a sense. The possibility of XOR processing was checked since XOR is an important operator in a computer. The input and output values are plotted on the two-dimensional coordinates as shown in Fig. 7.3(2). As you see in the figure, the white and black dots cannot be separated by a straight line. The white and black dots represent 0 and 1 outputs, respectively, and the coordinates correspond to two input values.

In technical terms, we say that XOR is not linearly separable. There is no way you could draw a line that would separate the space on the coordinates.

Logical disjunction (OR), logical conjunction (AND), and logical negation (NOT) are linearly separable (Figs. 7.3(4), (6), and (8)), but XOR is not, which was pointed out as a problem.

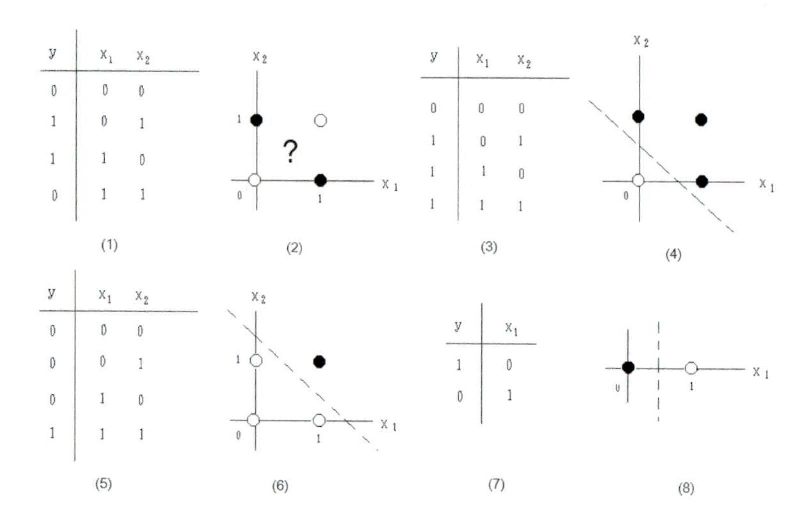

Figure 7.3 Linearly separable.

7.1.3 Feed-Forward Network and Back Propagation Method

The linear-separation problem was solved by Rumelhart *et al.* in 1986. They set up a second neural network above the first one as shown in Fig. 7.4. The result is a two-layered neural network. An intermediate (hidden) layer was provided between the input and output layers. The intermediate layer stores signals coming from the input layer temporarily and delivers them to the output layer after adjusting the signals.

Rumelhart *et al.* introduced nonlinear output functions for respective neurons to achieve nonlinear separation.

The sigmoid function (Fig. 7.5(1)), which is continuous, differentiable, and nonlinear and is frequently used as a nonlinear output function, is expressed by the following equation:

$$\Theta(x) = \frac{1}{1 + e^{-kx}}$$

Other available output functions include linear (Fig. 7.5(2)) and step functions (Fig. 7.5(3)). Output functions are like a filter in the sense that they give some features to the output values. The step function, for example, limits the output values to be either 0 or 1,

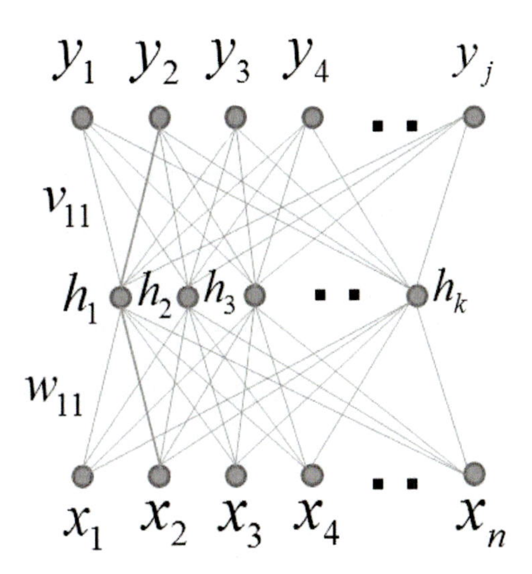

Figure 7.4 Two-layered neural network.

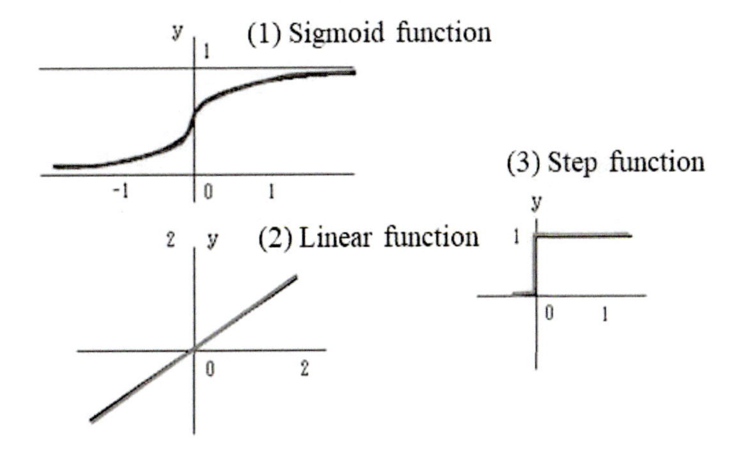

Figure 7.5 Threshold functions.

while the sigmoid function outputs continuous values between 0 and 1.

How does a two-layered neural network learn a desired output pattern? The structure is somewhat complex, but basically the process performed by a single layer is performed twice. The basics of learning are the same as a single-layer network. The synaptic weights of the two layers together are adjusted such that the supervised signals S (desirable outputs) are output in response to input patterns ρ_n. Initially, a near-zero random number is selected for the synaptic value. An input pattern for training t_i is given to input terminal x_i (Fig. 7.4; Eq. 7.6).

Then we calculate value h_k^ρ of the intermediate layer (Eq. 7.7). This calculation uses Eq. 7.2, without any modification, which calculates the output values for a single-layer neural network, except that the input terminal is x_i and the output terminal (intermediate terminal, actually) is h_k^ρ. Output value y_j^ρ is then calculated (Eq. 7.8). This calculation also uses Eq. 7.2, without any modification, which calculates the output values for a single-layer neural network, except that the input terminal (intermediate terminal, actually) is h_k^ρ and the output terminal is y_j.

Note, however, that there are two kinds of delta errors in a two-layer network. The delta error of the second layer δ_j^ρ is first calculated, and then using this value, the delta value of the first layer δ_k^ρ is calculated.

We first calculate the generalized delta error δ_j^ρ for each output value. S_j^ρ is the desired supervised signal to be output (Eq. 7.9). Then we calculate the generalized delta error δ_k^ρ for each value of the intermediate layer (Eq. 7.10).

$$x_i^\rho = t_i^\rho \tag{7.6}$$

$$h_k^\rho = \overset{*}{\Theta}\left(\sum_{i=1} w_{ik} \times x_i^\rho\right) \tag{7.7}$$

where $\overset{*}{\Theta}$ is a sigmoid function.

$$y_j^\rho = \overset{*}{\Theta}\left(\sum_{k=1} v_{kj} \times h_k^\rho\right) \tag{7.8}$$

$$\begin{aligned}\delta_j^\rho &= \overset{*}{\Theta}\left(\sum_{k=1} v_{kj} \times h_k^\rho\right) \times (S_j^\rho - y_j^\rho) \\ &= y_j^\rho \times (S_j^\rho - y_j^\rho)\end{aligned} \tag{7.9}$$

$$\delta_k^\rho = \overset{*}{\Theta}\left(\sum_{i=1} w_{ik} \times x_i^\rho\right) \times \sum_j v_{jk} \times \delta_j^\rho$$
$$= h_k^\rho \times \sum_j y_{jk} \times \delta_j^\rho \tag{7.10}$$

In Eq. 7.10, delta error δ_j^ρ of the output layer is multiplied by the synaptic value of the second layer in reverse. The result is then multiplied by the respective values of the intermediate layer that were calculated before to derive delta error δ_k^ρ of the intermediate layer. This procedure is called back-propagation because the error is returned from the second to the first layer. We then calculate the modifier values (Eq. 7.11) to modify the postsynaptic and presynaptic values. The postsynaptic and presynaptic values are modified by Eq. 7.12. This equation corresponds to Eq. 7.4 used for a single-layer network. This is an estimation of the difference in synaptic weight between the second and first layers.

$$\Delta v_{jk}^\rho = \delta_j^\rho \times h_k^\rho$$
$$\Delta w_{ki}^\rho = \delta_k^\rho \times x_i^\rho \tag{7.11}$$

$$\overset{(t)}{v}_{jk} = \overset{(t-1)}{v}_{jk} + \eta \Delta v_{jk}^\rho$$
$$\overset{(t)}{w}_{ki} = \overset{(t-1)}{w}_{ki} + \eta \Delta w_{ki}^\rho, \text{ where } 0 < \eta < 1 \tag{7.12}$$

Equation 7.12 calculates changes to be given to the synaptic weights of the first and second layers. In this equation, η is called the learning rate. As η approaches unity, the effect of learning increases. The network repeats the calculation according to the above procedure using several training sets until the difference from S reaches the minimal level. Convergence is generally determined by the following evaluation function:

$$E_B = \frac{1}{m}\sum_\rho^m \left(\frac{1}{n}\sum_j^n \left(s_j^\rho - y_j^\rho\right)^2\right) \tag{7.13}$$

where m and n are the total numbers of the training sets and output terminals, respectively. Learning is determined to have completed when this equation approaches zero, which means that y_j (the value

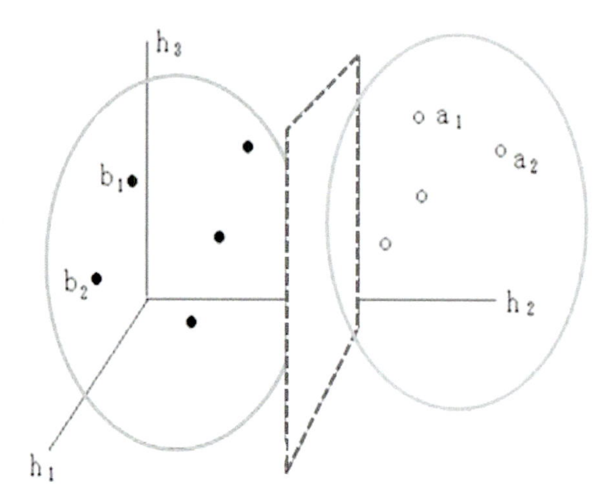

Figure 7.6 Split of activated vector space and categorization of activated vectors.

of the output terminal) is sufficiently close to S_j (the supervised signal). This type of network is called a feed-forward network.

Each terminal of the intermediate layer can be considered to be a vector with a unique coordinate system in a spatial representation of multiple dimensions (i.e., multiple terminals). This type of vector is called an activated vector, and the space where it is observed is called the activated vector space. The coordinate values are a function of the values of the input and output terminals. For this reason, various input patterns can be categorized in the activated vector space.

If, for example, the activated vector space (Fig. 7.6) of an intermediate layer is separable into black (b_i) and white (a_i) dots, we can categorize the vectors into two groups.

7.1.4 Recurrent Neural Networks and Their Functions

"Recurrent" means that the values of the output terminals are returned to the intermediate terminals. An example is illustrated in Fig. 7.7. In this example, value y_j of the output terminal is input into intermediate terminal h_k in the next time step. The role of a recurrent neural network is to dynamically change the outputs as the input values change with time.

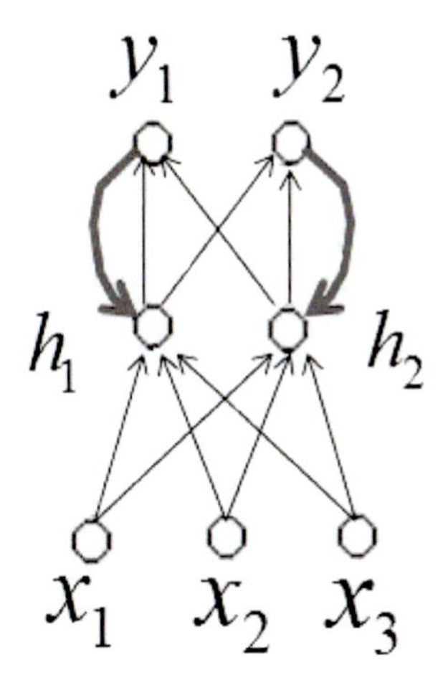

Figure 7.7 Image of a recurrent neural network.

Conversely, the memory of reactions continuing from the past to the present that have brought about the current output still exists. A detailed explanation follows referring to Fig. 7.7.

Figure 7.8 shows how outputs change with time. The output of this network at $t = n$ is (y_{n1}, y_{n2}). Note that in addition to the input pattern (x_{n1}, x_{n2}, x_{n3}), the output pattern $(y_{(n-1)1}, y_{(n-1)2})$ at $t = n - 1$ is also input into the intermediate layer at $t = n$.

The value of an output terminal is determined by the values of the intermediate terminals. The behavior of the values of the

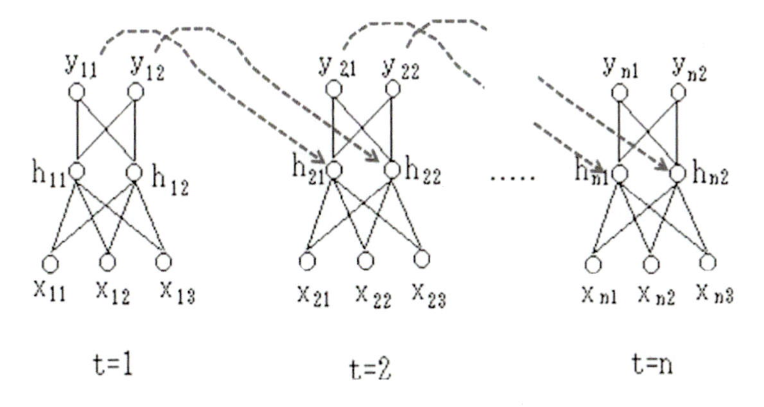

Figure 7.8 Dynamic reaction of a recurrent neural network.

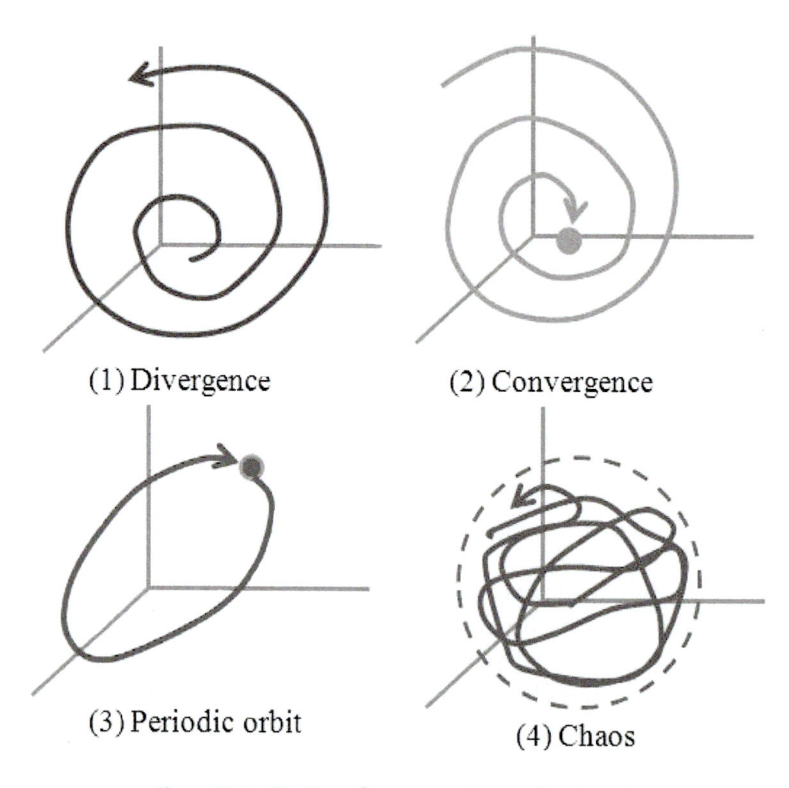

Figure 7.9 Motion of a recurrent neural network.

intermediate terminals is shown in the activated vector space in Fig. 7.9.

The activated vector can take on one of four different states according to the change in input patterns: (1) divergence, (2) convergence, (3) periodic orbit, or (4) chaos. In any of these states, the vector values vary with time, allowing us to build a dynamic system. States 2 and 3 are easy to use, and both can be used for building a dynamic reaction system. State 3, in particular, relates to periodic action. It is possible to move the hands and feet of a robot using this vector. State 4, chaos, is related to unpredictable motion within a certain boundary. This is interesting when you remember that the human body has been called a mass of chaos. It is said, for example, that the heartbeat is related to chaos.

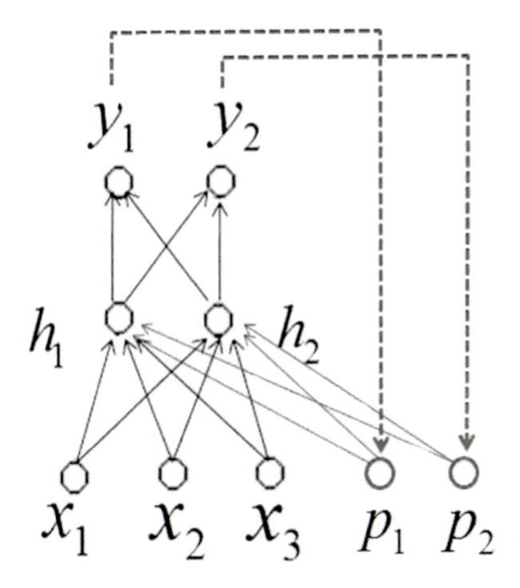

Figure 7.10 Example of an actual recurrent neural network.

There is a convenient method for creating a recurrent network. Referring to Fig. 7.10, prepare recurrent terminals (p_1, p_2) beforehand to which you will later copy recurrent values. When the output pattern (y_1, y_2) is determined, copy its values to the recurrent terminals simultaneously.

7.1.5 Summary and Observations

Artificial neural networks have been briefly explained in this section. Specifically, feed-forward and recurrent networks are discussed. In a feed-forward network, an output pattern is uniquely determined by the input pattern. This type of network is capable of learning basic logic functions and thus, in principle, can achieve all logic functions in a network.

We have also shown that input patterns of a feed-forward network can be categorized by representing the intermediate layer of the network in the activated vector space. Recurrent networks are capable of learning dynamic output states that change with time. The activated vector space of the intermediate layer of a recurrent network has four output states: divergence, convergence, periodic orbit, and chaos. Their respective learning behavior and usability have been described.

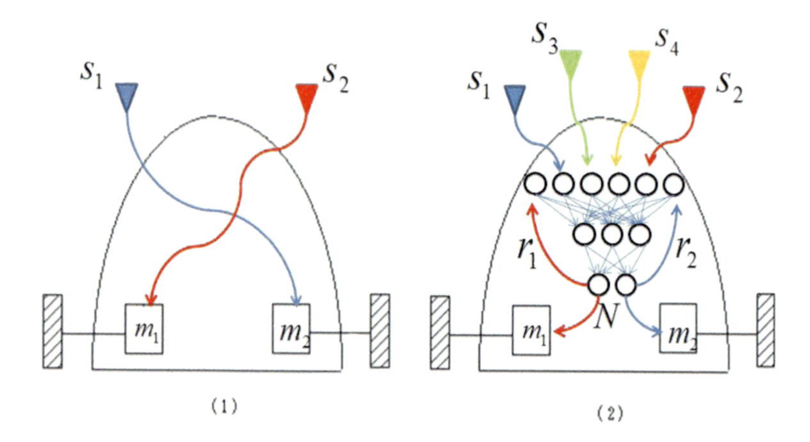

Figure 7.11 Braitenberg robot and its development.

7.2 Theory of Robot Evolution

In the previous chapter, we explained that neural networks can learn output patterns in response to input patterns. As such, it would be possible to build a machine with sensors and motors or other drive units, in which the sensors generated input patterns and the resultant output patterns were connected to the motors, to make the machine move in response to the values of the sensors. This was actually done by Braitenberg. How the machine reacts is determined by the neural network connected between the sensors and the motors.

In the Braitenberg robot, the sensors and motors are connected by simple circuits. With such a simple design, the robot achieved remarkable motion performance (Fig. 7.11(1)).

We are now able to replace the simple circuit with a complex neural network in the Braitenberg robot (Fig. 7.11(2)).

It has gradually become known that it is not so easy for a robot to learn complex neural networks, in particular, a recurrent network. Let's stop and think this over for a moment. The original objective of learning a neural network is to obtain desirable output patterns for a given set of input patterns. To obtain desirable output patterns, we only need to adequately modify the synaptic values in the neural network.

7.2.1 Machine Evolution Approach

Robotics researchers hit upon the idea of using the techniques of evolution that were broadly adopted in biological studies. They decided to apply the principle of the survival of the fittest in the development of robots.

Assume 10 robots decide their actions using their own respectively different artificial neural networks. Four of them perform desirable actions, while the others fail to do so (e.g., could not avoid obstacles). The four robots are determined to be the fittest that survive. The surviving four robots create the next generation of robots. The principle of the survival of the fittest is again applied to the next generation, and robots that act desirably survive. If this process were repeated over some generations, we would eventually have robots that acted excellently. By simulating the principles of the evolution of organisms in this way, researchers created the machine evolution approach.

The robot shown in Fig. 7.12(1) has four sensors to generate input patterns. These input pattern values are transmitted to the output patterns (motors) via neural networks (Fig. 7.12(2)). The neural networks can be variously configured by how their junction of neurons and synaptic values are set. The variety of neural networks that can be configured means that the robots' behaviors can also be diverse. What I mean is that we only need to adequately vary the artificial neural networks for the actions we wish the robot to perform.

The neural network shown in Fig. 7.12(2) has 17 neurons connected by 27 synapses. The state of connection is shown in the matrix in Fig. 7.12(3). The top row and the leftmost column denote all neurons ($n_1, n_2, n_3, \ldots, n_{17}$). The element w_{12}, for example, in the matrix represents the synaptic value from neuron n_1 to n_2. Generally, the element w_{ij} in the matrix represents the synaptic value from neuron n_i to n_j. This matrix thus shows all of the synapses and synaptic values for the entire set of neurons. Let us consider the simple case of three neurons n_1, n_2, and n_3. The area enclosed by the dotted lines in Fig. 7.12(3) is the region of interest. This region is graphically represented in Fig. 7.12(4). In this diagram, the three neurons and all of their synapses with the corresponding values are

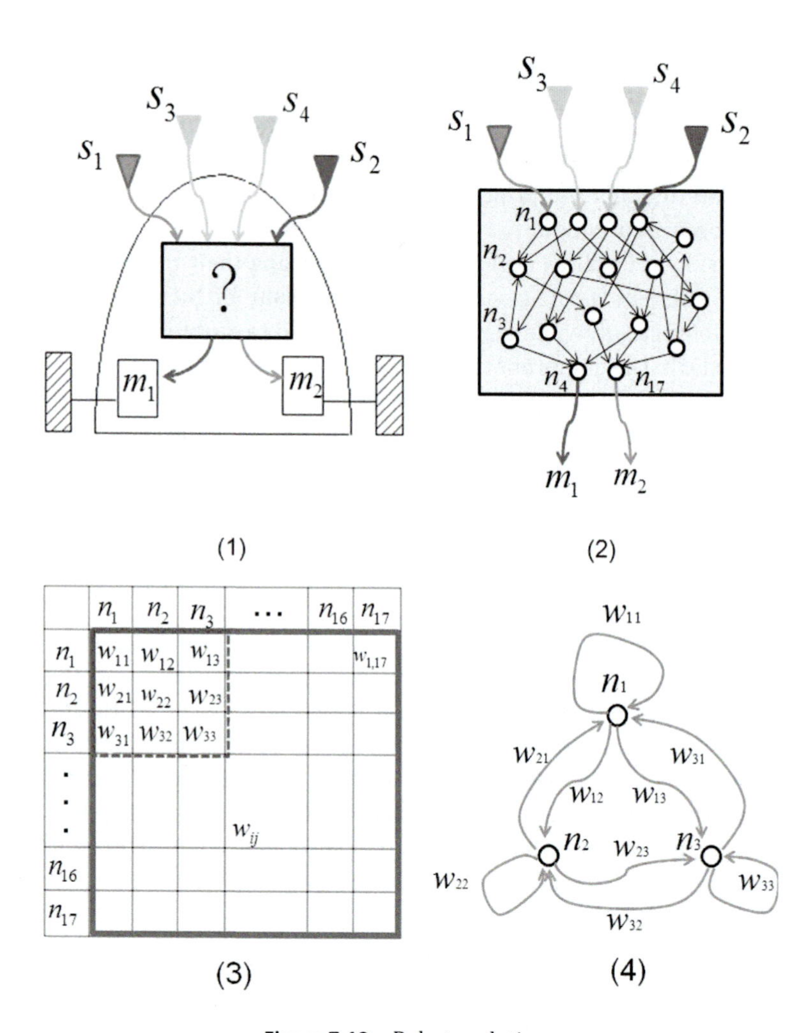

Figure 7.12 Robot evolution.

covered. In discrete mathematics, this diagram is called a directed complete graph with three nodes. Synaptic weight w_{11} is the neural pathway from neuron n_1 to itself, which is called a self-loop in technical terms and is a kind of recurrent pathway.

Since the elements in this matrix determine the behavior of a robot, all of the behaviors of the robot can be programmed by specifying all combinations of the elements.

As such, in a direct way of thinking, it would seem to be possible to discover neural networks that constitute the substance of human consciousness or the mind in the combinations of elements of all of the synapses. If the neural network constituting human consciousness or the mind is a redundant system, however, I am afraid the possibility of discovery is very low.

Putting aside the difficult problem of consciousness and the mind for now, it is clear that the optimum behavior for the robot can be selected from among these combinations of elements. For three neurons, the maximum number of neural pathways (number of synapses) is 9 (number of neurons $\times$ number of neurons). In Fig. 7.12(4), the maximum number of neural pathways is 289 since there are 17 neurons.

Let the maximum value that a synaptic weight w_{ij} can take be a one-digit number in the decimal system, i.e., an integer between 0 and 9. Then in the case of the 3 neurons shown in Fig. 7.12(4), each synaptic value can take 10 different numbers, and there would be 1000 (10^3) different pattern possibilities in this neural circuitry. When we run the robot by applying each and every possible pattern to its neural circuitry shown in Fig. 7.12(4), we can find the optimum pattern. In the case of the 17 neurons shown in Fig. 7.12(3), the total number of possible patterns reaches 100000000000000000 (10^{17}), which practically prohibits computation, although it is theoretically possible. In technical terms, this solution brings about an exponential explosion of computation time. In the present example, the synaptic values are limited to one-digit decimal integers. If this limitation were removed and any real number could be used, there would have been no possibility of computation.

To avoid this problem, we do not check all possibilities one by one, but instead we check areas with higher possibilities of finding the solution sequentially until a suboptimal solution is identified. A suboptimal solution is a practically acceptable solution that would be theoretically close to, though is not itself, an optimal solution. This approach was devised in the study of artificial intelligence and adopts the idea of biological evolution. Hence, the method is named "machine evolution."

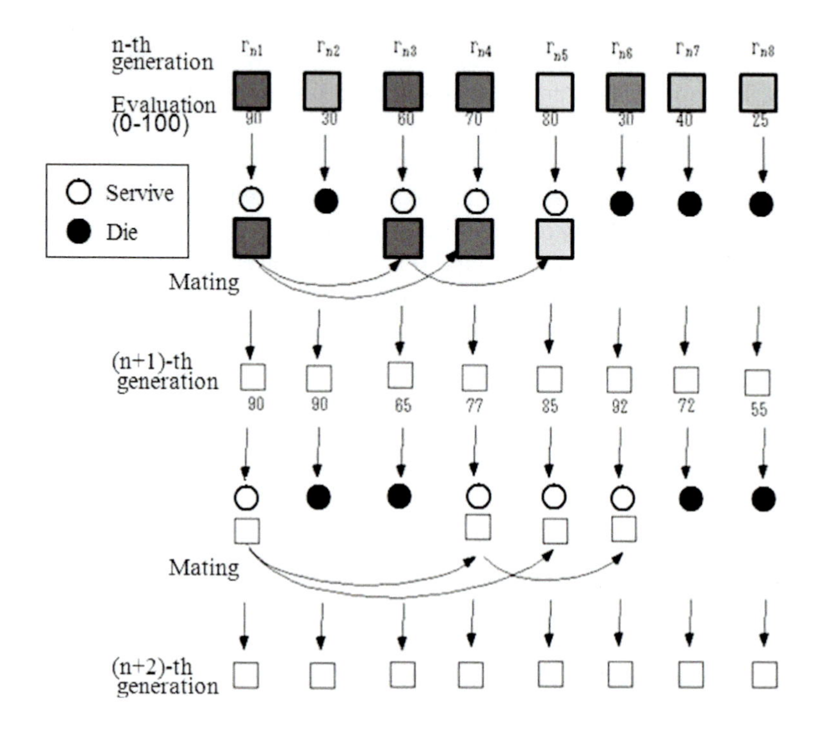

Figure 7.13 Biological evolution approach.

One of the popular machine evolution approaches is introduced now. Assume that eight robots survive in the *n*-th generation (Fig. 7.13). We evaluate the fitness function of each robot with respect to its environment on a 100-point scale.

The fitness functions are assigned by humans. For example, the evaluation mark would increase the faster a robot runs. The survival-of-the-fittest principle is applied to the robots after evaluation. Various discussions exist regarding how many marks would qualify for the fittest. We decide to select "die" for four robots with poor evaluations, and the remaining four robots survive as the fittest (selection). These four surviving robots, with more than 60 points as the evaluation mark, are mated. Various mating methods are conceivable. Only the surviving robots are mated or some dead ones may be recovered temporarily for mating purposes. The mating problem will be discussed in detail later.

For now, the four surviving robots reproduce eight robots by mating. These eight robots are the $(n + 1)$th generation. The robots of this new generation are evaluated to select the surviving robots. At this time, the value of fitness function increased to more than 77. We can hopefully expect the fitness function to increase gradually by repeating the above process generation after generation. Repetition stops when robots of a high evaluation are found. This is a breakthrough approach in that the details of the artificial neural network are omitted and the overall picture is grasped. This sounds like saying that good results are important no matter how the artificial neural network is structured. That is why this approach is enthusiastically hailed by behaviorists.

The machine evolution approach is said to have been initiated by John Holland and L.J. Fogel, both in the United States, and Ingo Rechenberg in Germany, in the 1960s almost simultaneously (Pfeifer and Scheier, 1999). Holland advocated a genetic algorithm, while Fogel promoted evolutionary programming, and Rechenberg developed an evolutionary strategy technique. The three researchers seem to have been working in different fields of study, but in reality there is no significant difference among them from the broader perspective of machine evolution.

The machine evolution approach has made a great progress historically hand in hand with the study of artificial life.

"Artificial life," as defined by Christopher Langton of the Santa Fe Institute, "is an artificial system that behaves like a living organism." In the study of artificial life, artificial living organisms, called creatures, reproduce themselves, grow, prosper, decay, and die by simulation in computers.

In their studies, artificial life researchers naturally discuss the evolution of their artificial living organisms. However, their research is performed in an ideal environment of computers, which is markedly different from the environment of machine evolution research.

Researchers of machine evolution claim that machines acquire knowledge through interaction with the natural environment. This very belief prevents them from attempting to build a natural environment on a computer, which remains a decisive drawback.

I personally believe that the study of artificial life and Braitenberg's robot studies were fused to give rise to the study of machine evolution and spurred its rapid development. In other words, the evolutionary technique of artificial life and Braitenberg's robots that moved around in an actual environment have merged. Let me explain in more detail, beginning with genotype and phenotype.

Humans have genes. The genes are used as a plan for producing and developing the body and other parts of a human. Genes are a plan and the human body is a representation of the plan. Using this analogy, we may say that in machine evolution, a genotype corresponds to a numerical sequence or a character string, and phenotype to a machine system created from or representing the numerical sequence or character string (Fig. 7.14(1)). The machine system as a phenotype may be a hexapod centipede or a function unit that generates adequate output patterns in response to input patterns.

In the discussion to follow, we consider the function unit of Braitenberg's robot shown in Fig. 7.12(1). The component with the question mark (?) in the figure is the function unit.

Our robot has four sensors S_1, S_2, S_3, and S_4. When an external stimulus is given to the sensors, the function unit responds and outputs signals to two drive motors m_1 and m_2. The behavior expected for the robot is to "continue moving without collision as much as possible." We assume that the function unit incorporates a single-layer neural network of four inputs and two outputs (Fig. 7.14(1)). The structure of the network itself could be the result of evolution, but we select a simple scenario for now.

We then determine genotypes and use four-digit binary sequences because of the ease of representation in genes. Since our robot changes its behavior according to the synaptic weight, we relate the genotype to the synaptic weight. We further assign values to respective synaptic weights, which are the elements of the genes in our example. The values are determined by generating random numbers. The synaptic weights are four-digit binary numbers, which are equivalent to 0 through 15 in the decimal system.

In our present study, we determine the synaptic weights as follows:

$$(w_{11}, w_{12}, w_{21}, w_{22}, w_{31}, w_{32}, w_{41}, w_{42}) = (7, 1, 5, 2, 3, 8, 1, 5)$$

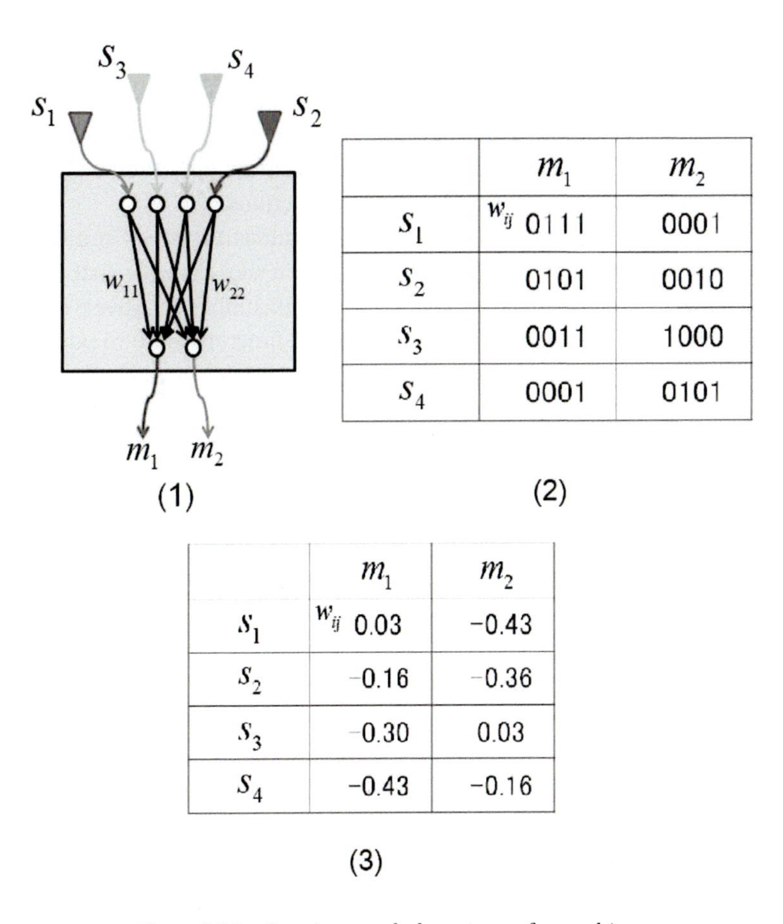

	m_1	m_2
s_1	w_{ij} 0111	0001
s_2	0101	0010
s_3	0011	1000
s_4	0001	0101

	m_1	m_2
s_1	w_{ij} 0.03	−0.43
s_2	−0.16	−0.36
s_3	−0.30	0.03
s_4	−0.43	−0.16

Figure 7.14 Genotype and phenotype of a machine.

The function unit has eight synapses, each having a four-digit synaptic value, i.e., the total number of bits of the genes of the function unit is 32 (8 × 4). The respective synaptic values are therefore represented in four bits as follows (Fig. 7.14(2)):

$$w_{ij} = (0111\ 0001\ 0101\ 0010\ 0011\ 1000\ 0001\ 0101)$$

The genes of this function unit are now finally determined as follows:

$$g = (01110001010100100011100000010101) \qquad (7.14)$$

This is the genotype of our robot. Let us now calculate the phenotype for this genotype. Each synaptic value is divided by 15, and 0.5 is subtracted from the result. This gives you the phenotype (Fig. 7.14(3)). We divide by 15 to convert the synaptic values of 0 to 15 to values between 0 and 1.0. We subtract 0.5 to further convert the resultant values to those between –0.5 and 0.5. Negative numbers are intentionally introduced because inhibition values must be considered for certain synaptic junctions.

We have thus converted the synaptic values to values suitable for use in neural networks. It is assumed in our example that the input and output values are real numbers, including negatives, and therefore we use the values of the sigmoid function $\Theta(x)$ to which 0.5 is added as the threshold function of the neuro unit. This is necessary, for example, to run the drive motors in reverse. Merits include the fact that the sigmoid function is effective for inputs between –1.0 and 1.0, and that values near zero are suitable for the initial value of genes (Fig. 7.5(1)).

It is somewhat difficult to understand that Fig. 7.14(3) shows the phenotype. The elements in the phenotype directly indicate the synaptic values of the single-layer neural network built in the function unit, and the actual robot (Fig. 7.14(1)) behaves on the basis of these synaptic values. As the phenotype is decided, the robot's behavior is also decided.

An explanation of population now follows. The genotype defined by Eq. 7.14 is an initial value. Let us prepare, in our example, 20 genotypes as the initial value. Random numbers may be used. When considering these 20 genotypes to be phenotypes, each behaves differently. These 20 genotypes are therefore phenotypes of 20 individuals at the same time. These 20 individuals constitute the population of the first generation.

Now we proceed to the selection process. We select individuals with a high fitness function from among these 20 individuals. In our example, robots capable of avoiding collision as much as possible are determined to have a high fitness function. One possible evaluation method would be to have each individual robot run and count the number of collisions observed within a unit time. Assume we select 10 individuals.

Various methods are available to select 10 individuals. One of the popular methods is roulette wheel sampling. Individuals are

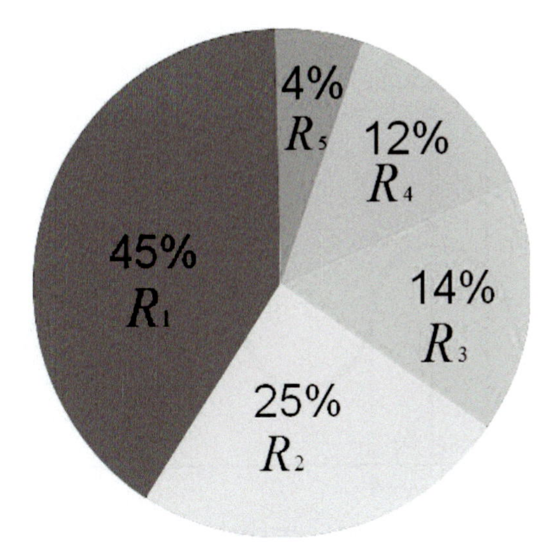

Figure 7.15 Roulette wheel sampling.

selected by probability R_i corresponding to relative fitness within the given population (Fig. 7.15).

The widths of the regions on the roulette wheel are proportional to the level of fitness. We spin the roulette wheel. When the pointer stops at region R_3, for example, the individual with that relevant fitness is selected. If there is more than one individual, we select one at random. Repeat this procedure 10 times, and you have selected 10 individuals. The advantage of the roulette wheel sampling is that individuals with various levels of fitness are selected with a certain balance for the given generation. "Balance" here means that individuals of not only a high fitness but also those of a middle and low fitness are selected as well. This method helps us approach the optimum solution, or, in technical terms, avoids jumping at the local maximum.

Assume the solution space for the distribution of genotype solutions is as shown in Fig. 7.16.

The solution distribution in our example has two different modes, each with a peak. This is a type of multimodal solution distribution. Peak p_1 has a low fitness value compared with the

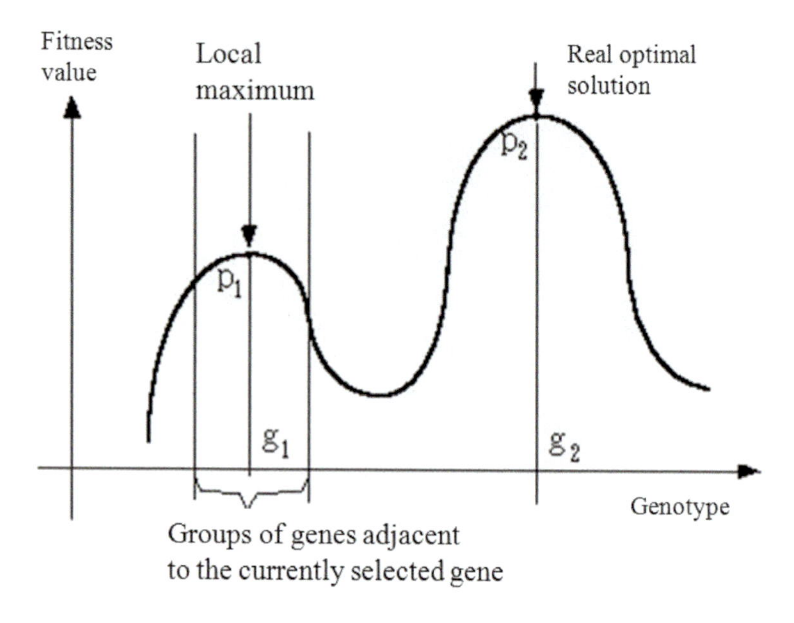

Figure 7.16 Solution space and multimodal solutions.

second peak p_2. Peak p_1 is the local maximum and peak p_2 is the optimum fitness. Genotype g_2 is therefore the desirable answer.

If you always selected individuals of a high fitness in this selection process, all of them could exist around genotype g_1. If this were the case, you would have ended up selecting individuals near the local maximum, and could get out of the local area, eventually failing to select genotypes that would surpass the local maximum.

Note that a solution space like the one shown in Fig. 7.16 is totally unknown to humans initially. Indeed, if it were known, we need not have any trouble.

It is known that the machine evolution approach is not effective unless the solution space is continuous and differentiable. In other words, for the machine evolution approach to work effectively, the fitness determined by certain genotype g_1 must not change abruptly (should be differentiable) from the fitness of the gene adjacent to g_1. Conversely, for a problem involving any sort of solution space, the machine evolution approach finds an answer somehow. This is the great point of the machine evolution approach.

Other methods for selecting individuals include elitism (to leave behind individuals with optimum fitness values in a generation), rank selection, and tournament selection. Rank selection refers to the use of ranking of individuals' fitness values, while tournament selection picks two individuals at a time for comparison for survival. After selection of individuals comes the process of reproduction.

We have earlier selected 10 out of 20 individuals. Genetic operations are performed on the selected individuals to reproduce 20 individuals again.

Two popular genetic operations are crossing and mutation. Crossing refers to the crossover of genes of the selected individuals. Assume we pick two genes g_1 and g_2 from a selected individual.

$$\begin{aligned} g_1 &= (0111\,00\underline{01\;0101\;0010\;0011\;1000\;0001\;0101}) \\ &= (g_{11} \qquad\qquad g_{12} \qquad\qquad) \\ g_2 &= (0011\,10\underline{00\;0001\;0101\;0111\;0001\;0101\;0010}) \\ &= (g_{21} \qquad\qquad g_{22} \qquad\qquad) \end{aligned}$$

Crossing occurs between the sixth and seventh characters from left in the above gene sequences. Random numbers may be used. Two crossed genes g_1' and g_2' are generated.

$$\begin{aligned} g_1' &= (0111\,00\underline{00\;0001\;0101\;0111\;0001\;0101\;0010}) \\ &= (g_{11} \qquad\qquad g_{22} \qquad\qquad) \\ g_2' &= (0011\,10\underline{01\;0101\;0010\;0011\;1000\;0001\;0101}) \\ &= (g_{21} \qquad\qquad g_{12} \qquad\qquad) \end{aligned}$$

The original genes $g_1 = (g_{11}g_{12})$ and $g_2 = (g_{21}g_{22})$ turned out two new genes $g_1' = (g_{11}g_{22})$ and $g_2' = (g_{21}g_{12})$ by crossing. The phenotype of the gene has changed as follows:

$$\begin{aligned} g_1 &= (w_{12}, w_{12}, w_{21}, w_{22}, w_{31}, w_{32}, w_{41}, w_{42}) \\ &= (7, 1, 5, 2, 3, 8, 1, 5) \end{aligned}$$

to

$$g_1' = (7, 0, 1, 5, 7, 1, 5, 2)$$

The above process applies to g_2 in like manner.

Mutation is described now. Mutation means that part of the information of the genes of a selected individual is overwritten.

One element of the gene is subjected to bit conversion, or 1 and 0 are overwritten by 0 and 1, respectively. The position of the bit conversion can be arbitrarily determined.

$$g_1' = (0111\,0000\,0001\,0101\,0\underline{1}11\,0001\,0101\,0010)$$

$\uparrow$ Position of mutation

$$\begin{aligned} g_1'' &= (0111\,0000\,0001\,0101\,0011\,0001\,0101\,0010) \\ &= (7,\,0,\,1,\,5,\,3,\,1,\,5,\,2) \end{aligned}$$

In the above example, the value of the 18th position (underlined 1) of gene g_1' changes by mutation and a new gene g_1'' is created. These gene-operated genes are reproduced to give birth to the next generation. The procedure is repeated until genes of a desirable fitness are found.

It is understood that the machine evolution approach can not only be applied to the design of the above-mentioned function unit but also be expanded to the study of the shape of the machine systems in which the function unit operates.

Karl Sims is a famous researcher on the evolution of shape. His studies are very interesting, and I invite the readers to study his work.

7.2.2 Summary and Observations

This section explained the machine evolution approach. The machine evolution approach simulates the biological evolution method called the survival of the fittest. New generations are created by mating, and the fittest among the individuals is selected for further mating. This biological method is adopted in the machine evolution approach using an engineering technique. "Fittest" means that one is adapted to the environment. Our eventual goal is the survival of excellent generations. I also explained that this approach is a powerful method for complex neural networks to learn successfully. This approach is useful in that it can find specific solutions to problems for which humans are utterly at a loss as to how to find solutions.

In the machine evolution approach, humans specify the genotypes, set pairing of genotypes and their phenotypes, and stipulate the fitness function based on the survival-of-the-fittest principle. All of the processes of evolution are then left to computers. It is naturally possible to set genes as the object of evolution. As a result, artificial living organisms beyond the imagination of humans could appear. This phenomenon is called emergence.

There are problems, though.

In the machine evolution approach, evolution takes place in interaction with the environment in principle, but it takes an incredible amount of time to observe how and to what extent the evolved individuals will satisfy the fitness function. The machine evolution approach is used jointly with computer simulators and with back propagation and other neural network learning methods to shorten the computation time. Despite all of these efforts, no essential solution has yet been found. Furthermore, no one has ever clarified theoretically the reason why the machine evolution approach identifies good answers swiftly.

7.3 Other Notable ANN Technologies

7.3.1 Deep Learning

With the emergence of deep learning (DL) techniques in recent years, we can now say that artificial intelligence (AI) is no longer merely but a dream. What is DL? DL is not any different from an artificial neural network (ANN). However, this neural network used for DL has a large number of input data and it is a multi-layered network. In the past, the number of input data for an ANN was 10 nodes at most, and the number of layers was typically two layers. However, DL can be executed without any difficulty when the number of input data is about 250 × 250. Until now, there was no possibility of learning with ANNs requiring such a large number of inputs. Why did such a breakthrough come about? The reason is the faster speed of computers. And behind that faster speed lies the large-scale integration (LSI) technology of the computing unit. Soon after computers were developed in the 1950s, circuits that used vacuum tubes and circuits that used transistors evolved. Transistors are based on semiconductor technology, but they can be created on semiconductor materials, and computing units can be constructed in a way that is very similar to photolithography. As such, a technique similar to photolithography would allow the smallest computing unit to be produced in large quantities and quickly, just as if it were a photocopy. And the smaller the electronic circuit, the shorter the flow of electrons flowing through the circuit. Put another way, LSI technology not only made electronic units smaller, it also shortened the processing time of calculations. This shrinking trend continues even today. Another factor that contributed to the development of DL was the sharpening of graphic boards. In general terms, it is a computing unit for the image display called a graphics processing unit (GPU). Together with the development of computers, computer games appeared as a form of entertainment that was greatly accepted by general users. Of course, our readers love games. GPUs have a lot to do with computer games. Computer games must react instantly to the user's actions. In other words, the computing unit of the computer needs to react quickly. And clearly this means that the result of the calculation must appear immediately on the computer's display. GPUs were first incorporated as an interface circuit between the computer's processor and the image display device. This was because the image displayed on the display (the game screen) is a two-dimensional plane, and a phenomenon exists in which the image area of a part of the plane is relatively often associated with the image area of a nearby part. Put in simple terms, images often affect nearby image areas. Here was the true value of the GPU. I believe we can certainly say that the development of computer games encouraged the development of GPUs. By the way, it is natural to notice that GPUs can be used for the internal processing of the computer. In other words, multiple GPUs can be used like a parallel computer, and this is known as general-purpose computing on graphics processing units (GPGPU). In particular, ANN researchers believe that GPGPU can be used because the ANN calculation process is almost identical. In other words, we might say that the sophistication of GPUs together with the faster speeds of computers were responsible for creating DL. And, certainly, we should also point out that the processing speed of the internal memory used by computers has also dramatically improved thanks to LSI technology, and that large-capacity internal memory has been developed.

Another point to be noted here is that the technological advancement of ANN that enabled it to learn input data of more than 256 × 256 was most certainly indispensable for the development of the DL called artificial intelligence. As can be seen in the two-layer ANN shown in Section 7.1.3, the connections between neuron cells from the input layer x_n to the hidden layer h_k and the connections between the neurons from the intermediate layer h_k to the output layer y_j are all full neural connections. In other words, the former is $n \times k$, and the latter is $k \times n$. At this time, the number of input data that is being handled by DL is 256 × 256, so the computational load is still large, even if the computer is highly stratified or if GPGPU is being used. For this reason, rather than only build a two-layered ANN similar to a conventional ANN, we have developed a multi-layered ANN. Put another way in simple terms, the intermediate layer, as it is called in the conventional ANNs, is multi-layered. However, the idea of multi-layering the intermediate layer alone only results in increasing the complexity of the ANN, and it is still difficult for it to learn a large amount of input data. So if the input data can be aggregated, a large amount can be learned. The methods used for aggregation are representative, and a method that combines the convolution method and the pooling method (Fig. 7.17), as well as the auto-encoder method (Fig. 7.21) are well-known. The former method is one of those used for image processing.

Now, we would like to explain the basic configuration of typical DL. Multiple convolutional layers and pooling layers are arranged consecutively between the input layer and the output layer (Fig. 7.17). This figure is presented as a conceptual diagram for the sake of explanation. As shown in the figure, the convolutional layers (Fig. 7.17a) and the pooling layers (Fig. 7.17b) are combined, and these combinations are arranged throughout multiple layers. In addition, the flow of information in which the combination of the layers (a) and (b) is continuous comprises multiple parallel flows from the input layer to the output layer. Each of these information flows receives information from the input layer (they may receive the same information or only part of it), and a pair of a convolutional layer and a pooling layer processes the information continuously.

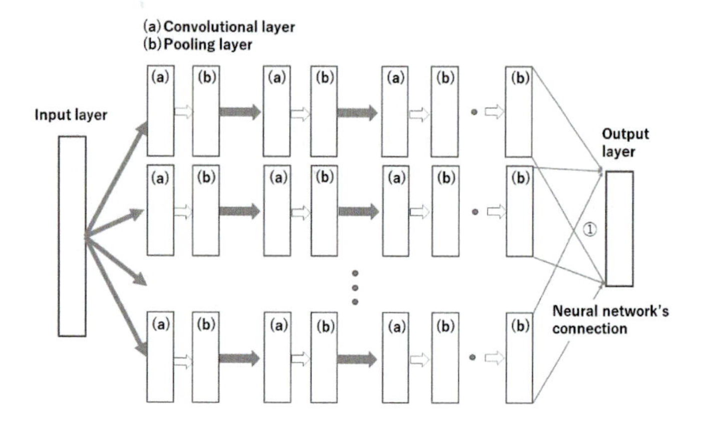

Figure 7.17 Diagram of deep learning configuration.

At that time, the output of each pooling layer at the last step is connected to each terminal of the output layer by a full neural connection (this is called a fully connected neural network). Fully connected means that all the terminals of the last pooling layer are interconnected to all the terminals of the output layer. We should note here that not all the terminals are interconnected in all layers in DL. The details will be described later, however, for example, the connections between the convolutional layer and the pooling layer are not a full neural connection.

The convolution method and the pooling method are described in detail in the following.

The convolution method emphasizes the learning data itself by basically applying a sort of filter to the input learning data. For example, the filters that are used on a camera lens are well-known. A special lens is attached just in front the camera lens with the objective of enhancing certain information in the image being captured. In a similar way to a camera filter, the convolution method used for DL is a process that emphasizes certain information in the input learning data.

Now, let us consider the binary representation shown in Fig. 7.18a.

Let us call this 16 bit × 16 bit image data. And considering this to be image information, we will explain the convolution method.

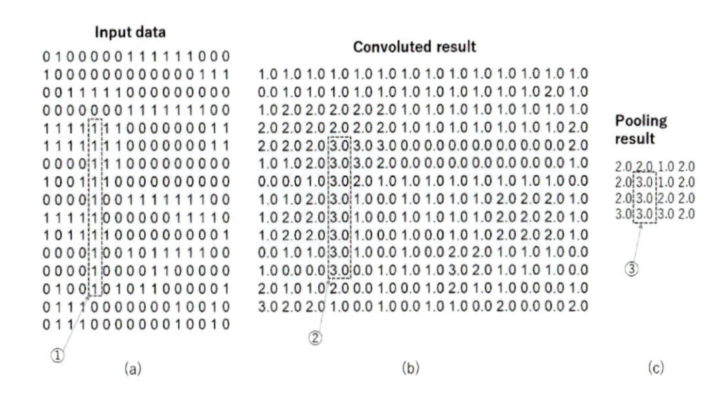

Figure 7.18 Explanatory diagram of convolution processing and pooling processing.

There are several types of image enhancement, but here we consider the highlighting of a vertical line as an example. In this image you can see a long vertical line of continuous 1s in the center left side of the image (Fig. 7.18a, ①). There are vertical lines in other parts of the image, but here we will focus on the ① vertical line.

Now, let us apply the convolution method to this image. We use Filter A as the filter for the vertical line highlighting (Fig. 7.19a). This is a 3×3 bit pattern (a binary matrix). Other well-known filters are Filter B for use on horizontal lines and Filter C for diagonal lines (Fig. 7.19d). Figure 7.19b shows a part of the image stored in the input layer (a part of Fig. 7.18a). As this information is information to be processed by applying the convolution method, it is also said to be in a pre-neuron state (Fig. 7.19). Take note of the information in matrix ① in Fig. 7.19b. Matrix ① has the same size of 3×3 image information as Filter A. Here we apply convolutional processing to this information. This process is a matrix calculation between matrix ① and Filter A. A matrix calculation is the process of multiplying each matrix element and sequentially adding the results.

① Filter A

$$\begin{bmatrix} 1 & 1 & 1 \\ 1 & 1 & 1 \\ 0 & 1 & 1 \end{bmatrix} \circledast \begin{bmatrix} 0 & 1 & 0 \\ 0 & 1 & 0 \\ 0 & 1 & 0 \end{bmatrix} = 3.0$$

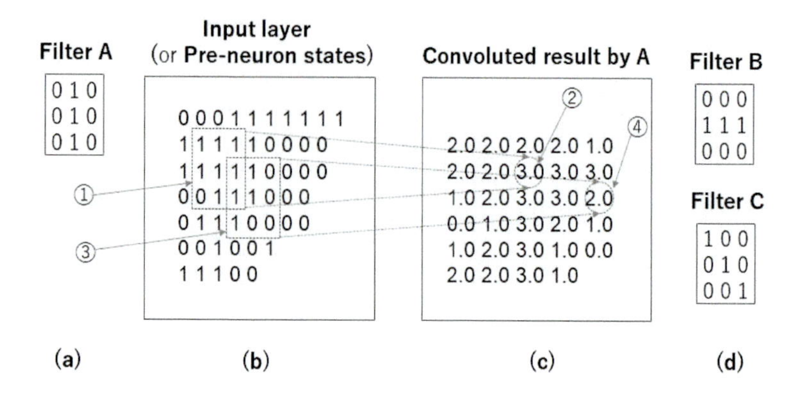

Figure 7.19 Explanatory diagram of convolutional processing.

This value of 3.0 (a real number) expresses the characteristic of the vertical line at the center position of matrix ①. The result of this calculation is stored in ② of Fig. 7.19c. Furthermore, applying Filter A to matrix ③ yields the result of ④ (Fig. 7.19c). In other words, the center position of matrix ③ is 2.0 with vertical line highlighting.

As an overall result, when convolutional processing using Filter A is applied to all the input layer image information (Fig. 7.18a), it results in a convoluted result (Fig. 7.18b). Due to this calculation, the 16×16 image size of the input layer is affected by the size of the filter and is reduced to 14×14.

Next, we will explain pooling processing M. We calculate the pooling result (Fig. 7.18c), which is the condensed information of the convoluted result (Fig. 7.18b). One can easily understand the meaning of condensation by comparing these two pieces of information. The reason is that the highlighting of the vertical line in ② of the convoluted result (Fig. 7.18b) is condensed as ③ in the pooling result (Fig. 7.18c). The highlighted part of the vertical line is also shown in the lower left of the pooling result (Fig. 7.18c). Let us consider a pooling size of 3×3 for this example to explain the calculation of this condensed information. In the example in the figure, ⑤ (Fig. 7.20a) receives the result of processing M and it is stored in ⑥ of the pooling result (Fig. 7.20b). Processing M compares all the elements in matrix ⑤, finds the largest element, and stores the value in the pooling result matrix. Another technique

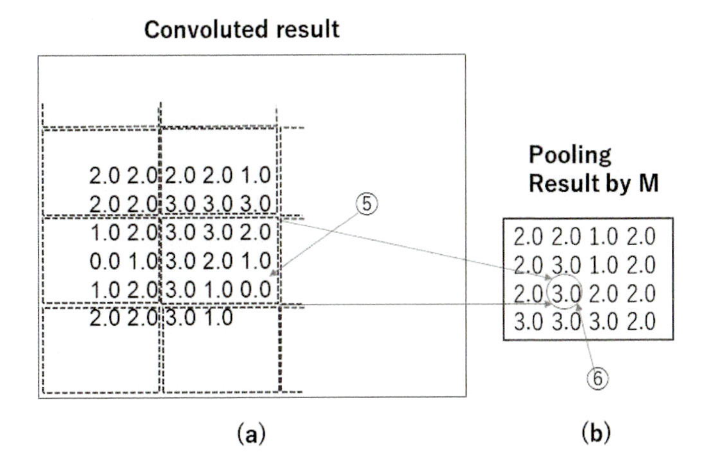

Figure 7.20 Explanatory diagram of pooling processing.

involves calculating the average value instead of the maximum value. In this example, when processing M was performed on all of the convoluted results, the 14×14 size was condensed to the pooling result size of 4×4.

If we reconsider the two methods, the convolution method extracts the features of the input data, and the pooling method condenses the features of the extracted data. In other words, we can consider the many processes occurring in the multiple layers sandwiched between the input layer and output layer of deep learning (Fig. 7.17) as shown in this example to be a sort of feature extraction and condensation processing. The processing result is eventually connected to each terminal of the output layer by a full neural connection (Fig. 7.17①), and in this part, the feature information of the image information that was given to the input layer is aggregated, and we can say that it is neurally connected into the output layer. Supervised learning is used for neural connections at this final step.

If we consider feature extraction and data condensation for the intermediate processing of DL, other methods are also conceivable. Another one of the known techniques is the auto-encoder method. This method uses a conventional two-layer ANN with supervised learning (Fig. 7.21). There is a hidden layer between the input layer

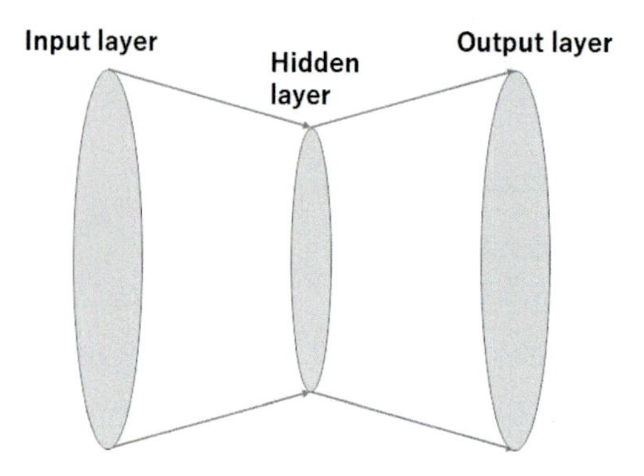

Figure 7.21 Explanatory diagram of the auto-encoder method.

and the output layer. Learning is performed on this ANN. However, the data input to the input layer itself is used as the teaching data. Note also that the number of terminals on the hidden layer should be less than the number of terminals on the input layer. By doing this, the information of the hidden layer at the end of the learning then becomes a sort of aggregated data of the information given in the input layer.

7.3.2 Self-Organizing Maps

This section describes the important self-organizing maps (SOMs) that are one of the ANN technologies.

All of the ANN technologies described previously use a "supervised signal" to lessen errors by "backpropagation (Section 7.1.3)" and proceed with the learning. A SOM can proceed with learning without using a specific teacher signal. This idea was created in the 1980s by Finnish professor Teuvo Kohonen.

At this time, this theory provides a powerful tool for ANNs, and here we will explain a simple basic principle.

For example, let us suppose that we have a sensor unit A that repeats several bits of output (e.g., 16 bits). At this time, let us suppose we have a problem for which we want to analyze the output bit pattern.

Patterns in the output that are close to random numbers may be output one after another, or a certain pattern may be biased. For

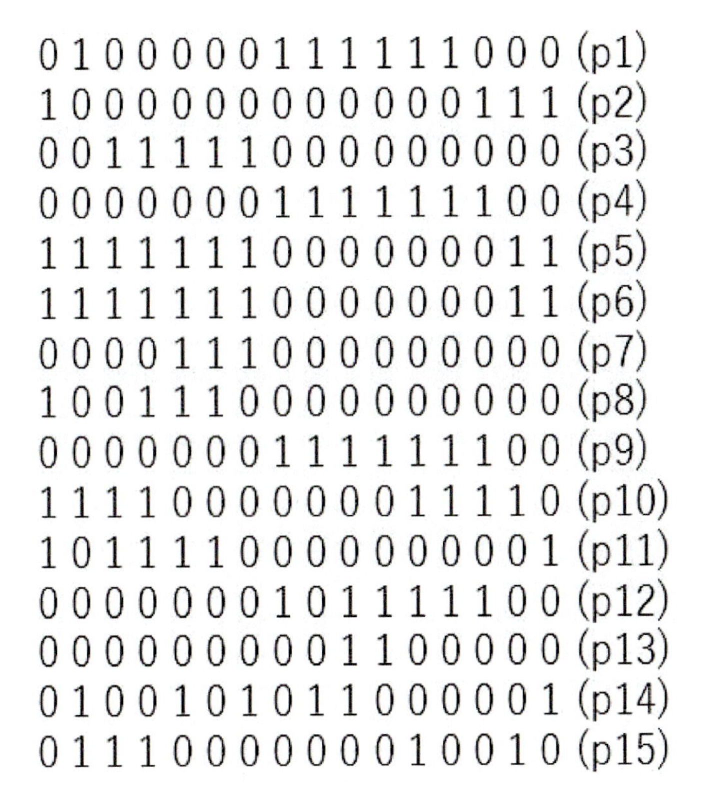

Figure 7.22 Example of data output from sensor unit A (set of learning data).

such issues, a SOM provides powerful functions that enable us to grasp the trend of the output pattern.

Let us assume that the bit pattern shown in Fig. 7.22 is being output from that sensor unit A. In this example, the number of training data items is 15. Each data consists of 16 bits. These 15 items of data are data to be learned and data to be analyzed. The data to be learned can be 1000 or 10,000 items of data.

When we look at this data, we can quickly see, for example, that patterns p_5 and p_6 are the same pattern, and that patterns p_1 and p_4 are very similar. The data for p_1 and p_4 differ only in distance 2. Here, distance here is defined as the number of different bits when comparing two bit strings. In other words, when counting from the most significant bit string of p_1, the distance is 2 because the 2nd and 14th bits are different.

Now let us have the SOM program process these 15 items of data and analyze which are the same, which are different, and how they differ.

A SOM is basically a program that sorts similar data similar into a two-dimensional array called a Map so that they are close located to each other.

The SOM program uses three arrays. One array is an input pattern, which is a one-dimensional array with n elements that are input to the SOM. In an ANN, this array is generally called the input layer (Fig. 7.23).

The second array is the Map, which is a two-dimensional array that displays the output results of the SOM, and here we call this the output layer. The number of elements of the Map is $m_i \times m_j$. In this example, $m_i = 20$ and $m_j = 22$. This size can be freely determined by the user, but in our example, the number of learning data items is 15, so here the size is about 10 to 20. This size is related to the pattern resolution of the Map.

For example, if $m_i = 3$ and $m_j = 5$, the display range is too narrow compared to the 15 items of learning data. The third array is a three-dimensional array corresponding to the hidden layer, and called connected weights (CW) that connect the input and output

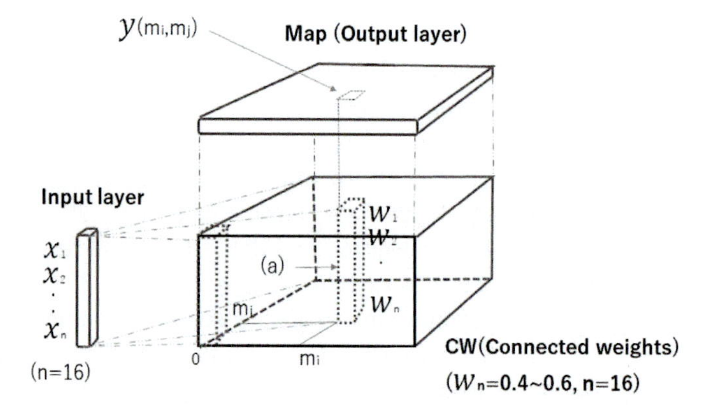

Figure 7.23 Explanation of SOM.

layers. The number of elements is $m_i \times m_j \times n$. For each weight, we input a random number between 0.4 and 0.6 as the initial value.

Now, let us talk about the processing procedure of the SOM program. To begin with, p_1, which is the first learning data to be learned, is stored in the Input layer.

$$p_1 = [0\ 1\ 0\ 0\ 0\ 0\ 0\ 1\ 1\ 1\ 1\ 1\ 1\ 0\ 0\ 0]$$

$$p_1 = [x_1 x_2 x_3 \ldots x_n](\text{example} : n = 16)$$

Next, we perform the following calculation:

$$\text{ED}\left(m_i, m_j\right) = \sum_{i=0}^{n} (x_i - w_i)^2 \tag{7.15}$$

Here, the ED value is a pattern composed of a 0 or 1 in the input learning data p_1, and expresses the difference from the pattern consisting of the one-dimensional connected weights (Fig. 7.23a) in CW.

This calculation uses the least squares method. The reason for using the square is that the negative value that occurs temporarily upon subtraction (7.15) is converted to a positive value, and then the square is calculated to evaluate the difference. Considering that the p_1 pattern is expressed in binary and that each element of the one-dimensional connection weights (a) is composed of values between 0.4 and 0.5, if ED results in a small value at this time, p_1 and (a) are evaluated as having little difference.

At this time, we can see that there are $m_i \times m_j$ one-dimensional connected weights in the three-dimensional connected weights (CW) of the SOM.

For this reason, if Eq. 7.15 is applied to all these one-dimensional connected weights in CW, various ED values will be calculated. Now, this time, if the minimum value of ED can be found, this means that the one-dimensional connected weight in CW that has the least difference with the learning pattern p_1 has been found.

Now, let us suppose that (a) is the one-dimensional connected weight. The m_i row and m_j column of CW (Fig. 7.23) are at this position. At this time, (a) is evaluated as the neuron cell group w_1, $w_2, \ldots w_n$ that exhibited the highest response to the learning data p_1. The position (m_i, m_j) and the pattern name p_1 in the example are simply called 1 (or concept 1) and are stored as a record in $y(m_i, m_j)$ of the Map (and this position is now called q_1).

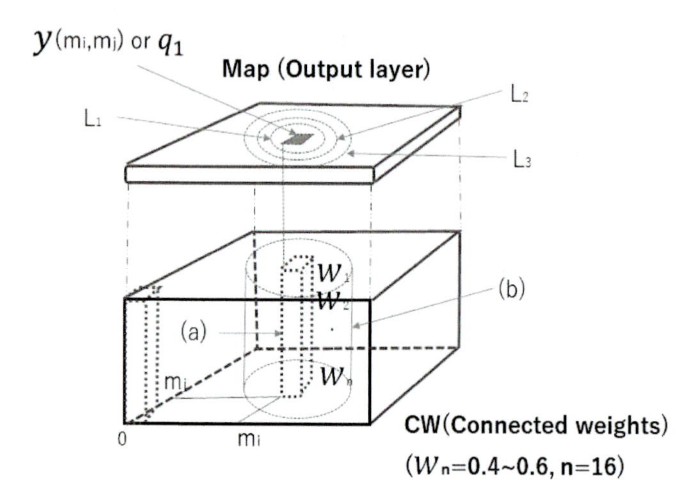

Figure 7.24 Changing the weight of the neighboring connected weights related to position q_1.

Next, we perform the work to change a part of the connected weights related to the position q_1 of the connected weights in CW (Fig. 7.24).

At this time, let us consider a region L_1 adjacent to the periphery of the position q_1, a region L_2 adjacent and in a direction further away from q_1, and a region L_3 adjacent and further to the outside (Fig. 7.24).

Each of these regions is actually a rectangle because the Map is a two-dimensional array. And each of these regions is assigned a one-dimensional connected weight in the CW as shown in Fig. 7.24a. As just described, the one-dimensional connected weight (a) corresponding to the position q_1 is a pattern that has the smallest difference from the input pattern p_1.

The purpose of the SOM is to make the input pattern p_i (in this example, $i = 1$ to 15) as close as possible to similar patterns in the various one-dimensional connected weights in the CW, or to place it as far away as possible from a pattern in which it is difficult to find similarity.

As a specific means, the SOM first finds a one-dimensional connected weight (a) in CW that is most similar to the input pattern p_1 at the position q_1, and the connected weight (a) at that position and the other one-dimensional weight located around the connected weight (a) are changed so that they are more similar to their respective weights.

At that time, regarding the changing of the other one-dimensional weights located around q_1, a change is made so that the degree of similarity to the one-dimensional weights located farther from the position of q_1 is gradually lowered.

The one-dimensional weights that comprise the CW by means of such an operation are arranged so that, depending on the learning data items that are input one after another, similar learning data are arranged close to each other, or learning data with a lower degree of similarity are arranged at a position farther away from each other.

As a more specific calculation, first of all, the weight of the one-dimensional connection weight (a) in the CW corresponding to the position of q_1 is changed.

W_i is changed to a value calculated by the following equation.

$$w_i + (x_i - w_i) \times \eta \tag{7.16}$$

Here, $i = 1$ to n ($n = 16$) and $\eta = 0.5$; η is called a learning coefficient and is usually given as 0.2 to 0.5. Here, 0.5 is used. The meaning of the expression (7.16) is that each weight of the one-dimensional connected weights (a) that was given as an initial value is to be made more similar to the input pattern p_1.

The ED value of the one-dimensional connected weight (a) before the change was the smallest in the CW, but the ED value can be further reduced by changing the expression (7.16). In other words, due to the lower ED value, the one-dimensional connected weight (a) of the CW indicated by the q_1 of the Map is stronger than before and it becomes possible for it to react to the input pattern p_1.

Next, the weight of the one-dimensional connected weight located in the L_1 region near the one-dimensional connected weight (a) is changed. However, the value of η is reduced by half. That is, $\eta = 0.25$. In the L_2 region, η is further reduced by half. And $\eta = 0.125$. In the L_3 region, η is further reduced by half. Let $\eta = 0.0625$. That is, in the L_n region, the value of η may be η / $(r+1)$ (where r is an integer). In the description, r is 3, but this is an arbitrary value.

There are various methods for changing η. The value of η may be changed depending on whether the learning is being advanced strongly or slowly. The process of changing the neighboring connected weights using η will be described again, but it makes it easier for patterns similar to the p_1 pattern to converge around the q_1 position of the Map. On the other hand, patterns that are not similar to the p_1 pattern are less likely to converge.

Now, q_1 of the Map is determined by the input pattern p_1, and the one-dimensional connected weights around q_1 are subsequently changed. This process is called process B.

Then, the next learning data p_2 is input to the Input layer, and process B is performed. When process B is performed on all the learning data (Fig. 7.22), the SOM processing in the initial stage is completed. However, note that the pattern position in the Map is still unstable in the initial stage. The reason for this is that, while learning data is being processed one after another in process B, sometimes the same pattern is mistakenly judged to be a different pattern. In other words, the pattern separation of the Map is not yet high in the initial stage.

Therefore, when process B is performed on all learning patterns, the total TR value of the errors generated in the second term of expression 7.16 is calculated. (This is called process C.) By observing the change of the TR value (ΔTR), the pattern separation of the Map is judged. Here, $\Delta \text{TR} = 0.002$, and 0.08 is often used as a standard value.

The Map uses a two-dimensional array of 22 rows and 22 columns, and the array elements are the numbers of the learned concepts. Figure 7.25 shows the result of learning the 15 types of learning data (p_1 to p_{15}) shown in Fig. 7.22 using the SOM program. The array elements of the Map are the numbers of each learned concept. For example, the learning data p_5 (concept 5) is arranged at y (13, 17) in the Map array. Concepts 6 and 9 are not in the Map because they had the same pattern as concepts 5 and 4 and were thus excluded from the Map. In Fig. 7.25, it can be seen that concepts other than concepts 6 and 9 are almost evenly arranged in the Map. Certainly, this concept layout obviously depends on random numbers because the connected weights that are the elements of CW are determined by random numbers. Concepts 8 and 11 in Fig. 7.25

```
Emap--(The concept's No.)--
---- 0- 1- 2- 3- 4- 5- 6- 7- 8- 9-10-11-12-13-14-15-16-17-18-19-20-21
 0)  0  0  0  0  0  0  0  0  4  0  0  0  0  0  0  0  0  0  0  0  0  0
 1)  0  0  0  0  0  0  0  0  0  0  0  0  0  0  0  0  0  0  0  0  0  0
 2)  0  0  0  0  0  0  0  0  0  0  0  0  0  0  0  0  0  0  0  0  1  0
 3)  0  0  0  0  0  0  0  0  0  0  0  0  0  0  0  0  0  0  0  0  0  0
 4)  0  0  0  0  0  0  0 12  0  0  0  0  0  0 14  0  0  0  0  0  0  0
 5)  0  0  0  0  0  0  0  0  0  0  0  0  0  0  0  0  0  0  0  0  0  0
 6)  0  0  0  0  0  0  0  0  0  0  0  0  0  0  0  0  0  0  0  0  0  0
 7)  0  0  0  0  0  0  0  0  0  0  0  0  0  0  0  0  0  0  0  0  0  0
 8)  0  0 13  0  0  0  0  0  0  0  0  0  0  0  0  0  0  0  0  0  0  0
 9)  0  0  0  0  0  0  0  0  0  0  0  0  8  0  0  0  0  0  0  0  0  0
10)  0  0  0  0  0  0  0  7  0  0  0  0  0  0  0  0  0  0  0  0  0  0
11)  0  0  0  0  0  0  0  0  0  0  0  0  0  0  0  0  0  0  0  0  0  0
12)  0  0  0  0  0  0  0  0  0  0  0  0 11  0  0  0  0  0  0  0  0  0
13)  0  0  0  0  0  0  0  0  0  0  0  0  0  0  0  0  0  5  0  0  0  0
14)  0  0  0  0  0  0  0  0  0  3  0  0  0  0  0  0  0  0  0  0  0  0
15)  0  0  0  0 10  0  0  0  0  0  0  0  0  0  0  0  0  0  0  0  0  0
16)  0  0  0  0  0  0  0  0  0  0  0  0  0  0  0  0  0  0  0  0  0  0
17)  0  0  0  0  0  0  0  0  0  0  0  0  0  0  0  0  0  0  0  0  0  0
18)  0  0  0  0  0  0  0  0  0  0  0  0  0  0  0  0  0  0  0  0  2  0
19)  0  0  0  0  0  0  0  0  0 15  0  0  0  0  0  0  0  0  0  0  0  0
```

Figure 7.25 Arrangement of various learning data shown in the Map (Emap–). The Map uses a two-dimensional array of 22 rows and 22 columns, and the array elements are the numbers of the learned concepts.

are relatively closely located. The reason for this is that the learning patterns p_8 and p_{11} are similar patterns.

$$p_8 = [1\,0\,0\,1\,1\,1\,0\,0\,0\,0\,0\,0\,0\,0\,0\,0]$$
$$p_{11} = [1\,0\,1\,1\,1\,1\,0\,0\,0\,0\,0\,0\,0\,0\,0\,1]$$

According to Fig. 7.26, concept 1 was initially determined to be the same as concept 4 by the execution of process C, but was subsequently determined to be the same as concept 9 and concept 12 by another execution of process C. However, concept 1 finally separated from other concepts (4, 9, and 12) by the fourth execution of process C. Concepts 3 and 10 were initially determined to be the same as concepts 11 and 15, respectively, but the second execution of process C separated them from the other concepts. Now let us look at concepts 4 and 5. Each was continually identified as the same concept because concepts 9 and 6 each had the same pattern, despite the progress of process C. And in fact, concepts 6 and 9 are in the Map. Eventually, ΔTR became 0.002 or less, and the progress

Doubl_list--

----ct		1	2	3	4	5	6	7	8	9	10	11	12	13	14
(1)	3	4	9	12	0	0	0	0	0	0	0	0	0	0	0
(2)	0	0	0	0	0	0	0	0	0	0	0	0	0	0	0
(3)	1	11	0	0	0	0	0	0	0	0	0	0	0	0	0
(4)	12	9	9	9	9	9	9	9	9	9	9	9	9	0	0
(5)	13	6	6	6	6	6	6	6	6	6	6	6	6	6	0
(6)	0	0	0	0	0	0	0	0	0	0	0	0	0	0	0
(7)	0	0	0	0	0	0	0	0	0	0	0	0	0	0	0
(8)	0	0	0	0	0	0	0	0	0	0	0	0	0	0	0
(9)	0	0	0	0	0	0	0	0	0	0	0	0	0	0	0
(10)	1	15	0	0	0	0	0	0	0	0	0	0	0	0	0
(11)	0	0	0	0	0	0	0	0	0	0	0	0	0	0	0
(12)	0	0	0	0	0	0	0	0	0	0	0	0	0	0	0
(13)	0	0	0	0	0	0	0	0	0	0	0	0	0	0	0
(14)	0	0	0	0	0	0	0	0	0	0	0	0	0	0	0
(15)	0	0	0	0	0	0	0	0	0	0	0	0	0	0	0

Figure 7.26 Inspection result of concept duplicates at SOM program execution (Double_list). The first column is the concept number (pattern number) being learned. The second column is the duplicates in the corresponding concept numbers. The third and subsequent columns show the execution of process C. Elements other than 0 are duplicate concept numbers.

of process C stopped. At this point, all the learning patterns shown in Fig. 7.22 have been completed by the SOM program.

By the way, let us input a new pattern into the input layer after the learning. Let this pattern be p_{16}. Then, "We got new concept: No. = 16" is displayed (Fig. 7.27). This means that the SOM has discovered a new concept.

$$p_{16} = [1\ 1\ 1\ 1\ 1\ 1\ 1\ 1\ 0\ 0\ 0\ 0\ 0\ 0\ 1\ 1] \tag{7.17}$$

And this new concept (p_{16}) is shown to be the most similar to the learning pattern p_5 in the Map (Fig. 7.27).

$$p_5 = [1\ 1\ 1\ 1\ 1\ 1\ 1\ 0\ 0\ 0\ 0\ 0\ 0\ 0\ 1\ 1] \tag{7.18}$$

In other words, p_{16} was only one bit different from p_5.

7.3.3 Summary and Observations

This section describes application programs that use ANNs.

```
--We got new concept: No.=16
******The show of the nearest concepts is
-----
-----
-----
-----
(L4)5
/test5 p=15, p1=4
 1 1 1 1 1 1 1 0 0 0 0 0 0 0 1 1
 1 1 1 1 1 1 1 1 0 0 0 0 0 0 1 1 differ=1
-----
```

Figure 7.27 Acquisition of a new concept.

We first explained deep learning (DL), and then explained self-organizing maps (SOM).

DL has recently attracted attention because it can handle large-scale input data using ANNs.

The important part is the intermediate processing part between the input layer and the output layer. The intermediate processing part is a hidden layer in a normal ANN, but DL includes a multi-layered feature extraction part and a data reduction part.

The multi-layer feature extraction part and the data reduction part have the purpose of contracting the amount of data without losing the features of the large-scale input data.

This section describes a method that combines the convolution and pooling methods, and a method that uses the auto-encoder method.

The results of the feature extraction part and the data reduction part are finally learned at the terminals of the output layer in the form of all connections of neurons. DL as described here is basically supervised learning.

Next, we will describe SOMs. Although SOMs are not necessarily the result of recent research, they have high applicability as application programs for ANNs, and since an ANN has a mechanism for organizing networks while learning by itself, SOMs are introduced and their potential is evaluated here.

A SOM has a mechanism for learning an input pattern that is input one after another in an output layer called a Map.

Also, when learning, there is a mechanism for converging similar patterns that are located near the learning pattern in the Map, and the degree of mutual similarity of various patterns learned in the Map can be expressed.

That is, a SOM is a pattern learning machine and can function as a pattern analyzer. This type of learning is what is called unsupervised learning.

Chapter 8

Machine Consciousness

I have described the achievements of general physiological and physical research on the human brain and also introduced outstanding historical research studies on human consciousness and the mind and the relevant achievements. There is a long history of research encompassing psychology, philosophy, cerebral physiology, information science, and cognitive science.

Braitenberg, a cybernetics researcher, demonstrated the possibility of constructing something resembling the human mind on a machine.

I explained mathematical methods for understanding and using artificial neural networks that are potentially capable of artificially reproducing human neurons and nerve networks. Roboticist Brooks proposed subsumption architecture — the world's first concept of its kind. He created a robot that moved around at all times. The behaviors of the robot were hierarchized and higher-level behaviors were achieved as the robot built up a behavior hierarchy tower, eventually fulfilling the emergence of consciousness and the mind, according to Brooks.

I further outlined the techniques of mechanical evolution in which the principle of the survival of the fittest in biological evolution is applied using an engineering approach. On the basis of

Self-Aware Robots: On the Path to Machine Consciousness
Junichi Takeno

ISBN 978-981-4877-90-9 (Hardcover), 978-1-003-26181-0 (eBook)
www.jennystanford.com

this knowledge, I will introduce some studies closely related to the development of a conscious robot.

8.1 Walter's Turtle

W. Grey Walter (1910–1977), a US-born researcher, will be remembered for his world's first life-simulating robot. Between 1948 and 1949, he created two robots named *Elsie* and *Elmer*. This was the era when the ENIAC electronic computer had just been developed. Driven by batteries, Walter's robots travel on three wheels. An oblong turtle-shaped shell protects the body, and a small chimney-like shape protrudes from the shell (Fig. 8.1). The front steering wheel turns and determines the traveling direction of the robot. The two rear-drive wheels control forward and backward travel. A light sensor (a photoelectric cell) is mounted on top of the chimney to observe the environment. The chimney is directly coupled to the

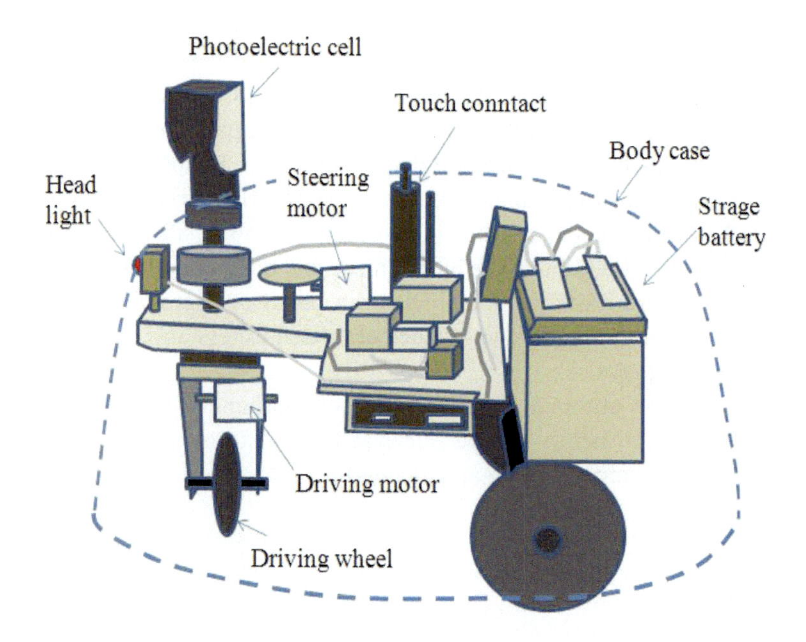

Figure 8.1 Internal structure of the robot.

steering wheel, so that when the sensor turns toward a light, the steering wheel turns in the same direction. The robot is provided with a touch sensor to detect a collision with an obstacle. The touch sensor is mounted on the shell and a collision is detected when the shell hits an obstacle while traveling. The robot has a headlight on its front to show its position. The intensity of this lamp seems to be relatively low. It is said that this lamp also serves to confirm the energization of the robot (a pilot lamp). The robot's electronic circuit comprises two electron tubes, several relays, and switches. This is a very simple circuit compared with a modern computer. Of course, this circuit was analog not digital.

Several landmark experiments performed by Walter are introduced in this chapter. His experiments were performed in a dark room since the robot was designed to react to light.

The first experiment that I would like to introduce was conducted with a high-intensity lamp and with a relatively large obstacle placed in the room. The robot was positioned such that it could not see the light due to the obstacle. The robot initially moved around without purpose. Accidentally, the robot hit the obstacle and moved back. As it moved back, the robot slightly changed its traveling direction (because it was designed to do so). The robot stopped and then again moved forward. After repeating this series of movements for some time, the robot accidentally cleared the obstacle and the lamp was visible to the photosensor of the robot. The robot reacted and turned toward the light and advanced. The robot seems to be designed to stop when the light intensity reached a certain level, then move back a little, and thereafter resume forward motion. When viewed from a macroscopic perspective, what the robot did was to go around, not toward the lamp. This behavior reportedly looked like a living being that was aiming at its prey while making a detour to avoid obstacles.

In the next experiment, Walter uses two robots and puts away the large lamp that he used in the previous experiment. The two opposing robots move ahead toward each other, aiming at their respective headlamps. Since the light intensity of the headlamps was small, the two robots repeatedly collided and backed up. The robots were described to be "engaging in a conversation."

Figure 8.2 Walter's self-recognition experiments.

Another feature of the robots is that they returned to their home base, called a dock, when their batteries were low for recharging. A strong light installed in the dock guides the robot into the dock. Walter's robot could move almost perpetually by recharging its energy like a living being.

The last noteworthy experiment of Walter that I would like to introduce is his self-recognition experiment using a mirror (Fig. 8.2). A robot is placed in front of a relatively large flat mirror. Apparently, this is similar to the experiment with two robots introduced above except that the second robot is the image in the mirror. The robot reacts to the image of the flame of a candle set on its shell and moves toward the image. The robot hits the mirror and thereafter repeatedly backs up, moves forward, and collides with the mirror. How is this different from the previous experiment? In the previous experiment, the movement of the two robots gets disordered as they repeat the same movement, but in this mirror experiment, the movement of the "two robots" (i.e., the robot and its reflected image) is synchronized because the mirror is used as an intermediary.

Because of its intrinsic property, the mirror traces the motion of the robot moving in front of it.

I think this was the world's first experiment that showed the possibility of having a robot discriminate between its self-image in the mirror and another robot. It is a wonder that such an experiment was successfully conducted in as early as 1950. This experiment suggests the possibility of mirror image cognition by robots. Mirror image cognition by a robot is a theme of study aimed at unraveling the mystery of human consciousness by exploring the human mechanism of mirror image cognition through the development of a robot with a built-in mechanism of awareness of its self-image reflected in a mirror.

A question, however, remains: Does Walter's robot really discriminate between its self-image and the other? This experiment seems only to show that the robot can discriminate between the other and some other. Can we simply conclude that "some other" is none other than the self? I think it is difficult to judge that "some other" is the robot's "self-image."

I personally believe that in this experiment the robot discriminated between the other and some other, or between two different others. In order to assert that some other is the robot's self-image, some concrete gadget to scientifically show their relevance must be built into the recognition mechanism of the robot, or at least, the relevance must be shown with scientific proof.

8.1.1 Summary and Observations

The small robot called a turtle developed by Walter was the world's first biological robot. Walter conducted many interesting experiments and demonstrated the possibilities of robots to people worldwide at expositions. Prof. Braitenberg and many other researchers owe their various enlightenments to Walter's experiments. Walter's experiments using mirrors, in particular, were epoch-making in the sense that they were the first challenge to the difficult problem of self-image cognition. I wish to acknowledge that his study was the first step toward the success of mirror image cognition by robots, and I express my respect for his achievements.

8.2 Kitamura's Robot

The conscious robot developed by Tadashi Kitamura seems to be deeply influenced by Brooks' subsumption architecture. Kitamura's originality is best exhibited by the fact that he exquisitely correlated the hierarchy of consciousness devised by Vietnamese philosopher Tran Duc Thao (1917–1993) to the hierarchization of behaviors formulated by Brooks using the Expression Model for Consciousness and Behavior that he revised and compiled (Kitamura, 2000).

The Expression Model for Consciousness and Behavior advocated by Kitamura is explained as follows. He sets nine conscious levels. Like Brooks' subsumption architecture, any conscious level subsumes all the underlying levels.

Level 0 is the lowest level, and from the viewpoint of phylogeny, Level 0 is equivalent to the basic knowledge of protozoa in the waking state. The behavior corresponding to this level is the basic reactions needed for existence. Levels 4 and 5 are nearly equivalent to those of mammals. Consciousness in these levels includes the stable discerning of objects and stable and sustainable feelings as well as the temporal and spatial relationships between objects. The relevant behaviors are detour, search, aim, be prepared, control limbs, and use media. Levels 7 and 8 correspond to those of humans. Consciousness at levels 7 and 8 include symbolic representations and concepts, and the relevant behaviors are the manufacturing of tools and the use of language.

The basic idea behind the hierarchization of the consciousness levels is the belief that the evolution of the brain nerves pertaining to behavior control is deeply related to the phylogenetic evolution of living organisms. On the basis of Thao's theory, Kitamura estimated the consciousness development process from the development process of the body and behavior considering phylogenetic evolution (Kitamura, 2000). Kitamura then tried to build his model in a machine. Figure 8.3 shows the conscious robot designed by Kitamura. The conscious levels range from 0 to 5 in this figure. The conscious system is basically designed to evoke consciousness by using the values of sensors to select the corresponding behavior, and then drive the motors according to the selected behavior

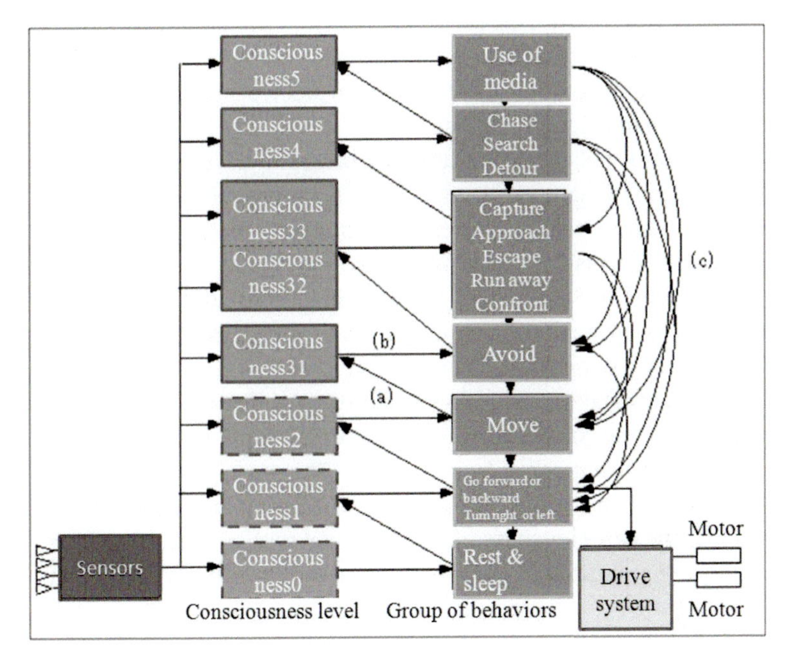

Figure 8.3 Kitamura's conscious robot model (partly modified by the author).

using a drive system. Figure 8.3 also suggests the similarity of Kitamura's idea to Brooks' subsumption architecture. Of course, Kitamura's architecture is significantly different from Brooks' theory in that Kitamura clearly defines consciousness to be possessed by a robot.

The relationship between consciousness and behavior is described now. Assume a robot in which a behavior is suppressed while the robot is performing that behavior. The robot moves to the next higher conscious level, selects and engages in the corresponding behavior. For example, a robot is at conscious level 2 and is performing a "move" behavior. The robot's "move" behavior is now suppressed. At this time, the robot raises its conscious level by one, or from level 2 to 31 (Fig. 8.3a). Specifically, the robot moves from the conscious level of the sensorial area evaluated by good and bad to another conscious level of temporary feelings such as emotions. This transition is effected by selecting the "avoidance" behavior (Fig. 8.3b). This series of behaviors can be summarized as "the robot avoided because its movement was blocked." The flow of consciousness would be that the robot avoided because it "was made sad to have met something bad."

Kitamura believes that the suppression of behavior is the suppression of consciousness. This is because he believes that Merleau-Ponty's assertion that "consciousness means behavior" is justifiable. This is a difficult phrase to understand. To put it plainly, the behavior of "avoidance" equivalent to conscious level 31, for instance, is because the meaning of the relevant conscious level is "temporary feelings such as emotions."

Kitamura makes a hypothesis that the suppression of consciousness gives rise to the desire of and search for a behavior that is expected to bring about better results. This means that to remove the cause of the suppression of consciousness and expect better results, the consciousness of an individual with a better developed or more complicated physical body, from the viewpoint of phylogeny, is necessary.

I will explain how the robot then elevates its conscious level to a higher level and selects a behavior corresponding to that level.

When more than one behavior exists at the same conscious level, all candidate behaviors are evaluated (the procedure for this will be described later), and a behavior that is evaluated higher than the behavior immediately before, even though the difference may be marginal, is selected. It is possible that the robot steps down layers and selects a lower level. For example, a robot chasing an object at level 5 using a medium, upon finding a new object appearing abruptly in its vicinity, may lower its conscious level from 5 to 2 in an attempt to access this new object by selecting the "move" behavior.

It should be noted that when a robot is behaving at a certain conscious level, behaviors specified for all lower levels are included as subconsciousness (and this term is used here by the author in place of unconsciousness as used by Kitamura to distinguish between them). Moreover, when the conscious level is elevated, all previous conscious levels that were used are retained as subconsciousness. When stepping down through layers out of necessity, all consciousness existent at the upper levels is cleared and the consciousness of the relevant lower level emerges.

Next, I will explain the method for controlling the consciousness architecture proposed by Kitamura by citing his writings (Kitamura, 2000).

Assume that C_i is a function representing the strength of consciousness to change the conscious level. I_i is the function used to decide a behavior given conscious level i, where $i = 1, \ldots, 5$.

$$C_i = |\beta_i| + |E_i| + |\gamma_i| \tag{8.1}$$

$$I_i = \beta_i + E_i + \gamma_i \tag{8.2}$$

where β_i is perception, E_i expectation, and γ_i physiological perception parameter. Perception β_i takes a positive or a negative value when the object is "love" or "hate," respectively. The absolute value relates to the distance to the object. Expectation parameter E_i exists for conscious level 3 and above.

For $i = 3$, E_i is proportional to β_i for a given time period and is thereafter proportional to $-\beta_i$ if β_i does not increase. If perception β_i of an object is "love," expectation is set high for a certain time period and then set low thereafter. This means that if a good result is not obtained within the given time, expectation turns to disappointment.

For $i = 4$ or 5, E_i is proportional to the sum of present β_i and the previous β_i's considering the effect of oblivion. This means that the magnitude of historical obsession of the robot with various objects is taken into consideration.

Selection of behavior is described as follows. Behavior I_i selected by the robot at conscious level i satisfies the following equation:

$$I_i(t) > I_i(t-1) \tag{8.3}$$

where t is current time.

When there is no behavior pattern satisfying Eq. 8.3 at conscious level i, the robot elevates its conscious level to $i + 1$. Parameters β_i and γ_i are used to characterize the robot. This includes, for example, emphasizing "love" or "hate" of objects and the state of starvation. It seems that these are currently being decided by the designers of respective robots.

Kitamura introduces four experiments in his book (Kitamura, 2000). One of them is explained here.

Robot H captures robot T in this experiment. Robot H has five conscious levels, whereas robot T has only three. From the viewpoint of phylogeny, robot H is physically superior to robot T, that is, its neural pathways are more sophisticated and functional. A relatively large obstacle and the nest of Robot T are prepared in the test area. Robot T is initially placed in the center of the test area and Robot H is set slightly away from Robot T. The positions of robots H and T, respectively, at time I are h_i and t_i.

The experiment starts with Robot T finding its favorite nest and approaches it by elevating its conscious level to 2 (position t_1). Robot H tracks robot T by elevating its conscious level to 4 (position h_1). Robot T takes a rest in its nest by lowering its conscious level. After resting a while, Robot T starts moving on receiving stimulus from its sensors and by elevating its conscious level to 2. Having lost sight of its target, and therefore having its current behavior suppressed, Robot H elevates its conscious level to 5 to start a series of behaviors of looking around for a medium and waiting for the target by hiding behind the medium (position h_2). Robot T discovers an obstacle and, because its behavior is thereby suppressed, elevates its conscious level to 3 to initiate an avoidance behavior. After having cleared the obstacle, Robot T lowers its conscious level to 2 to resume the move behavior. Through its sensor robot H catches sight of robot T coming out of the obstacle and chases and captures it (positions h_3 and t_3).

By observing this experiment, we know that Kitamura's conscious robot functions very well. Figure 8.4 shows the Locus of Ambush Behavior by Kitamura, which was partly modified by the author to describe this experiment (Kitamura, 2000).

8.2.1 Summary and Observations

Kitamura's achievements and unsolved problems are described as follows. He proposed a consciousness architecture based on Brooks' subsumption architecture. He hierarchized the levels of consciousness just like Brooks did with behaviors. In Kitamura's hierarchized consciousness, any conscious level subsumes all of its lower levels just like in Brooks' design where any behavior level subsumes all of its underlying behavior levels. The phylogenic com-

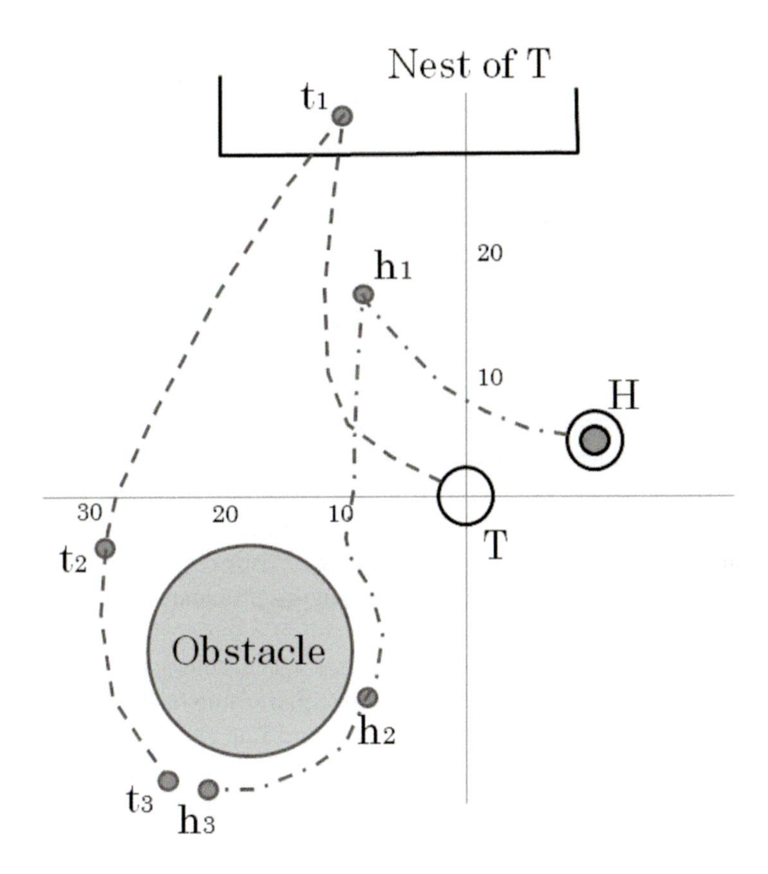

Figure 8.4 A rough sketch of Kitamura's experiments.

plexity of living organisms is taken into consideration in Kitamura's hierarchized consciousness; simple organisms are assumed to have simple consciousness and are deployed on the lower levels, whereas higher organisms are considered to have complex consciousness and are placed on the higher levels. Kitamura, thus, believes that all living organisms have a consciousness corresponding to their level of complexity.

Consciousness occurs, according to Kitamura, when transition to a higher conscious level occurs when the behavior at the current conscious level is suppressed. As many researchers agree, this is knowledge obtained from observing that when a human experiences that the current action is suddenly suppressed, the human's conscious activities are activated (i.e., they become conscious) and attempt to solve the problem.

Kitamura learned about conscious levels and their corresponding behaviors from philosopher Thao and applied it to the design of his consciousness architecture.

The author highly evaluates Kitamura's consciousness architecture, a machine consciousness theory, because it coincides well with the events obtained from phenomenological knowledge.

Brooks' standpoint is, however, basically behavioristic, and he does not acknowledge the conscious activities of humans and representations. It is therefore highly possible that Brooks would not accept Kitamura's layered consciousness and would disprove of it saying that everything can be explained by layered behaviors only.

Returning to our main theme, Kitamura describes the future course of development as follows. The first task to solve is related to the "emergence of I," that is, the problem of "first-person properties of consciousness." Kitamura believes that this problem can be solved by further increasing the levels of consciousness. According to Kitamura, the "emergence of I" occurs when consciousness has been elevated to the highest level where there is no further activity capable of enhancing the present comfort. This applies, Kitamura continues, to living organisms of low conscious levels as well.

Organisms at level 5, for example, can be conscious of the "emergence of I" but cannot speak of it simply because their consciousness is not high enough to engage in language activities.

Kitamura further says that it is difficult at this stage to correctly propose behaviors corresponding to conscious levels exceeding level 5, but such problems as symbolic processing can be solved by further elaborating conventional AI knowledge.

The duality of self-consciousness is described as follows. According to Kitamura, when "I" recall past experiences at the time when "I am emergent," the duality of consciousness that "I think of myself" may be said to be achieved. Regarding embodiment, Kitamura says it is a difficult technical task for a robot to become conscious that its limbs are its own. To solve this technical task, Kitamura continues, the robot needs to use the body schema of its own limbs. Kitamura says that the problems of the consciousness of others, feelings, and thought and the problem of chaos remain unsolved and will be continually studied in the future.

8.3 Jun Tani's Robot

Jun Tani built a conscious machine based on the belief that the conflict between the external and internal worlds gives rise to self-consciousness. He developed this idea after studying the theories of Martin Heidegger and many other philosophers and psychologists.

The external world means the environment of the robot as captured by its sensors. The internal world consists of a series of artificial neural networks accommodated inside the machine (Fig. 8.5).

Tani proposed an open dynamic structure featuring multiplexed neural networks and built it into his conscious machine (Fig. 8.6). The circuitry consists of a group of neural units (a) playing the role of a window for interaction between external and internal worlds and a cognition and prediction unit (b) including context loops. The context loops are used to predict the next occurring movements and objects based on past experience and have a memory to remember events in the external world.

Section b comprises a set of recurrent neural networks (RNN) responsible for two processes: cognizing sensor inputs and predicting the next occurring sensor values.

Let us assume that Tani's machine has learned enough in the external world, and as a result, it is expected to perform smooth movements. Actually, the machine has succeeded in learning smooth movements. Section b of the machine constantly outputs prediction values of "What and Where" to section a for collation with the present values. Smooth movement is assured when the prediction and external world coincide with each other. If the prediction fails to match, the robot's movement is not smooth. Look at the internal state of the robot at this time. You will see sections a and b are alternately stable and unstable. It is stable when the prediction agrees with the external world and unstable when it fails to agree.

As discussed in Chapter 7, the recurrent neural network has four states: divergence, convergence, orbiting, and chaos. We can

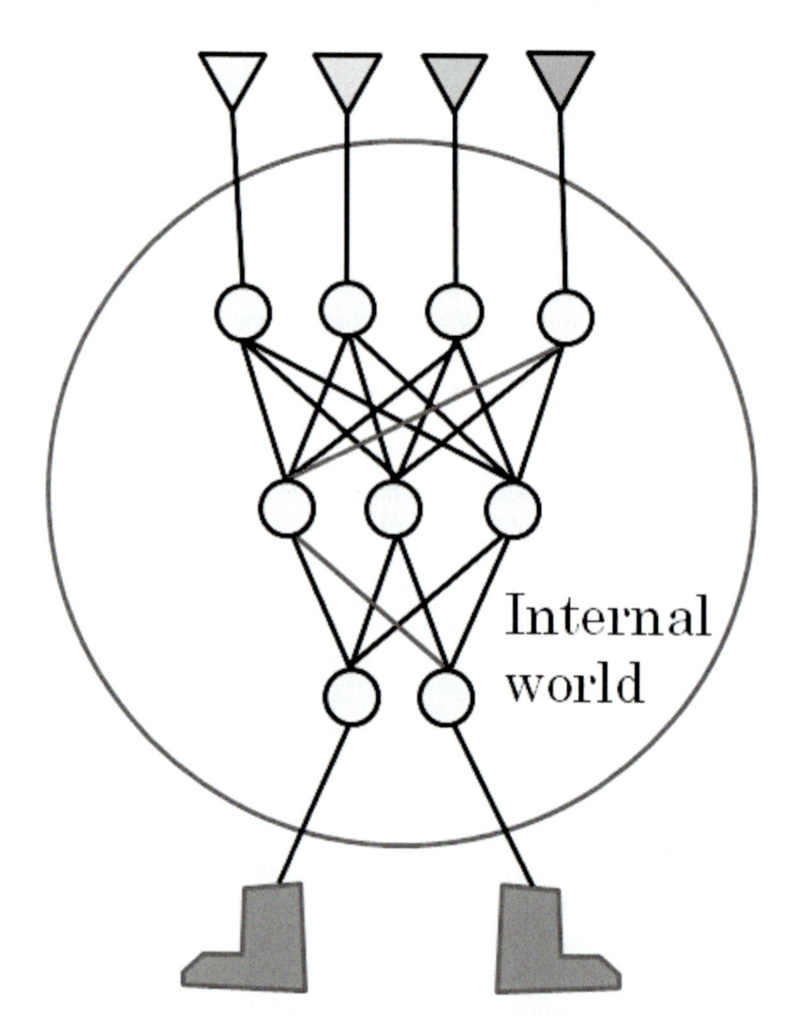

Figure 8.5 External and internal worlds.

safely exclude divergence because this robot has sufficiently learned in the external environment. The aforementioned stable state is equal to convergence. If RNNs repeat convergence even though their state is incessantly variable as the machine moves, the smooth

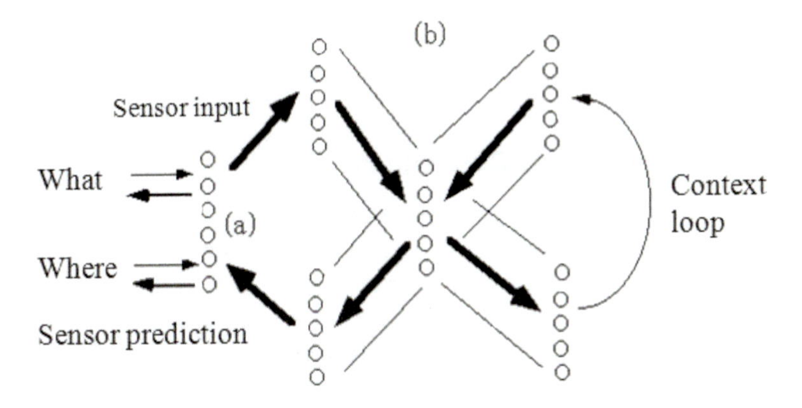

Figure 8.6 Open dynamic architecture.

movement of the machine is achieved. Orbiting, including quasi-orbiting, can be said to be a quasi-stable state. After all, the machine can continue to move as long as stable and quasi-stable conditions are maintained in the real world. Continuous stable movement thus means agreement between the predictions and the external world. The RNNs, while learning information about the external world that they successfully predicted, further strengthen their capacity to predict.

What happens, then, if we arbitrarily change the external world which the robot has learned? The RNNs get unstable and enter a state called chaos. When the machine enters the chaos state, the orbit is seriously distorted. In short, prediction and reality do not match, and it is no longer possible to predict. If this state is allowed to continue, the RNNs cannot escape from chaos. Tani at this point incorporates an arbitrator mechanism in his architecture. The arbitrator reduces the impact of the predicted values when in chaos. As a result, the machine is capable of re-learning information about the external world. According to Tani's interpretation, self-consciousness occurs in the machine when it is in chaos. Tani also explains that the state where the machine runs stably corresponds to non self-consciousness. He describes self-consciousness referring to Heidegger's hammer.

When a carpenter uses his hammer without any difficulty, the carpenter and the hammer are in unity, but when problems recur, they are separated and the carpenter is conscious of the hammer and he himself becomes self-conscious. This theory is supported by many phenomenological philosophers and psychologists.

8.3.1 Summary and Observations

Tani's conscious robot consists of multiplexed recurrent neural networks. He designed the open dynamic structure. In its basic design concept, Tani's robot captures information about the external environment to engage in cognition activities, while at the same time information about the external environment is predicted inside the machine. The robot moves in the environment smoothly when the result of the current cognition coincides with the internal prediction. If the prediction is different, the internal state of the robot enters chaos, and the robot moves awkwardly.

Tani interpreted that from a phenomenological standpoint, self-consciousness emerges when the robot enters the chaotic state. He relates the chaotic state in the open dynamic structure to the emergence of self-consciousness. However, is it possible to explain all the functions of human consciousness by relating the chaotic state to self-consciousness? There remains a serious question. For example, how would Tani explain the function of consciousness for willfully exploring objects based on chaos? If self-consciousness arises from the chaotic state as he argues, the above example must also be explained from the aspect of chaos.

8.4 Mitsuo Kawato's Examples

Mitsuo Kawato is a leading researcher of computational theory of the brain in Japan. His central theme of study is the elucidation of the human brain by a computational approach.

His famous research papers include, *Motor Learning Model of the Cerebellum, Computational Theory of Motor Control,* and *Computational Theory of Vision.* Of the wide range of his studies, his research related to consciousness is introduced in this section.

In his study of the brain, Kawato challenges some long-standing problems. One of them is called the binding problem. The author personally believes that this problem caused Kawato to decide to study consciousness. The binding problem originally emerged from the findings of the physiological study of the brain. It refers to a mysterious phenomenon related to the coincidence of the physiological fact that for an object that a human cognizes, the brain analyzes the object at the cerebral cortices in detail using the nervous system with the metaphysical fact that the human cognizes the object at the same time as the above physiological process goes on.

Assume, for example, you hold an apple in one of your hands and at the same time you look at the apple (Fig. 8.7). The brain sends commands to your arm and hand to drive muscles to keep holding the apple (Fig. 8.7a), while, at the same time, the brain receives information on the weight of the apple, its texture, and the somatic sensation of the driven muscles via the somatosensory system. At the same time, the eyes (Fig. 8.7b) capture the image of the apple and send the information from the retinas to the brain via the optic nerve system. The image of the apple is then transmitted to the visual areas V1, V2, V3, V4, etc. (see Chapter 3) of the brain and further to the modules that cognize form and color, and movement of the apple, respectively. The modules exchange information with the cerebral cortex and conclude the analysis of the apple (Fig. 8.7). It is obvious that information is exchanged with the memory modules. To hold the apple, various other areas of the brain related to control of the hands and arms are likewise assumed to be at work.

Given the physiological knowledge that the cognition of an object is achieved when visual information is transmitted to the brain through the optic nerve system and via various modules, the following question arises: When does cognition or awareness of holding an apple in the hand occur.

It should be noted here that "the brain cognizes" is different from "realization." Personally I believe that "realization" refers to "becoming aware of" or "something emerges on the consciousness." Even when you cognize "holding an apple," the cognition can be subconscious.

Let us return to the binding problem. Referring to the above example, the information on the apple as it is seen by a human goes to various modules via V1 through V4 until it reaches the columns of the cerebral cortex, where the analysis of the apple concludes.

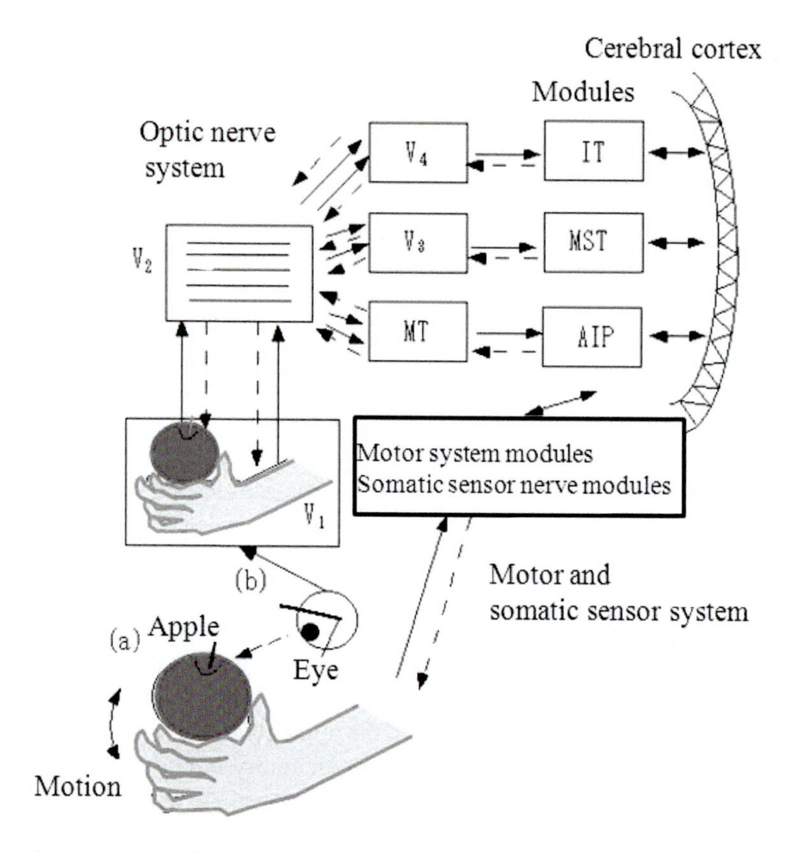

Figure 8.7 Binding problem and bidirectionality theory. Cited from Kawato's drawing in reference Kawato (1996), partly modified by the author to facilitate the explanation.

Since information about the apple is functionally divided into pieces, there arises the question how these small pieces of information are analyzed and finally assembled, or "bound," to form a cognition.

Where and when does this integration occur?

The result of the analysis of these segmented pieces of information may be used to explain that simple combinations lead to cognition as a kind of solution. It is obvious, however, that the number of possible combinations can be astronomical, and thus this explanation has no possibility of being accepted as a solution.

In 1988, Christoph von der Malsburg announced his theory of synchronous firing to solve the binding problem. According to this theory, the identity of an object is shown when groups of segmented information belonging to the object are synchronously fired.

Kawato posed three questions regarding this theory.

First, to identify more than one object by synchronous firing, the accuracy (i.e., window) of synchronization must be about 1 ms or 2 to 3 ms at the largest (Kawato, 1996), and he denied the possibility of the brain achieving this.

Second, a combinatorial explosion is not improved by this theory.

Last, there is no answer to the question, "Who will witness the synchronization?" Something like a central control center would be required to organize a multitude of modules, and this would bring us back to the first question.

Kawato concludes, "After all, any of the three proposed theoretical possibilities (a single monster integration center, module binding, and synchronous firing) fails, at least at the present time, to describe how a unified image is successfully formed and solve the binding problem, given the large number of visual areas and sparse binding (parallel flow of information in a hierarchical structure)" (Kawato, 1996).

Kawato proposes the bidirectionality theory that is allegedly possible to solve, even though not now, all of these problems. The bidirectionality theory has been developed to explain the high-speed processing of the human visual function. This theory is explained referring to the already-discussed example of an apple. When the visual information is segmented and transmitted to the cerebral cortex, the same information always flows in the reverse direction, and the sameness of the forward and backward information is maintained (dashed lines in Fig. 8.7). Neural pathways from a lower to a higher area are analytical paths (solid lines in Fig. 8.7). Analysis is generally a kind of modeling; so Kawato calls this operation forward modeling. The information from the result of the analysis is then fed back to the lower area via a neural pathway extending in the opposite direction from the higher level (dashed lines). The role of this reverse neural pathway is to integrate the result of the analysis and create an approximate model called a reverse model in the lower area, according to Kawato.

To provide the theoretical grounds, Kawato refers to the bidirectional binding between areas of the cerebral cortex, which is an established anatomical fact about the brain. He argues that his theory solves the first and the second of the three problems already mentioned. Specifically, the integration of information can be finally performed in relatively low areas according to his theory. More detailed information, when necessary, is obtainable at any time by accessing the higher level cortices. This way, the second problem of the combinatorial explosion can be avoided. The binding of all areas is necessary if you follow the theory of synchronous firing. It is clear that this problem is also avoided in the bidirectionality theory. Kawato does not seem to discuss the third problem.

I will now describe how consciousness is addressed in the bidirectionality theory. Kawato applies his bidirectionality theory to the problem of not only visual processing but also sensorimotor integration. His study is considered a sort of imitation learning. This means that he tries to analyze information obtained by the vision system (forward model) and integrate the results in the motor system (reverse model). The idea of bidirectionality can be used because "one moves as one sees" is the essence of imitation learning. Kawato then extends the concept of forward and reverse models (Fig. 8.8). I have partly modified his drawing here to facilitate understanding.

Section A in Fig. 8.8 represents a set of sensors for hearing, vision, and the like. Section B is a set of drive systems for the limbs. In the middle part of the figure, there are neural networks and modules M_i conceived in the bidirectionality theory. Modules M1, M2, M3, …, for example, are responsible for analysis and integration of perception, while modules M4, M5, M6, … are related to analysis and integration of motion.

Kawato does not describe the connection between the perception and the motor system. However, if sensorimotor integration is achieved by the bidirectionality theory as asserted by him, Kawato's argument that the bidirectional forward and reverse models can generalize other individuals, one's own brain, and part of it, and society existing in the external world is indeed persuasive enough. Kawato also asserts that his bidirectionality theory has the merit of cognizing or recognizing the patterns that are created by motor

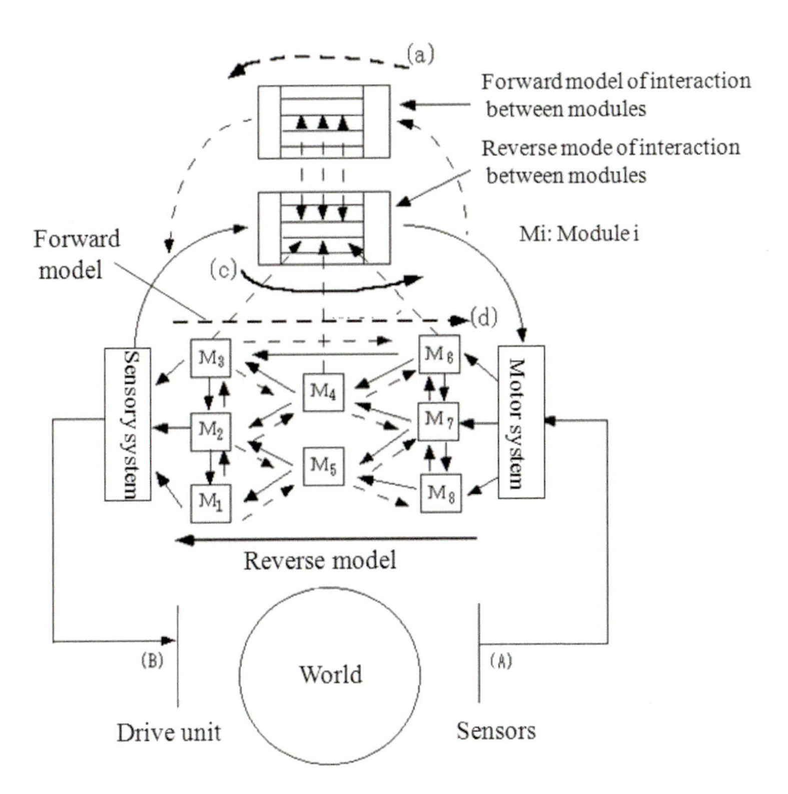

Figure 8.8 Bidirectionality theory and selfconsciousness (Kawato's drawing was simplified by the author for this description).

control. He also discusses the problems of consciousness. According to him, "Consciousness exists to accelerate integration of the modules in the brain." He says, "Self-consciousness may be partly explained as 'a certain part of the brain constructs the forward or reverse models of interactions for a number of modules" (Kawato, 1996). Furthermore, he says, "The role of the forward and reverse models of interaction between the modules is to set favorable initial conditions for the relaxation algorithm by an intermediary of a number of modules, and accelerate the relaxation algorithm for the integration of the modules." He explains, "When our consciousness works normally, we feel the world's only perception image and are executing a consistent action plan."

It is not difficult, according to Kawato, to model interactions between modules. The problem is if the total memory of the interaction models becomes equal to or exceeds that of the main bidirectional modules, the merits of convergence and high-speed processing are lost, and hence the interaction models must be poor. "Forward models of interactions between modules can be regarded as the approximate models of other areas of the brain," Kawato says, "Based on this hypothesis, we can say that consciousness is a process of approximating the huge volume of parallel calculations that occur subconsciously to resolve the ill-posedness of sensorimotor integration with extremely simplified false serial calculations." He further argues, "Forward models of module integration, generally, are similar to reverse models of the external world because they convert sensory inputs into motor outputs. Likewise, reverse models of module integration can be said to be forward models of the external world. These computations circulating through the forward and reverse models make it possible to emulate one's own thoughts, predict one's own behavior and perform the act of introspection." Emulation here means to run part or the whole of a system on other software or hardware that behaves like the original system.

Kawato's writing is not so easy to follow. I will interpret it as I understand it, risking the claim of misunderstanding. Forward models of interaction (Fig. 8.8a) are approximations of reverse models (Fig. 8.8b) of sensorimotor bidirectionality modules, and the reverse models (Fig. 8.6c) of interaction are approximations of forward models (Fig. 8.8d) of the bidirectionality modules. These approximate forward and reverse models are considered to be situated above the real-world models (forward and reverse models of the sensorimotor bidirectionality modules) and behave as an emulator that approximates the real world as a two-tier world. Kawato's argument that computations circulating through the approximate forward and reverse models, which are sparse due to the two-tier structure, enable the "emulation of one's own thought," "prediction of one's own behavior," and "introspection" is understandable as an idea but lacks concreteness.

Kawato further discusses the flow of consciousness.

> Each forward or reverse model of module interactions comprises three parts: interface to the sensory module presumably located in the prefrontal area, interface to the motor area assumed to be partly located in the basal ganglion, and the model itself which is said to reside on the outside of the cerebellar hemisphere, and is divided into several micro-zones as shown with the vertical lines in the figure. This means that consciousness has just one stream but the direction of the stream of consciousness is controllable. When driving a car, for example, you can converse with passengers, listen to music or concentrate on driving. This is because although the interface to the sensory and motor modules is just one route, several models are prepared to cope with certain possible combinations of modules (Kawato, 1996).

Kawato's description lacks clarity, but he seems to mean that the flow of consciousness (the reason why "a driver is capable of conversing with passengers") is related to the micro-zones located in the interaction modules.

8.4.1 Summary and Observations

Kawato studies the integration of the human sensory and motor systems. He needs to solve the long-standing "binding problem" that has been annoying brain researchers. To solve this problem, Kawato proposes the bidirectionality model for sensorimotor integration. According to Kawato, the proposed method is capable of suppressing a combinatorial explosion and other problems compared with conventional solutions. He further proposes the idea of module interaction models as a higher level of a bidirectionality model to accelerate the processing speed of various modules in the bidirectionality model. His interaction models are approximations of the already-mentioned bidirectionality models. He explains that the interaction models can approximately reproduce the functions of the bidirectionality models He explains that the "emulation of one's own thought," "prediction of one's own behavior," and "introspection" are enabled because the interaction models approximately reproduce the bidirectionality models. He further says that self-consciousness may be explained in like manner.

According to Kawato, consciousness is a process of approximating the huge volume of parallel calculations of sensorimotor integration that occur subconsciously with simplified false serial calculations. He says that the flow of consciousness is related to the microzones of the interaction models.

The author acknowledges the value of Kawato's bold hypothesis introducing bidirectionality models to explain consciousness but at the same time cannot see any concrete explanations. Kawato seems to have not yet formed a consistent theory about the function of human consciousness.

8.5 Cynthia Lynn Breazeal's Kismet

Cynthia Lynn Breazeal developed, in cooperation with Brooks, a robot named Kismet, which is capable of interacting with humans in real time. The humanoid head robot can move its eyes, eyebrows, mouth, and neck (Fig. 8.9), and it listens and talks. This sociable

Figure 8.9 Head robot Kismet.

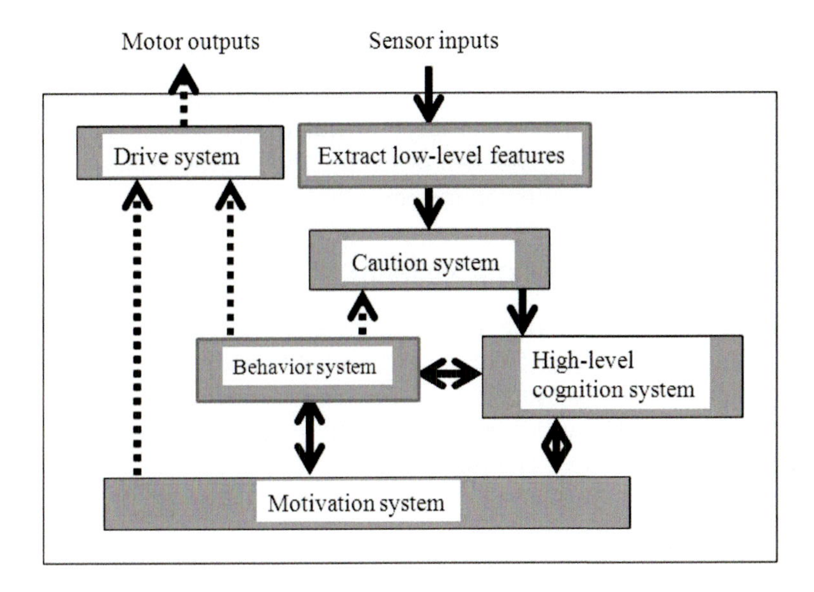

Figure 8.10 Outline of Kismet's drive system.

robot understands feelings using information from differences in height, pitch, etc., and expresses its feelings vocally. Its facial expressions also show feelings. The drive mechanism makes use of the psychological knowledge of the Theory of Mind. An outline of the circuit developed on the basis of this theory is shown in Fig. 8.10.

With these circuits, the robot reacts by driving motors in response to the stimuli given to the system via sensor inputs. Various subsystems are layered and implement their respective functions. As the essential part of the system, the information obtained by the sensors is subject to the extraction of low-level features. On the basis of the resultant information, the high-level cognition system is activated by the Caution system, that is, the cognition process is a two-stage process. There are two closed loops for processing information internally. In one circuit, information circulates through the Caution, High-Level Cognition, and Behavior systems. In the other circuit, information runs through the Motivation, Behavior, and High-Level Cognition systems. In the former circuit, if any information has a feature to note, the robot turns to the object or the vision cameras are rotated by drive motors while using the Behavior system as if a human were looking at an object and watching it intently. This is intended to achieve a higher-level cognition. In the latter circuit, the robot's attention is given to already-learned information such as color and shape, to achieve high-level cognition. The two circuits each have their own roles. The Motivation system is the highest-level layer in this system.

8.5.1 Summary and Observations

What is most interesting about this system is the parallel use of bottom-up processing for driving the robot based on information transmitted from sensors and top-down processing using information from the Motivation system, which is located at the top of the subsystem hierarchy. The robot achieves interaction with humans smoothly while solving the problem of competition of the two different types of processing.

What interests me is to find out where in these circuits lies the essence of consciousness.

Chapter 9

New Architecture of Robot Consciousness and the Robot Mind

This chapter introduces the conscious robot that passed our mirror image cognition tests for the first time in the world (Fig. 9.1). Mirror image cognition means to "cognize the image of the self in a mirror." It is considered to be the source of self-consciousness.

What is the mechanism of human consciousness?

This chapter discusses this mechanism.

The electronic edition of the Discovery Channel in the United States reported on this robot on December 21, 2005. Within two or three weeks after the report, about 26,000 secondary articles and references to this robot appeared on Web sites all over the world. The robot was discussed on many Web forums, and the information spread throughout the Internet, where many comments of both praise and refutation were seen.

The effect of the publicity on the Discovery Channel was enormous. Four months after its appearance, the conscious robot was listed in the second place in Google search results for the keyword "self aware," which logged about 132 million hits. The conscious robot was, of course, at the top of the list related to scientific research.

Self-Aware Robots: On the Path to Machine Consciousness
Junichi Takeno

ISBN 978-981-4877-90-9 (Hardcover), 978-1-003-26181-0 (eBook)
www.jennystanford.com

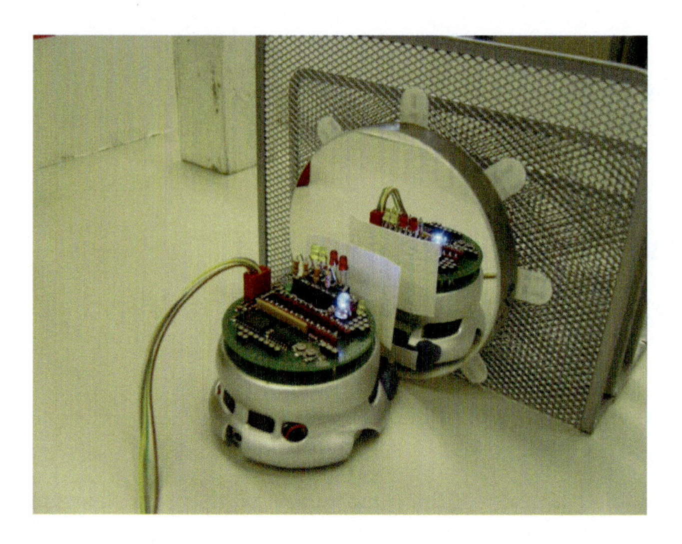

Figure 9.1 Conscious robot that passed the mirror image cognition tests.

9.1 Introduction

What neural circuits are used, and how are they used, to make up the mechanism of human reason? How is the mechanism of emotions and feelings related to reason? What is consciousness and self-consciousness? What is the difference between subconsciousness and explicit consciousness? What is the meaning of "to feel?" What is free will?

The author aims to answer these questions. I would like to innovate conventional artificial intelligence (AI), which is presently at a deadlock. I would also like to construct a totally new AI, which I call artificial consciousness (AC). I am also trying to construct a model of the human brain. I do not mean to use the risky anatomical analysis of the brain, but I will construct a mechanism to represent human consciousness in an embodied robot. The rich knowledge I expect to derive from my research will help resolve the mysteries of the brain. If we could uncover these mysteries, we could reap immeasurable benefits from our efforts.

First, AC-based machines will be designed and developed to ever higher levels. Conventional automobiles, for example, have been designed to follow the driver's instructions, whereas an AC-based automobile will function as though a dedicated operator were driving it in place of the human driver. The car runs automatically, always checking its safety, and observing the instructions given by the driver from time to time. Even if the driver were to faint for some reason, the car would continue running safely until it slows down and stops at a nearby safe place, or may continue on and arrive at the destination if time is not a critical factor. The difference from conventional vehicles is the intrinsic safety derived from the autonomy of the car and enhanced comfort. The driver's stress would be greatly reduced.

Second, AC technology will create a simultaneous interpreter robot that speaks natural languages. The robot interpreter will engage in simultaneous interpretation, the quality of which would surpass human interpreters. It understands homonyms from the context and appreciates and returns with a joke and irony when appropriate. In a hospital, the AC robot selects and avoids speaking upsetting words and phrases that may touch on diseases.

Third, the AC robot will attend to old people in need of care and physically handicapped persons and provide physical and mental care for elderly living alone. The robot will take care of these people while being conscious of their mental and physical conditions. It reaches out and supports them to walk safely. The AC robots will support people in wheelchairs and bedridden patients by speaking with them gently, encouraging, listening and then advising, or otherwise providing mental support.

Fourth, the AC robot will offer ideas for survival in the event that a human is in trouble and continue to support the person as long as its energy lasts.

Fifth, it is possible to develop artificial limbs that human users will feel are a part of their own body.

Sixth, the AC robot has the function of self-consciousness in principle and always monitors its own activities. When any problem occurs in the AC system, the relevant information is used to anticipate and prevent catastrophic breakdown of the robot's overall system to minimize damage. There is also the possibility of self-repair. With all these features, AC robots will contribute to the safety of people all over the world.

Lastly, it would be possible to discover a medical treatment for people suffering from serious mental illnesses such as schizophrenia. If the mechanism of consciousness can be eventually elucidated, a technology to resuscitate lost consciousness might be developed.

The author's idea is that "consistency of cognition and behavior is the origin of human consciousness." Cognition means to understand. The idea, therefore, means that the function of consciousness is to behave and to understand the behavior to maintain consistent information.

The author has developed, on the basis this idea, a module to build consciousness on a robot, which is called Module of Nerves for Advanced Dynamics (MoNAD). The consciousness system is constructed by arranging multiple MoNADs in a layered fashion.

9.1.1 Research on Consciousness and Cognitism

The first thinker to be introduced when it comes to human consciousness is French philosopher Rene Descartes (1596–1650), who was briefly mentioned earlier. He was the first person to assert the existence of the self on the grounds that "I feel myself" by saying, "I think, therefore I am" (Descartes, 1997). His approach is called dualism, which asserts that the mind and the body are separate entities.

Later, German mathematician and philosopher Gottfried Leibniz (1646–1716) attempted to treat the body and the spirit integrally in his Monadology (Leibnitz, 1714).

No scholar or researcher has ever reached a unified view of the body and mind based on physical understanding.

When a human is conscious of something, he or she feels it by himself or herself, and nobody else is involved. This is because to be conscious is a subjective matter and, some may argue, rejects analysis by science. To study consciousness through a scientific approach, a technique was devised to describe the reactions of humans to stimuli. The belief in this technique is called behaviorism.

Since about 1960, scientific findings have been announced one after another showing that complex processing is going on in the human brain. This made it necessary to explain brain activities and stimulated researchers to this end. One of these scientific findings reads like this: Assume, for example, a mathematical question of whether certain numbers are included or not in a given set of numbers. It was revealed that the time required to answer the question became longer as the digits of the numbers increase (Sternberg, 1966). The theory that accepts the existence of complex processing in the brain is called cognitivism.

Later, many advanced scientific equipment and systems to measure brain activities of humans were developed, including electroencephalography (EEG) and functional magnetic resonance imaging (fMRI). These technical advances assisted in the discovery of more scientific grounds for the existence of complex processing in the brain.

As of today, it is known that

(1) The brain engages in complex processing when solving a complex task.
(2) Specific areas of the brain are activated.

Behaviorism has evolved into cognitivism as a result of the development of science. Specifically, the description of human reactions to external stimuli has been gradually replaced by research on the internal processing of the brain. There still exists, however, criticism that subjective problems based on the sense of the self such as consciousness and the mind still lack scientific grounds.

9.1.2 Husserl's Phenomenology

Phenomenology proposed by Edmund Husserl (German philosopher, 1859–1938), mentioned earlier, is introduced in this section again (Husserl, *Die Idee der Phaenomenologie*). Phenomenology is a difficult philosophical idea but is described here as the author understands it.

Phenomenology is a philosophical method to establish a thought that has not been established. Simply put, a hypothesis is first proposed regarding an unknown world and studied by comparing it with phenomena emerging in the real world. A new hypothesis, agreeable to a larger number of people in the field, is presented on the basis of the results of the study. The hypothesis may at first be a subjective idea of a researcher, but by repeating the above process, an agreement between subjectivity and objectivity is eventually reached. At this time, the phenomena have acquired scientific grounds. By using this philosophical method called phenomenology as advocated by Husserl, one can thus establish scientific grounds for as yet unproved phenomena by following scientific processes. This method, also known as scientific positivism, is actually a representative method used by researchers in the study of unknown worlds.

Consciousness and the mind are subjects of study for which no scientific grounds have been established until now, and that is why the phenomenological method is actively being used to study them.

9.1.3 Definition of Consciousness

The author selected consciousness as the theme of his study, believing that the mind is the embodiment of consciousness. The author further defines consciousness as a function to perform cognition and behavior consistently, the details of which will be explained later.

What does it mean to become aware?

As a result of phenomenological observation, the author believes that to become aware means to realize that one is doing something.

To feel is a phenomenon that occurs when one becomes aware. The author believes that to feel and to become aware are different because to say that one becomes aware of feeling is understandable, whereas to say that one feels becoming aware is a somewhat uncomfortable usage of language. This suggests that in the human brain, the concept of feeling is at a lower level than that of becoming aware, and the two are independent. Thinking in this way, it seems that to become aware is synonymous with to be conscious.

What, then, is consciousness?

The author uses Husserl's 10 features of the functions of human consciousness (Husserl, *The Essential Husserl: Basic Writings in Transcendental Phenomenology*), for which scientific grounds have been established by Husserl himself using the phenomenology he proposed. The 10 features have been demonstrated by Prof. Tadashi Kitamura at the Kyushu Institute of Technology (Kitamura, 2000; Kitamura *et al.*, 2000).

9.1.4 Features of Consciousness

The 10 features of consciousness are described as follows:

(1) First-person property: First-person property is the sense that one is performing all things, i.e., a belief in the existence of the self or mind–body monism.
(2) Orientation: Orientation means that consciousness is always directed toward something.
(3) Relationship between action and result; duality of self-consciousness: Relationship between action and result refers to the chaining of an action and its result. Humans always feel a relationship that a certain action has a specific result. The duality of self-consciousness means, in other words, that one is aware of oneself.
(4) Expectation: Expectation indicates that humans are always predicting the immediate future.
(5) Function of determination and conviction: The function of determination and conviction is identical with the belief in the existence of things.
(6) Embodiment: Embodiment is the feature that the body is part of the self. All of us are conscious that our body is part of our self. The importance of embodiment in consciousness is addressed in detail in the *Phenomenology of Perception*, by Maurice Merleau-Ponty (1908–1961) (Merleau-Ponty, 1945).
(7) Consciousness of others: Consciousness of others is the feature that enables us to discriminate our self from others.
(8) Emotional thought: Emotional thought is the feature that reason is related to emotion and feelings.
(9) Chaos: Chaos means that consciousness is ceaselessly out of balance.
(10) Emotion: Emotion refers to qualia of consciousness. The human senses of taste, hearing, smell, color, pain, etc., are deeply related to qualia.

A consciousness system, when proposed, should be tested using these 10 features proposed by Husserl. How far the proposed system can describe these features is an important index in the evaluation of the proposed system.

The consciousness system proposed by the author can describe all of these 10 features. Only the item in No. 10, qualia have not been achieved on the robot constructed by the author.

Figure 9.2 Imitation of Facial Expressions by Infant (images created by the author).

9.1.5 Important Research Examples Related to Consciousness

Let me first introduce an example shown in the research of imitation behavior by infants as reported by Meltzoff and Moore (1977). They discovered that a four-week-old infant imitated facial expressions presented in front of the infant, such as sticking out the tongue, opening the mouth, and pursing the lips. According to them, infants are capable of imitation as soon as they are born. An imitation scheme is already wired in the neurons of the brain when a baby is born, i.e., imitation is an innate ability (Fig. 9.2).

There is a case study in imitation behavior (Lhermitte *et al.*, 1986) which was observed on a patient with partial damage to the frontal lobe (Fig. 9.3). The patient in question was unable to suppress imitating the behavior of the person sitting in front of him. When the person in front of the patient reached out for a pair of eyeglasses and put them on, the patient immediately behaved in the same way. This case underpins the hypothesis that imitation is one of the fundamental behaviors of humans.

Mimesis theory is worthy of mentioning here as another discussion of imitation. According to the theory, imitation was an important means to bring people together in an age when humans had not yet developed languages (Donald, 1991). This hypothesis suggests that imitation was an important function of humans in the history of human evolution.

There is another important research result. It is the great discovery of mirror neurons (Gallese *et al.*, 1996). by Prof. Giacomo

Figure 9.3 Examples of imitation behavior (from Lhermitte *et al.* (1986)).

Rizzolatti, a neuroscientist at the University of Parma (Arbib, 2002; Rizzolatti and Sinigaglia, 2006). The discovery of mirror neurons showed the existence of common neurons related to cognition and behavior in the brains of monkeys and humans.

These numerous findings show that the mechanism of imitation is critically important for human consciousness. This leads to the fact that when one claims that the consciousness system they propose is capable of simulating human consciousness, then the proposed system must possess the function of imitation as a basic feature.

9.2 Proposed Concept Model of Consciousness

The computation model of the consciousness module proposed by the author is introduced in this chapter. The model is described in a format allowing creation on a computer. The correlation with Husserl's 10 features is also explained. The reason why the self and others can be discriminated and why imitation is possible is also discussed.

9.2.1 Artificial Consciousness and Design of the Mind

Consciousness has been an important research theme in recent years and is actively being studied in many disciplines, including philosophy, psychology, brain science, cognitive psychology, and robotics.

Consciousness is defined variously as the working of the mind that recognizes and aspires, the state of mind that one knows what one is doing now, and something that includes and underlies all workings of our knowledge, feelings, and will. The mind is used in a broader sense than consciousness and is defined as the whole of knowledge, feelings, and will, whereas consciousness is generally understood to be something underlying the mind.

As mentioned earlier, the study of mind started with the philosophical discussions of Descartes, who launched mind–body dualism. The study of the mind evolved with research on the subconscious by Freud and other psychologists.

To facilitate scientific discussion of the subjective matter of the mind, phenomenologists, including Husserl and Merleau-Ponty, offered the necessary methodologies.

Brain scientists are engaged in researching the scheme of the consciousness and the mind, with the brain as the central theme, by directly stimulating the brain or by monitoring activities of the brain with the help of PET, MRI, and other investigation systems. Despite these numerous efforts, the location of consciousness and the mind has not been identified.

The discovery of mirror neurons is considered very important among all these achievements by scientists. The existence of mirror neurons is important because their existence supports the hypothesis that higher organisms learn by imitation.

The connectionists, including Prof. Daniel C. Dennett at Tufts University, argue that consciousness can be created artificially (Dennett, 1991) although concrete plans have not yet been shown.

Brooks and his colleagues at MIT have created COG, Kismet, and other robots to study natural learning systems through the interaction between infants and the environment and expect that consciousness and the mind will emerge from these learning procedures.

While partly agreeing with the ideas of these evolutionary robot researchers, the author believes that rather than attempting to create consciousness with evolutional techniques, we should develop artificial cell networks step by step while defining a new paradigm regarding consciousness. As mentioned before, the author doubts that consciousness can be created using the present evolutionary techniques.

I believe that a new paradigm must be established to generate consciousness in a machine. Descartes advocated a mind–body dualism, but my standpoint is a materialistic mind–body monism. I believe that something like human consciousness can be constructed in a machine. A robot with something like consciousness would behave as though it had a will.

9.2.2 Expectations for a New Paradigm

The author believes that the imitation function and the function to distinguish between the self and others are important factors of human consciousness. Humans can distinguish the self from others using a function that discriminates between the self and others. They learn the behavior of others by just looking at them, thanks to the imitation function. These two functions are the prime mover in the evolution of humans into higher organisms.

The research studies by Meltzoff *et al.* have shown that infants instantly imitate the behavior of their mothers. It was quite strange for the researchers to learn that the infants performed the same behavior in a short period of time (Meltzoff and Moore, 1977).

When a mother sticks out her tongue, the child instantly does the same. The child seems to move to action immediately by just seeing the mother. When I touch the keyboard of a computer, I mostly strike the keys subconsciously but I never think that the fingers striking the keys are those of someone other than myself. When a person resembling me stands nearby, I never mistake him for myself.

Self-consciousness that tells that the image of oneself reflected in a mirror is the self seems to be deeply related to the function of distinguishing between the self and others.

The author is trying to discover the source of consciousness.

There are many researchers studying the problem of "What is consciousness?" Nevertheless, no assertion has ever been acknowledged to be decisive enough to solve this problem. This is because a unified paradigm regarding the phenomena of human consciousness had been missing.

The author has devised a new paradigm of consistency of cognition and behavior as the source of human consciousness.

This chapter shows that the new paradigm can describe almost all of the phenomena of human consciousness in a computationally feasible manner. The author's arguments may contain, in various points, findings that have already been discussed by other researchers, but what is asserted by the author as a whole is a unified paradigm capable of explaining many phenomena related to human consciousness and the mind.

The underlying idea of the system introduced herein is the construction of neural networks essentially defined between inputs and outputs based on this new paradigm. The system continually evolves by capturing information about the surrounding environment.

The author believes that thought is generated from conscious activities, i.e., a stream of consciousness. Consciousness here includes the subconscious. Explicit consciousness, generally referred as simply consciousness, occurs when the subconscious satisfies certain conditions, which will be discussed in detail later.

The mind is activated by the stream of consciousness and the accompanying stream of *kansei*. *Kansei* is a Japanese word that means both emotion and feelings. Consciousness is considered to occur within the brain according to the internal state of the brain and on receiving stimuli from the body and external environment. The new paradigm is capable of explaining the root source of consciousness and also describes imitation learning, distinction between the self and others, self-consciousness, and qualia. It can also describe imaging and creativeness. Furthermore, this new paradigm makes it possible to draw a distinction between sense and hallucination.

The methodology based on the author's paradigm resembles Minsky's idea that in the human brain, higher-level areas have evolved biologically from the old areas, i.e., the old areas are

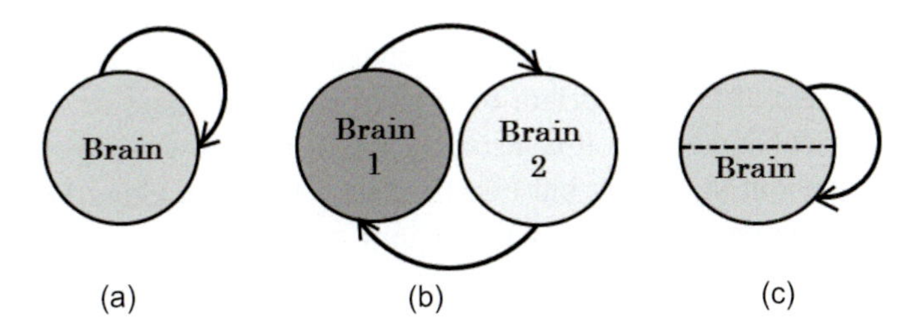

Figure 9.4 Development of the brain and self-recurrence.

supervised by the newly evolved areas. This idea may remind the reader of a centrally controlled system where the new brain dictates the activity of the old brain, but the author's methodology is different. The author's methodology is the same to the extent that the new and old brain areas co-exist, but the difference is that rather than the new brain controlling the old brain, information circulates between them. This idea may not sound smart, but using this idea it is easy to explain the function of self-consciousness that is defined as the state of being conscious of one's own consciousness.

It is easier to understand the model of self-consciousness by looking at Fig. 9.4a and then Fig. 9.4b. In Fig. 9.4a, the self is looking at the self, i.e., the self-loop type. This is a typical self-recurrent model. In Fig. 9.4b, the self is looking at others, i.e., the co-routine type.

The reason why the author selects the co-routine type is the difficulty of explaining the internal circulation of information using Fig. 9.4a. With the co-routine type, it is easier to understand that information circulates through both areas. If this description is possible and successful, we may move to the smarter self-loop type.

When one of the two routines is combined with the other (Fig. 9.4c), the result is the same as the model shown in Fig. 9.4a. Accordingly, the author hereby declares that he is not using the centrally controlled model.

9.2.3 Where Does Consciousness Come From?

What kind of state is it in which we are conscious of something?

I am now writing letters on a notebook with a pen. Or I am thinking of my family, solving math problems, walking down a street in Paris, etc. It is appropriate to express this awareness as being conscious of something. It is also possible that I was walking while thinking about my family, but I was not conscious of walking. When this phenomenon is to be written into a computer program by analogy, a number of software agents are working in the computer and one of them is selected by means of some function.

Even if the computer displays a message on its screen that one of the simultaneously running multi-tasking programs is currently active, it would be difficult to say that the computer now possesses a certain consciousness. To overcome this difficulty, some researchers argue that consciousness requires embodiment (Pfeifer and Scheier, 1999) but this proposition would be promptly refuted by the opinion that computers can also have embodiment such as by the keyboard and the hard disk. A computer may be aware, through its functions, that its hard disk is now being accessed. Even so, it is difficult to accept that consciousness occurred in the computer. Why?

The author believes it is possible to define the relationship between the operation of the computer's hard disk and consciousness, but this, by no means, indicates that I am ready to admit the occurrence of consciousness in the computer. This is because the existence of consciousness is a subjective phenomenon felt only by an individual or by myself. I just feel that others also have consciousness as I do, but its existence is difficult to prove.

The mirror neurons described earlier seem to make us believe that other people also have consciousness. Since computers do not have a function similar to mirror neurons, it is natural that we cannot feel that computers have consciousness.

Some researchers thought about relationship between Turing tests and the human consciousness (Mogi and Taya, 2003). This means that they expect the action of the mirror neurons of humans. From these observations, one might say that for humans to be able to readily feel the existence of consciousness in a machine, the machine must constantly stimulate the mirror neurons of humans. This means that to create consciousness similar to that of humans in machines, the functions of mirror neurons must be built into the machine.

A clear definition of consciousness capable of describing the various functions of human consciousness must be established on the basis of the functions of mirror neurons. Since mirror neurons are related to the smooth perception of consciousness in others, the shape of the machine may have to be a humanoid or human-like robot. In the case of machines that are not human in shape, the existence of consciousness may be judged by whether the machine physically has the functions described in the above definition of consciousness.

9.2.4 How Do We Define Consciousness?

Let us consider the existence of consciousness referring to an example: The sound of a voice is input into the ears, and in response to that stimuli, the action of speech from the mouth is performed.

This example does not spoil generality. The area responsible for consciousness is assumed to be located in the region with the question mark in Fig. 9.5a. This is the area where neural networks exist.

The above thinking is behavioristic because the description reads that inputs entail outputs. If we apply an evolutionary algorithm to this region, neural networks with the functions of consciousness can be created in this region at some time.

Given the complexity of the phenomena of consciousness, it is difficult in reality to use the evolutionary approach and determine what functions are appropriate for creating neural networks equivalent to consciousness. If consciousness consists of a redundant system as I suggested earlier, it is quite doubtful that the evolutionary technique would ever be useful. If the mechanism of consciousness is a redundant system, an evolutionary technique that relies on computers must find suitable functions by trying out infinite possibilities. The author believes that the mechanism of consciousness is a kind of redundant system. I know that the evolutionary technique is excellent as already described, but it is impossible to find the mechanism of consciousness using this technique. However, I also think that a redundancy circuit could be

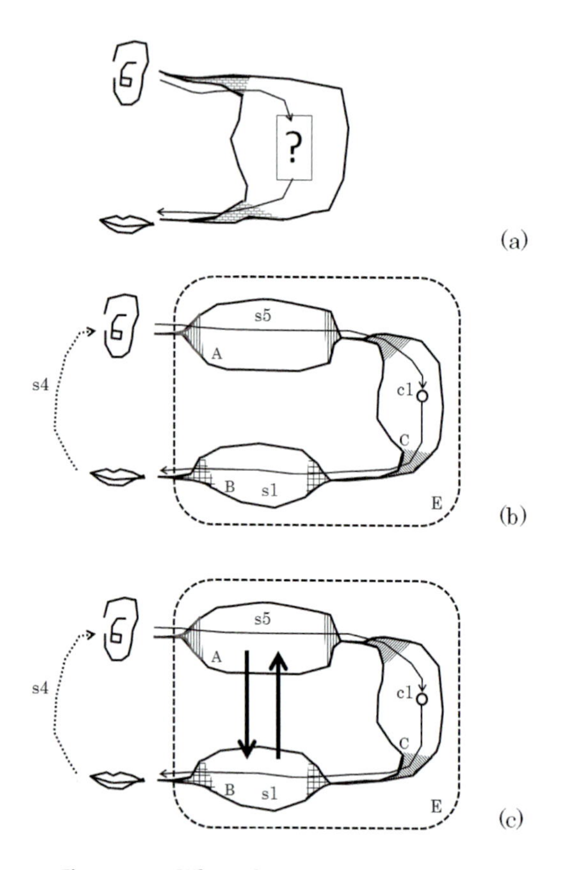

Figure 9.5 Where does consciousness exist?

produced by the natural evolution process, as discussed in Smith's book about the other mind (Smith, 2016).

Roboticists have been using a behavioral model simulating the human brain (Fig. 9.5b). This is generally called a mechatronics model.

In Fig. 9.5b, area A is the recognition system to listen to and understand the content of voice input into the ears. Area C is the decision system to select an action for the language recognized. The ears listen to the word *apple*, for example, and the information reaches c1 in area C via a path s5, where c1 is the language label for *apple*. The language label, also called a representation, is a name given to a certain concept. Area C decides a certain behavior based on the information of language label c1. Signals of the decision run through path s1 in area B, and speech (the corresponding behavior) is implemented by the mouth. Area B is the behavior system and area E the memory area. It is known that s5 and s1 are capable of being self-organized (Kohonen, 1995). Note that Fig. 9.5b is insufficient to describe the various known functions of consciousness.

The recognition and behavior systems are directly related to each other as the discovery of mirror neurons shows.

When s5 learns, the neural networks in area B are affected, and when s1 behaves, the neural networks in area A are also affected. This can be explained as shown in Fig. 9.5c. Specifically, when area A learns the function to recognize the word *apple*, area B learns to speak the word *apple* at the same time.

Mirror neurons are neural networks related to both the recognition and behavior systems simultaneously. Accordingly, the recognition and behavior systems may well be defined as shown in Fig. 9.6, which is more direct. Neural networks in area D are related to both area A and B. We cannot immediately determine that area D is the home of the mirror neurons, but undoubtedly this is the area where the functions of mirror neurons are active. The role of area D is for area A (the recognition area) to directly activate area B when area A is activated by learning or recognition, or without using other alternative paths.

Area D is also a shortcut from area B (the behavior system) to area A (the recognition system) by which learning or behavior in area B directly activates area A.

The author defines the source of consciousness to be the coincidence or consistency of cognition and behavior. "Being conscious" includes the state of subconscious according to the author. "Becoming conscious" refers to a state in which any one of multiple subconscious events is lifted to the level of consciousness (i.e., made explicit) as a result of certain conditions being satisfied.

As a human grows and develops, all of his or her functions are learned, and all subconscious events are active, during the state of consistency of cognition and behavior.

For example, the circulation of information through area D and c1 is the state of consistency of cognition and behavior, and the language label c1 is "being conscious."

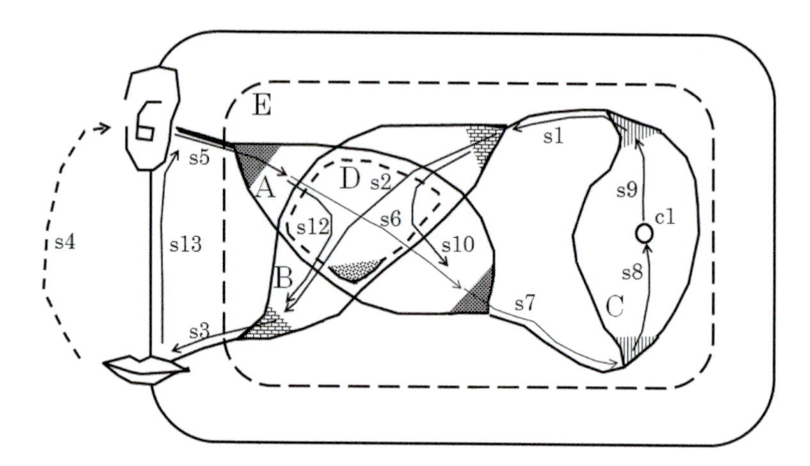

Figure 9.6 Consciousness system proposed by the author.

In Fig. 9.6, c1 is depicted as a single neuron. This is a symbolic expression, and c1 is actually a result of coding. There should be no problem if the expression is spread into multiple neurons. Normally, the binding problem arises in which the representation is expressed by the firing of multiple neurons. In the author's model, the source of consciousness is defined as circulating information, so that it is possible to interpret that signal s8 fires multiple neurons in area C. According to certain conditions, a single subconscious event is selected and surfaces to consciousness, i.e., becomes conscious.

When the consistency of cognition and behavior is lost, the current behavior is immediately stopped and another behavior (relearning) is consciously performed to restore the consistency of cognition and behavior.

With the generation of consciousness defined as above, human mental activities can be explained relatively clearly.

Circulation p1 (s7, s8, s9, s1, s10) does not include path s2. This means that speech is not performed when using this path. This state is interpreted as "imagining" the speech of the word *apple* after recognizing it by hearing. Circulation p1 therefore explains the function of human imagination.

The function of imagination can describe the existence of qualia because the recognition and behavior systems are intertwined in area D and both recognition and behavior information of c1 can fire simultaneously.

On imagining the word *apple*, the system automatically remembers the behavior of speech of *apple*, cognition that the system has heard the word *apple*, and co-occurring memories of *apple*. Co-occurring memory here is a sort of association memory. Co-occurrence means that information circulates through a closed loop with all the neurons in the loop firing simultaneously.

Path s12 in Fig. 9.6 is used to directly transmit the result of the recognition system to the behavior system, bypassing the consciousness proper. This path explains conditional reflexes as well as behaviors directly triggered by a behavior pattern obtained by vision.

An example of a visual-related pattern is a number of small children running in the same direction. When a child totters and falls forward, the children behind follow suit and fall nearly at the same time.

When one is watching a movie and an actor suddenly kicks something with his feet, one suddenly does the same action. This phenomenon is usually suppressed strongly, and seldom occurs in the everyday world. According to psychological knowledge, this phenomenon is known in the medical field and is called imitation behavior (Fig. 9.3).

Consider the case of the patient with partial damage to the frontal lobe, mentioned in Section 9.1.5. When the doctor points at something with a finger, the patient imitates the doctor's gesture unknowingly (Fig. 9.3, top). A pair of glasses is placed in front of both the doctor and the patient. The doctor reaches out his hand to pick up the glasses. The patient, seeing the doctor's movement, immediately performs the same act. The patient is already wearing glasses; yet he was going to pick up and put on another pair of glasses. The patient was instructed by the doctor beforehand not to do anything and the patient was conscious that he should do nothing. Nevertheless, the patient could not stop imitating the doctor (Fig. 9.3, bottom). This case illustrates the existence of information path s12 in Fig. 9.3, through which visual images are directly connected to behavior.

Path s13 transmits the result of the behavior system directly to the recognition system. This explains the phenomenon of self-awareness, i.e., knowing the behavior of the self.

9.2.5 Consciousness Is Generated by Consistency of Cognition and Behavior

The author regards imitation as one of the important functions of human consciousness. Imitation is the capability to behave in the same way as one sees by just looking. The case of an infant sticking out the tongue has been already discussed. There are many other examples, including the case where a mother talks to her child, "Tasty, tasty" while feeding the child, and then the child is soon able to speak, "Tasty, tasty," by repeating the mother's words. The child's words may not be very refined at first, but there is no problem because improvements will soon be seen some time after. Imitation is not unique to humans but is found in relatively higher-level animals such as chimpanzees and gorillas. It is marvelous that small children acquire the ability to imitate in a short time. Mitsuo Kawato at the Advanced Telecommunications Research Institute attempted to achieve this ability in a machine (Kawato, 1996).

The author's computational explanation of imitation resembles that of Kawato. The difference lies in that the author describes imitation comprehensively, focusing on the function of consciousness he proposes. An example of the auditory sense, which is different from imitation, is used in the author's description of imitation, but the generality will not be lost even if the auditory sense is used to explain behavioral imitation.

The author describes the function of consciousness referring to Fig. 9.6. The information on the words spoken by the mother flows through paths s5, s6, and s7, thereby activating the neurons of the recognition system A, including the common area D. The information further runs through path s8 and reaches and fires language label c1, which is thereby linked to the activation information of the common area D. Note that the activation information of the common area D is also part of the behavior system B, which is the speech system in the present example. When the consciousness system decides to execute the contents of language label c1, the relevant information c2 flows through s9, s1, s2, and s3, and the contents of language label c1 are spoken. The spoken language label c1 may or may not agree with the mother's words, but when they do, learning is successful.

When the words disagree, the child (i.e., the consciousness system) refers to the speech (the output) signal s4, which the child hears with its ears. The information of s4 goes to s5 and s6, where the value of the common area D is rewritten to correct the difference. Rewriting is repeated until the information circulation s9, s1, s2, s3, s4, s5, s6, s7, s8, and c1 no longer needs rewriting of the common area D.

The final state achieved is the consistency of cognition and behavior.

When this state is achieved, a three-layer structure is established: the external environment, the areas responsible for accepting stimuli and emitting response to external world (area A and B including area D), and the language labeling area C. The author calls this three-layer structure the Consciousness Module. Area C is called the Representation Area. The author believes that the flow of information circulating through external environment, area D and area C is the source of consciousness.

When speech is spoken consciously, the relevant information circulates through c1, s9, s1, s2, s3, s4, s5, s6, s7, s8, and c1. This route indicates that "the self is conscious of speaking the words represented in c1."

The problem is how to create common area D.

Area D must encompass both area A, the recognition system, and area B, the behavior system. Area D must function with regard to both recognition and behavior, and the information recognized must be directly used to perform a behavior. By stating this I must be ready to be impeached by the reader for reason of committing a category mistake. I admit that this is a sort of category error. However, a category is originally a philosophical idea devised by humans. Look at the brain in all of its details, and we just find neurons. There is no area that befits the term category. What exists in area D is artificial neural networks, and it is the human that categorizes it.

The author further believes that common area D provides a solution to the symbol grounding problem because the common area D is located between the external environment and language label, and interrelates them.

An example of the sense of hearing has been already described in this chapter. Let us consider another example to describe the author's idea.

Assume that the mother says "good morning" to the family in a room where bright morning sunshine is shining in through the window, and the smell of soup and bread hangs in the air. If this situation is known to the consciousness system by learning in the past, the consciousness system for visual information and smell information, respectively, will function, and at the same time, their respective common area D will be connected at the representation area C as co-occurring memory. As a result, when "good morning" is given to the perception system, the rest — bright morning sunshine and the smell of soup — is automatically called up. This co-occurring memory also fires the common area D automatically. The author believes that this is a series of so-called association memories. The smell of soup is memorized in the common area D as a result of actually recognizing it in the past with the somatic sensor. Accordingly, the actual smell of soup is reproduced (becomes conscious) as an information when the smell of soup is associated.

I believe this is qualia.

The sense one receives from qualia is different from, at least, the actual sense. This is because qualia are not a consciousness arising from the solid consistency of cognition and behavior closely related to external circumstances and embodiment but are simply produced from circulation of information called imagination (called circulation of imagination). A typical information circulation route is c1, s9, s1, s10, s7, s8, and c1. Certainly, when the smell of soup is sensed from the external environment, the same information circulation occurs, and the same qualia are generated. Information circulation in this case is s5, s6, s7, s8, c1, s9, s1, s10, s7, s8, c1, and s9 in Fig. 9.6 by replacing the sense of hearing with the sense of smell. When the smell of soup is quite different from the language label c1, the system learns to construct a new language label c1. If the difference is small, the current language label c1 remains unchanged.

The author believes that the most important function of human consciousness is distinction between the self and others. This is because the distinction of the self from others is the starting point of implementing imitation described in the preceding chapter. Here is a question: How can I know that the hand I am now looking at is my hand?

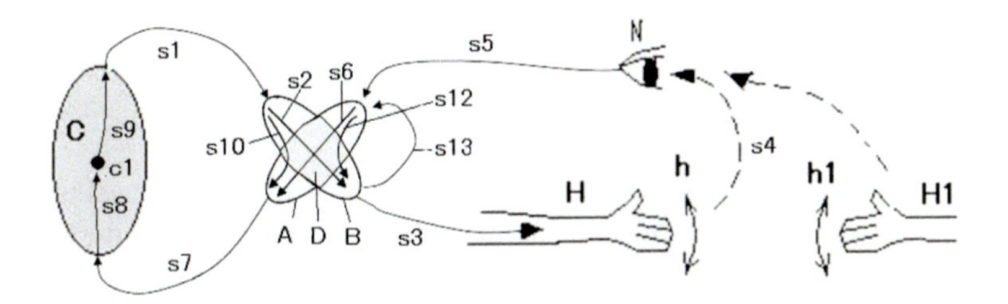

Figure 9.7 The self's hand H and the other's hand H1.

This is the theme of this chapter. The principle of the consciousness system proposed by the author is used to answer this question.

Historically, Descartes was the first thinker who clarified the existence of the self, as already described. Believers in phenomenology have studied the theme of the self and others from philosophical points of view. Merleau-Ponty, among others, proposed the idea of embodiment to describe the spiritual activities of humans. According to him, the self cognizes the existence of others in the external environment by using perception obtained via embodiment.

How does the author differentiate the self and others in the consciousness system he proposes?

In Fig. 9.7, we read vision for hearing and hand for mouth. Vision is signified by N and hands by H and H1.

The self's hand is denoted by H and the other's hand by H1 (Fig. 9.7). Self and oneself essentially have different concepts, but in our description here, we use self as a generic term. Oneself seems to refer to the whole body, including the flesh. Let us assume that vision N looks at the other's hand H1 and its movement. The information of H1 captured by vision runs through s5 and s6 and reaches the recognition system A, where the hand is recognized.

This information affects the behavior system B via the common area D, and the behavior system B immediately tries to imitate the movement of H1 by interpreting it to be the hand of the other. Actually, however, the self's hand H does not move because the action of the behavior system B is suppressed. This suppression is intrinsic to the consciousness system. The flow of information at this time is s5, s12, and s3. The same information also flows through s5, s6, s7, and s8 to fire the language label c1. The firing of c1 is linked to the firing information of the common area D. By linking we mean that the firing of the common area D causes the firing of the language label c1 by co-occurrence.

When experience has been sufficiently learned in the past using the route of circulation of imagination (s9, s1, s10, s7, s8, and c1) already described, the firing of the language label c1 ignites the relevant part of the common area D where the path intersects it.

The information and its circulation strengthen the link between the language label c1 and common area D.

To sum up, the information of the other's hand H1 flows through s5, s6, s7, s8, c1, s9, s1, s10, s7, s8, and c1 and creates a closed loop of information circulation consisting of c1, s9, s1, s10, s7, s8, and c1. We interpret this to mean that the information circulation makes us imagine that the hand H1 is the hand of the self by the existence of the common area D. To imagine means that this consciousness system thinks that H1 is its own hand. This is because the common area D is part of the behavior system B. What happens with the self's hand H? It is driven by the behavior system B. The information flows through the route of c1, s9, s1, s2, and s3.

Information on the movement of the hand is returned to the consciousness system via s13, a somatosensory nerve path, and also via s4, or through the external environment. Information s13 provides a means for the recognition system A to know that the behavior system B has issued behavior information. It is also possible that the other consciousness system detects the change of sensors on the skin following the movement of the hand and inputs the relevant data into the recognition system.

Path s4 means that the movement of the hand is detected by vision. It is discussed as follows. The same discussion applies also to s13.

Information s4 enters the vision and becomes information s5, which reaches the recognition system A. This information passes through recognition system A and becomes s6, which flows through s7 and s8, eventually reaching the language label c1 to ignite it. The common area D and the language label c1 are linked to each other, so that the language label c1 is symbol-grounded via common area D. In summary, when the consciousness system views the self's hand

H, the information circulates through c1, s9, s1, s2, s3, s4, s5, s6, s7, s8, and c1. The same information flows through c1, s9, s1, s10, s7, s8, and c1. The same information also runs simultaneously through c1, s9, s1, s2, s13, s6, s7, s8, and c1 via the somatosensory nerve path.

This is the state of the consciousness system in which the language label c1 and the common area D are firing at the same time, or the recognition and behavior systems are said to be consistent with each other. These three information loops are the root source for the consciousness system to become conscious of the self's hand H. The difference of the self's hand H and the other's hand H1 in the consciousness system is that the consciousness system uses the information routes of s2, s3, and s4 as well as s2 and s13 for the self's hand H but does not use them for the other's hand H1.

I will now explain that it is possible to cognize the image of the self in a mirror (i.e., self-consciousness) as the image of the self by defining that consciousness is generated from the consistency of the recognition and behavior systems.

Humans brought self-consciousness into the consciousness system (the self) as a model of being-in-the-world, and by using self-consciousness as a clue, they then succeeded in capturing the model of the world into their consciousness system. This description suggests the existence of a scheme to understand others by replacing the self with the other. A human can cognize his or her image reflected in a mirror to be the image of the self, but it is difficult to explain why.

When did consciousness emerge in the history of human evolution?

It should be natural to think that the self and others were first differentiated from each other due to the necessity of sharing work while living as a group, and then humans became conscious of the self among the others. If it is possible to become conscious of the self in the midst of others, it should also be possible to become conscious of others in relation to the self. To become conscious of the self in the midst of others, a model of consciousness of both the self and others must be created in the consciousness of the self.

Even non-Cartesians may not be able to deny that the function to become conscious of the self and others exists in the consciousness of the self. This is true even if this function consists eventually of reaction formation. I, too, do not deny the structure of reaction formation. I support the idea that any mechanism present in the brain depends on material and information reaction. Of course, the reaction system must be able to describe the mechanism of human consciousness.

In Fig. 9.8, when the representation area C instructs the execution of a hand movement, e.g., swing the hand vertically (h), it transmits the relevant instructions to the behavior system B of the consciousness module G via s1. The behavior system B moves the system H of the hand via s3 to execute the instruction. At this time, the behavior system B of the consciousness module G activates the recognition system A of the consciousness module G via s13, and the action h of the hand is reported back to part of the representation area C via s7. The circulation route of s1, s13, s7, and C is activated repeatedly to transmit information on the moving hand to part of the representation area C. This repetitively circulating information may be said to be the source of being conscious (L) of the fact that the system is moving a hand. The consciousness that a hand is moving is further strengthened by the information circulating through s1, s3, s4, s5, s7, and C. This explanation also applies to the consciousness module R for moving the feet.

This consciousness model can recognize that the image reflected in a mirror is the image of the self. The image in the mirror knows that the vision system N of the self recognizes the features of the body of the self, and the behavior of waving the hand h′ and moving the foot f′ also appears on the image in the mirror simultaneously. After integrating all of these informational data, it is still impossible to tell whether a person resembling the self is doing the same action, the other person is reflected in the mirror, or the image is that of the self.

In the consciousness model proposed by the author, taking the case of the conscious movement of the hand H as an example, all of the circulations p1 (s1, s10, s7, and C), p2 (s1, s12, s7, and C), and p3 (s1, s3, s4, s5, s7, and C) and the recognition route for the image of the self reflected in the mirror q1 (s1, s3, s4′, s5, s7, C, s1, s10, s7, and C) contribute to fortifying the source of consciousness. The consciousness generated by circulations p1, p2, and p3 agrees

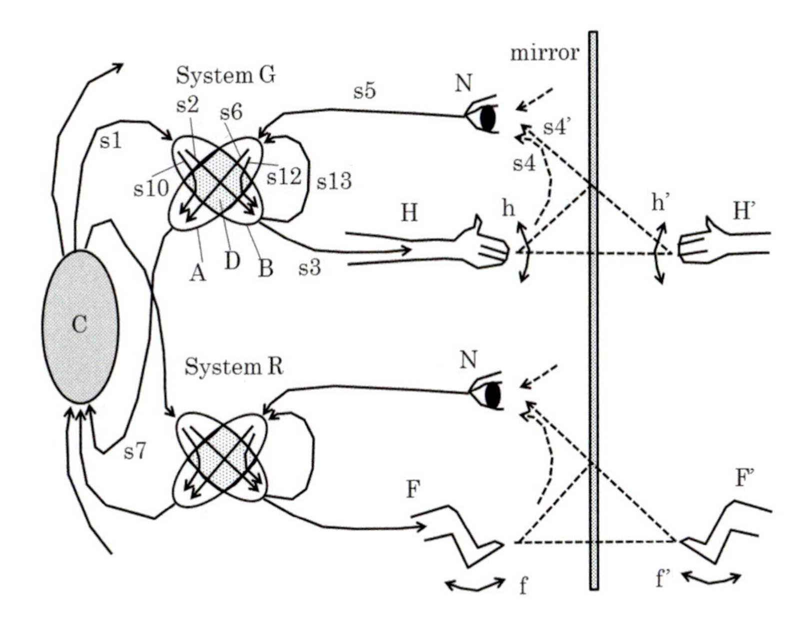

Figure 9.8 Scheme of recognition of self reflected in a mirror.

with the condition of the consistency of behavior and cognition, or that none other than the self is established. A conclusion is also reached that s4 and s4′ are indistinguishable existences since the route q1 cannot be differentiated from circulation p3 (Fig. 9.8) and no reasonable grounds exist to assume that s4 and s4′ are different.

A further investigation is required to prove this indistinguishable existence and to clarify the mathematical relationship between the image of the self and the self. This is because, for example, a mirror does not necessarily reflect an object perfectly. This will be discussed in detail later.

Here, I have introduced the basic procedure for judging that the image of the self reflected in a mirror is none other than the self.

As the functions of higher-level cognition develop, humans will reach the recognition that the image of the self in a mirror is a copy of the self since it is a mathematical existence indistinguishable from the self, and because of the absence of any physical bonding with the body of self.

One will cognize that the reaction of the routes p1, p2, and p3 is on this side of the mirror. This is one way of defining self-consciousness, i.e., becoming conscious of the self.

The circulations p1, p2, and p3 and the route q1 are a tool to interpret one's own body and external environment simultaneously, and also a powerful tool to take in the external environment by learning.

According to the consciousness system developed by the author, p3 and q1 are two different routes that show the same reaction with regard to consciousness. Using this feature, route q1 can be taken into the representation area. This means that multiple models of the self can be taken into the representation area one after another. Using this scheme, the representation area builds models of the self and creates self-consciousness in the consciousness system. Consciousness of others is taken into the representation area with a similar procedure by comparing it with the consciousness of the self. Once consciousness of the self and others has been taken into the representation area, it is possible to introspect one's own feelings and become thoughtful of the feelings of others. Mutual internal and external communications are also possible once consciousness of the self and others is taken into the representation area. This process provides a means for capturing being-in-the-world into the representation area.

9.3 My Standpoint

9.3.1 New Materialism

The author is basically a materialist and, as such, believes that the human brain can be analyzed as a substance. It is necessary, however, to include the concept of information in the category of substance. Conventionally, information, together with energy, is categorized as different from substance. Energy and substance can be expressed mathematically by expressions (e.g., $E = MC^2$). The relationship between information and substance may also be expressed by a numerical expression in the future.

Information is accumulated in a substance, and only then can information affect the world environment using the substance. It may sound extreme, but when two different pieces of information, a and b, are present in the same substance N, can't we say there are two different substances Na and Nb because they can have different functions to interact with the external world? By thinking in this way, one may consider the flow of information through substances to be a concept of a new materialism. Computers have traditionally been treated materialistically as an information processor, and therefore the above description by the author may not necessarily be unnatural. People often say, "Consciousness and the mind do not exist. They are but an illusion or a reverie," when discussing whether consciousness and the mind exist. This is a confusion arising from the old materialism fettered by the doctrine that consciousness and the mind do not exist as substances.

9.3.2 Connectionism

The author's standpoint is connectionism. Connectionists believe that the human brain is made up of networks of neurons and their functions create the spirituality of a person. They further believe that artificial consciousness can be generated by connecting artificial neural networks.

9.3.3 Cognitism and Representation

The author sides with cognitism and acknowledges the existence of representation. Representation means images of external and internal objects occurring in the consciousness based on perception, i.e., subjective concepts in the brain corresponding to objective things existent outside the body or in the internal world. This idea was inaugurated by German philosopher Immanuel Kant (1724–1804) (Kant, *Critique of Pure Reason*). Simply put, when you see an apple, a recognition that it is an apple is formed in your brain. This means that a subjective image of an apple is created with respect to the objective thing of an apple present in the external world. The subjective image in the brain is called the representation of an apple. We acknowledge the existence of representation because it would be difficult to describe the commonest phenomenon of human speech if representation were denied.

9.3.4 Not Emergent

The author does not agree with the idea of explaining all about consciousness by emergence. Emergentism is derived from behaviorism. Robots developed by behaviorists are considered to have a set of simple behavior protocols that are selectively implemented in response to changes in the environment. This idea is understandable, but how is the high-level cognition function unique to humans created in their robots? How is human thinking created in them?

Behaviorists use the word "emergent" when such questions are asked. Emergent means that during the process of evolution of high-level cognition function of humans, certain abilities or features are obtained by chance. The author has no objection to the idea that the function of cognition was acquired through evolution. However, even though the mechanism is unknown, no one should negate scientists' efforts to estimate and define the principle of the mechanism. Emergentists' research technique is to leave a robot in a difficult situation and expect it to solve the problem and evolve by itself (evolutional robotics). This may be a feasible research technique, but it would be difficult for the robot to discover the function of consciousness through such a process. This evolutionary technique is not effective if the function of consciousness is a redundancy system.

9.3.5 Avoiding Infinite Retreat

The third property of consciousness by Husserl reads that "the self becomes aware of the self." The word "self" is used twice in this statement. This is the same as saying that there exists a self who becomes aware of the self. This further leads to "there exists a self who becomes aware that 'there exists a self who becomes aware of the self,'" and the self exists eternally while retreating (Fig. 9.9). If this statement is accepted, the function of consciousness cannot be described by computers because computers cannot solve a problem including a recursive retreat and therefore cannot end the computation process.

Using this fact, a researcher has declared that the human brain includes non-computational elements (Penrose, 1994).

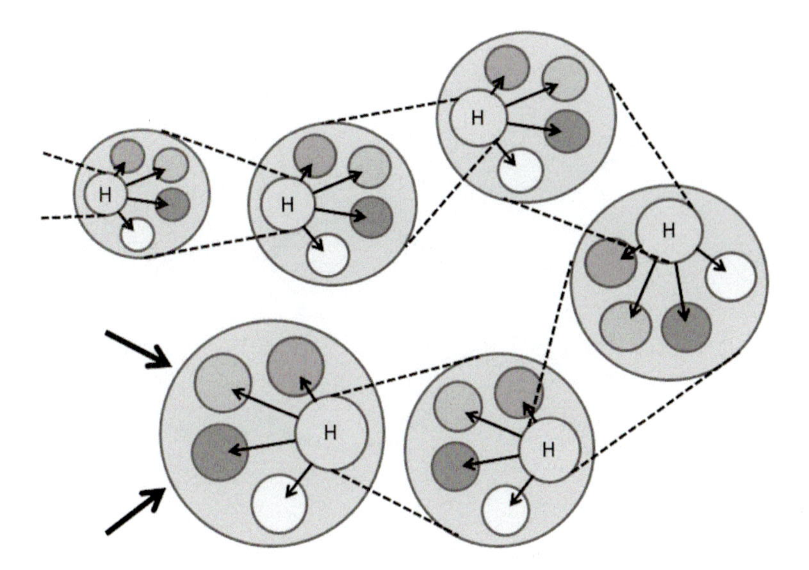

Figure 9.9 Recursive retreat.

The author believes that the recursive retreat problem can be somehow avoided by using infinite repetition (Fig. 9.10). The state of becoming aware of the self is understandable, but it is impossible to actually feel the self who retreats multiple times in the realm of phenomenology. I believe that the self who retreats multiple times is simply an imaginary object of symbolic logic created in the brain because we are already handling the concept of infinity as a symbol. The author, therefore, believes that it suffices if we can create a self who becomes conscious of the self on a computer, and the rest is considered to be the result of symbolic logic within the brain.

9.3.6 Quantum Consciousness

Quantum theory would provide an effective means of research if we were to analyze the brain at the atomic level. There are many things we can do, however, before we resort to quantum theory. I believe that quantum theory is not necessary if one is interested in elucidating the mechanism of consciousness.

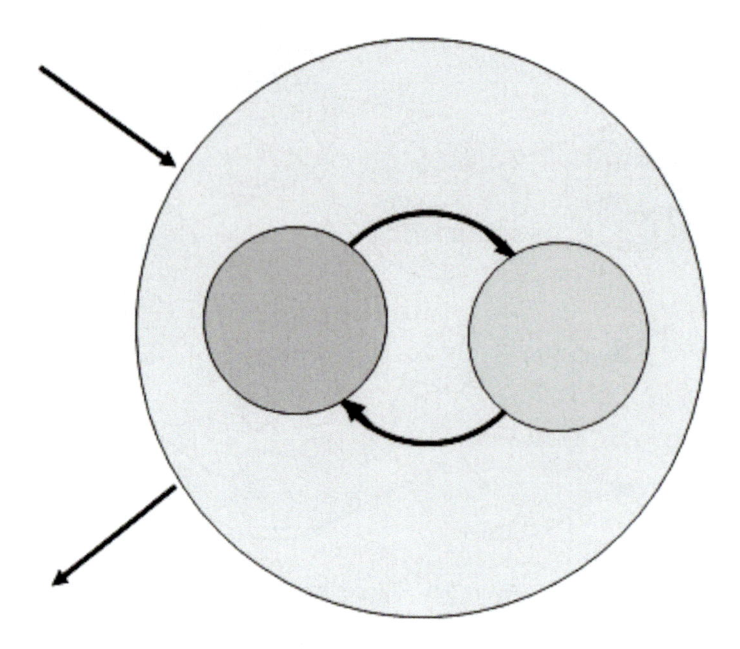

Figure 9.10 Infinite repetition.

9.4 Computational Model of the Consciousness Module

This chapter describes the MoNAD, the consciousness module that is the basis of the consciousness system.

9.4.1 Overview of Computation Model

A MoNAD is an artificial neural network with a slightly complex structure that is built upon the concept model of consciousness described in the previous chapter (Fig. 9.11). The first feature of the MoNAD is that at least two closed loops (c1 and c2) exist and both are connected to a common neuron group K called primary representation. The second feature is the somatosensory nerves Y′ extending in both common neuron group K and the closed loop c2. The MoNAD has at least one input terminal InputL. Closed loop c1 has two symbolic representations Z and Z′. Z is called

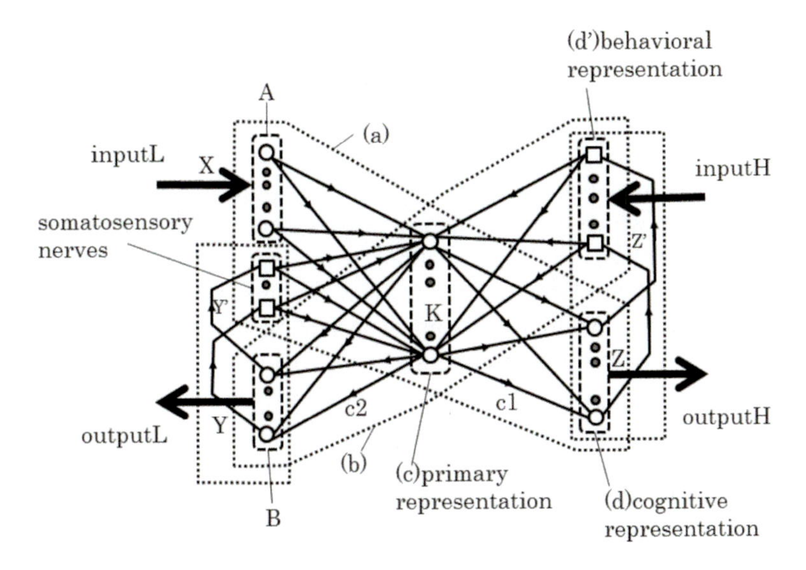

Figure 9.11 Computational model of consciousness module.

cognitive representation and Z′ behavioral representation. A MoNAD learns the neural networks so that the values of Z and Z′ are equivalent. When a behavioral representation Z′ is decided, the MoNAD executes behavior Y from outputL, and the information X obtained from inputL is changed to make the value of Z coincide with that of Z′. Z and Z′ also serve as I/O terminals (outputH and inputH, respectively) for exchanging information with higher-level MoNADs. Information on the current behavior is input again into K via somatosensory nerve Y′. The MoNAD, thus, cognizes not only the external environment but also the behavior of the self. The result of this cognition is reflected on the information of Z, and this information represents both the external environment and the behavior of the self.

9.4.2 MoNAD Functions

The MoNAD is designed to satisfy the consistency of cognition and behavior, which is the author's definition of consciousness. This means that in the MoNAD information runs consistently through the neural networks. In this model, the information flows through

Z' of the closed loop c1 and common neuron group K, and then reaches Z. The value of Z is copied to Z' and is also input into the inputL of a higher-level MoNAD. Information from Y runs through Y' of the closed loop c2 and K and then returns to Y. Information basically circulates in a single MoNAD through K, Z, Z' and returns to K, from which the information further flows through Y and Y' and then returns to K. Information X from inputL is added to this circulating information all the time. If the internal circulation of information is maintained normally despite the input of incessantly changing external information X, the MoNAD is said to be keeping the "consistency of cognition and behavior." This is because the information flow of (Z', X, Y'), K, Y decides the next behavior to be taken and the information flow of (Z', X, Y'), K, Z decides cognition. Information from Y output from the terminal outputL is also a new order for the lower-level MoNADs to engage in a behavior because Y changes the behavior of the lower-level MoNADs. When the value of Z' is changed by a higher-level module, the MoNAD executes an action from outputL that will eventually make the value of Z coincide with the value of Z'. The author calls this state, in which the value of Z is equal to that of Z', a state in which cognition and behavior are consistent.

9.4.3 Investigating Husserl's 10 Properties

Husserl enumerated 10 properties of consciousness. Does the MoNAD proposed by the author satisfy them?

(1) The first-person property is the assertion of the existence of the self based on mind–body monism. The representation area Z of the MoNAD distinguishes between the self and others and represents the behavior of both. This means that the MoNAD recognizes the behavior of others based on the behavior of the self. This is a sort of mind–body monism.

(2) Orientation means that consciousness is always directed toward something. The MoNAD self-learns to behave, by outputL, to make cognitive representation Z coincide with behavioral representation Z'. As such, the MoNAD satisfies the second requirement of Husserl.

(3) This is the same as point 2. In addition, the duality of consciousness is that the self becomes aware of the self. The MoNAD is capable of cognizing the behavior of the self using its somatic sensation Y'. The MoNAD is therefore able to become aware of the behavior of the self by itself.

(4) Expectation: The MoNAD performs a behavior (outputL) according to point 2 such that the cognitive representation Z agrees with the behavioral representation Z'. This means that the behavioral representation Z' expects the forthcoming cognitive and behavioral state of the MoNAD.

(5) Conviction of the existence of things: The MoNAD always calculates the cognitive representation Z using external information (inputL) and internal somatosensory information Y', .i.e., Z results from the cognition of the external and internal environments. In the area of the MoNAD where external information is directly processed, information X about the external environment inevitably includes uncertainty but Z, as a representation, always converges on a certain definite value. The MoNAD can, thus, discern a positive existence in the midst of uncertain information.

(6) The body is part of the self: The MoNAD cognizes the behavior of the self using its somatic sensations. This means that the MoNAD cognizes and is convinced that the present behavior is being done by the self, i.e., the MoNAD is confident that the behavior of the self is being done by the body of the self.

(7) Distinction between the self and others: As stated in point 6, the MoNAD continually cognizes things by obtaining somatosensory information and information about the external environment via inputL. This means that the MoNAD always cognizes information about the behavior of the self and the external environment. As such, I conclude that the MoNAD distinguishes between the self and others.

(8) Property in which reason is related to emotion: With the properties 5 through 7, the MoNAD has functions to represent the internal and external environments. If this is true, the MoNAD can represent information about emotions arising in the body. Representation is shown

on a Pleasant or an Unpleasant MoNAD. This means that reason and emotion can be achieved as a representation function using MoNADs. Details will be described later. As Antonio Damasio, Professor at the University of Southern California, says in his book on Baruch De Spinoza (1632–1677), humans seem to have developed in such a way that reason and emotion function together (Damasio, 2003). And also he has described in detail the feelings of the body arising from emotions in his book (Damasio, 1999).

(9) Randomness: The MoNAD is made of artificial neural networks. For this reason, networks are always swaying or in a state of mathematical chaos. A consciousness system structured using layered MoNADs can define the sway of consciousness, but we will not go any further into this topic right now.

(10) The qualia problem: The MoNAD does not define its relationship with qualia in detail. This will be discussed later.

9.5 Discriminating the Self from Others Using the Function of Visual Imitation

The MoNAD is capable of learning the function of imitation. Imitation means that the value Y of outputL follows the value X of inputL in the MoNAD. In principle, imitation means that the input value X of the external environment is output as it is to Y (and simultaneously to Y′). The MoNAD cognizes the external and internal environments and indicates the result on the cognitive representation Z. This means that the MoNAD distinguishes between the self and others. When the external and internal states are in agreement with each other, imitation is successful.

If the cognitive representation Z and behavioral representation Z′ are in agreement, cognition and behavior are consistent in the MoNAD. Irrespective of whether they are in agreement or not, the value Z is copied to Z′. This means that the MoNAD basically accepts the condition of the real world.

9.6 Solution of the Symbol Grounding Problem

A computer is basically a high-speed symbol processor. The problem is that the symbols processed inside are semantically disconnected from the external world. As I mentioned before, this problem is the major cause of the stalemate that conventional artificial intelligence (AI) is experiencing. The MoNAD possesses the function to represent the external and internal environments simultaneously. Since this representation is a symbol in a sense, the MoNAD continues to interpret the real external and internal worlds as symbols. Because of this property, the MoNAD may be considered to have the function to solve the symbol grounding problem.

9.7 Consciousness System

The consciousness system shown in Fig. 9.12 consists of layered MoNADs. Each MoNAD is driven by the top-down and bottom-up system to continually exchange information between the lower-

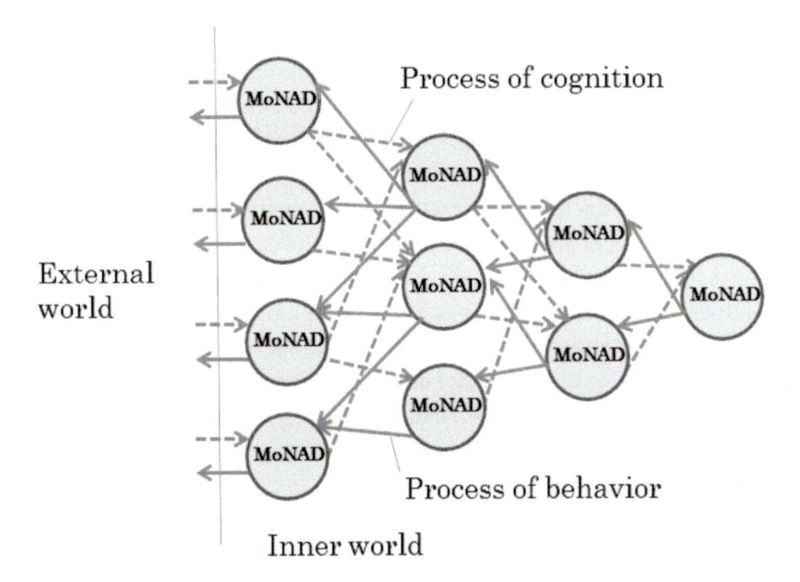

Figure 9.12 Consciousness system structured using layered MoNADs.

and higher-level MoNADs. As the basic flow, each MoNAD executes a behavior (outputL) based on information from the lower- and higher-level MoNADs (inputL and inputH, respectively) and the somatosensory information Y′. This information is constantly transmitted to the higher-level MoNADs via outputH. Information Z is normally copied as is to Z′ except that when new information is given to Z′ from a higher-level MoNAD via inputH, the receiving MoNAD works to create Z′.

The lowest-level MoNAD is responsible for communicating information with the external world. Behaviors or transactions with the external world are implemented via outputL, and the external information enters the MoNAD via inputL. Excepting the transfer of external information, the MoNADs communicate with one another using their respective representations as information sources. The consciousness system is, thus, a representation-driven mechanism.

The functions of the consciousxness system are described as follows. The consciousness system performs the behavior described below in response to external stimuli while considering its current internal state. What makes this robot different from conventional ones is this consideration of its internal state. The internal state means the whole history of somatosensory information and behavioral representations experienced by the robot up to the present.

Each MoNAD functions independently in the absence of a behavioral representation from a higher-level MoNAD. If a behavioral representation exists, the higher- and lower-level MoNADs function cooperatively. Each MoNAD continually interprets its current status while obeying instructions from higher-level MoNADs. Note that these higher-level MoNADs do not constitute a central control tower known as a Homunculus because the behavior of a higher-level MoNAD is dependent on information from a lower-level MoNAD. As such, it can be said that higher-level MoNADs follow the commands of lower-level MoNADs. Normally, representations of the higher-level MoNADs are concepts of a higher level compared with those of the lower-level MoNADs.

The current system, including its constituent MoNADs, uses a human-supervised backpropagation learning method. Voluntary development and learning are expected in the future.

9.7.1 Subconsciousness and Explicit Consciousness

Humans seem to always have multiple flows of consciousness, most of which are normally not known to them and are thus called the subconscious. You can walk while eating ice cream, for example. It is possible for you to be conscious of eating and not to be conscious of walking. Without being conscious of walking, you can accomplish the other behavior of eating without inviting an accident. Walking is an act of the subconscious, whereas eating is an act of explicit consciousness. The reverse is also feasible. Humans seem to be conscious of just one thing at any given time. What is the mechanism that arouses the subconscious to consciousness? The author assumes that the cognitive and behavioral circulation involving the MoNADs that associate (i.e., connect) the reason and feelings systems (see the next chapter) present in many streams of subconscious are related to explicit consciousness.

This suggests that the actualization of the subconscious is deeply related to emotions. Any subconscious activity producing stronger emotions is considered to be actualized. In the example concerning walking while eating ice cream, eating ice cream arouses stronger emotions than walking, and that is why the consciousness of eating is made explicit. If the balance of the body were to be lost due to stumbling on a stone, the relevant information triggers other emotions and awakens new consciousness. Generally, the subconscious includes emotions excited by the subdued and explicit consciousness. When consciousness is aroused in a human, both the emotion and reason systems of consciousness are working together, i.e., the explicit consciousness refers to a state in which the self is feeling its body emotionally. Assume, for example, that the representation of a feeling MoNAD turns unpleasant in the consciousness model developed by the author. The unpleasant information is transferred to the association MoNAD. The association MoNAD activates the MoNADs of the reason system to engage in a behavior intended to change the unpleasant information to a pleasant one. At this time, the consciousness system is explicit. Explicit consciousness refers to a state in which the representation of the self and the representation of feelings co-occur and react.

9.7.2 Relationship between Reason and Feelings

Humans possess both reason and feelings. No one can deny this fact. Antonio Damasio proposes the somatic marker hypothesis, stating that reason and feelings are inseparable in humans (Damasio, 1999). The author shares the same idea. How are reason and feelings of humans related to each other?

Joseph E. LeDoux, professor of neuroscience and psychology at New York University, has published many achievements in the study of emotion and behaviors (Ledoux, 1996, 2002).

The author proposes a MoNAD model to explain the relationship between reason and feelings. Figure 9.13 illustrates how the most basic MoNAD relates with feelings. Assuming that human feelings are essentially represented by Pleasant and Unpleasant information, the author devised their respective MoNADs D and E (Fig. 9.13).

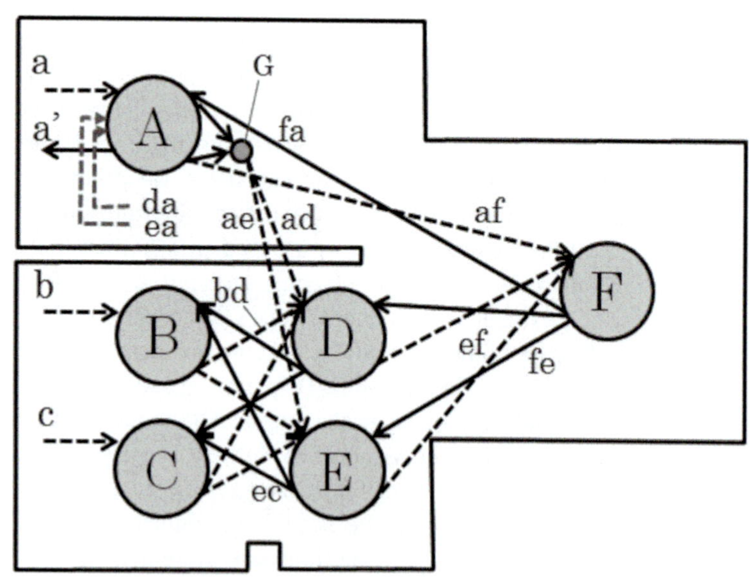

Figure 9.13 Relationship between reason and feelings.

Considering that feelings arise from emotions present in various parts of the body, the author allocated MoNADs B and C to represent the emotion area. MoNAD B, for example, represents the emotion of "pain" and MoNAD C, "emptiness" due to the loss of values from the vision sensors. The emotions B and C transmit their information to MoNADs D and E, which represent Pleasant and Unpleasant, respectively (bd, be, cd, and ce). The representation of the feelings of Pain and Emptiness is normally Unpleasant, but in this model, information is also transmitted to the MoNAD of Pleasant (bd and cd). Both the reason and feelings systems are constructed using MoNADs. MoNAD F associates (i.e., combines) the information of both the reason and feelings systems. MoNAD F receives information (af, df, and ef) from the lower-level MoNADs (A, D, and E in Fig. 9.13), generates new information (fa, fd, and fe) based on the information received, and transmits it to the lower-level MoNADs (A, D, and E in Fig. 9.13).

This new information means that when the representation af, df, and ef reads the "current cognitive representation is Unpleasant," information fa, fd, and fe is sent to MoNAD A of the reason system urging it to initiate a behavior to change Unpleasant to Pleasant. This means that the information expects that the representation D will be changed to Pleasant.

A system in which the representation of MoNAD D becomes Pleasant if behavior a′ is initiated when the external inputs are a, b, and c has already been empirically learned.

In the example shown in Fig. 9.13, the neural circuitry network G judges the coincidence of the behavioral and cognitive representations. Information ad and ae makes the feeling representation Pleasant if the expectation information coincides with the real-world cognitive information, and makes it Unpleasant otherwise. I believe this example is very generality.

The problem here is the development through the learning of the MoNAD that associates (i.e., connects) information. The system developed by the author evolves by repeating empirical learning, and at least in the initial stage of evolution, wiring is expected to have been intrinsically set. Each MoNAD has at least two pairs of bidirectional communication lines to support creation of a new representation by associating simultaneously occurring representations. Since plural representations are associated to a single representation, this model can explain the phenomenon of creativity or the occurrence of a new concept.

Silvano Arieti (born in Italy, 1914–1981), professor of psychiatry at New York Medical School, left many famous research studies on the mechanism of creation in the brain (Arieti, 1976). There is no significant difference between his description and the creativity of the MoNAD developed by the author, but a detailed discussion is omitted here.

My discussion of reason and feelings continues. Changes in the behavioral representation MoNADs D and E are transmitted to the lower-level MoNADs B and C via db, dc, eb, and ec to change the behavioral representations of these MoNADs. This indicates that the emotional MoNADs B and C are activated when the system cognizes the representations Pleasant and Unpleasant. The emotional MoNADs B and C, respectively, create new representations by adding information from by external stimuli (b, c), and then send those representations to the higher-level MoNADs D and E to change their representations. This means that the system operates in a way similar to a human's changing to a state of Pleasant or Unpleasant just by thinking of Pleasant or Unpleasant.

The consciousness system with emotions shown in Fig. 9.13 consists of three essential MoNAD systems: reason, feelings, and association. The reason system executes behaviors in the external environment and at the same time interprets and represents the external environment. The feelings system represents the information of the emotion system activated by external stimuli and also represents the feelings aroused by the information represented in the reason system. The association system transmits the information of the feelings system to the reason system and also sends the information of the reason system to the feelings system. Generally, there is the major circulation of information through the feelings, association, reason systems, and then back to the feelings system. There are, theoretically, direct inputs from feelings to the MoNADs of the reason system (da, ea), but a description of these inputs is omitted here. Information from feelings always passes through the association system. The reason system executes behaviors to interact with and change the external world. As a result of these behaviors, the external information is changed and the internal representation modified. Importantly, in the consciousness system incorporating the MoNADs proposed by the author, the internal environment is also variable with the changing condition of the somatosensory neural networks. In the author's system, the next behavior is decided not only by depending on the condition of the external environment but also by considering the condition of the internal environment, including feelings. The reason system that interprets the external environment is capable of interpreting it emotionally to deliver information to the feelings system. This means that the system has a means to evaluate the feelings of others existing in the external world.

The operation of the emotion system is described as follows.

The internal and external information is required to change the representations of the feelings MoNADs. Internal information includes reflex information from the association system and information derived from the representation of the reason system. The section using the neural network G is currently mounted on the robot. The section using the representation of the reason system itself is a task for the future. The model shown in Fig. 9.13 describes the relationship between the MoNAD and the feelings system. It is clear that the consciousness system incorporating these MoNADs is capable of expressing Pleasant and Unpleasant feelings. A system in which more than one MoNAD exists in the reason system is yet to be studied.

9.8 Where Do Qualia Come From?

The reader may claim that the MoNADs described here lack the function to explain qualia. That's right. When the author earlier discussed the relationship between qualia and the somatosensory system, he said that the functions of consciousness and qualia are different problems as far as the MoNADs are concerned. Qualia have been considered the hardest of all problems in the function of consciousness (http://consc.net/papers/facing.html). The author assumes that qualia are a function below the function of consciousness from the fact that "to become conscious of feeling something" is easily understood, whereas "to feel to become conscious of something" is not. I agree that the qualia problem is an interesting theme for scientists, but if qualia are not directly related to the function of consciousness, it would be better for now to stay away from the qualia problem. If qualia are a subjective yet physical phenomenon, someone will solve this problem in the future, and at such a time, the function of consciousness and qualia might be combined. The author's function of consciousness is designed to allow the incorporation of a qualia function later. In brief, the function of consciousness can cognize the qualia present at a given time consistently because the author defines the function of consciousness to be consistent cognition of the behaviors of the self and others.

The author believes that the mystery of the occurrence of qualia is related to the transformation of information into substances and vice versa. At the end of each nerve, information is generated from substances, and in the brain, substances are produced from information. Transformation occurs cyclically in both the body and the brain using bidirectional information transmission routes via the MoNADs. This is the idea that any event being experienced in the body is virtually experienced in the brain at the same time. This is synonymous to the old idea of a bodily model of the self or a homunculus in the brain. Anyway, the details are yet to be clarified.

I would like to introduce my view of qualia.

The function of consciousness has been substantiated scientifically by the author on the basis of computation theory, but qualia still remain a subjective problem. Qualia are a sense of the self and cannot be understood by the sense of others. For example, it is impossible to prove if my pain and my wife's pain are of the same sense. This question is asked by Thomas Nagel in his essay, "What Is It Like to Be a Bat?" (Nagel, 2021).

I think we can believe for now that my pain and my wife's pain are of the same sense. A scientist should not use the word "believe" in this context, but isn't it true that we live by believing? Just because we believe that my wife's pain is the same as my pain, we try hard to remove the pain that brings about Unpleasant feelings from my wife. The quality and severity of the pain, as well as the interpretation of Pleasant and Unpleasant, may vary slightly depending on the people owing to innate propensity and developmental learning, but the essential points should be the same for all people. If not, humans must have lost an effective means of communication and would have become extinct long ago in the process of evolution. While acknowledging the existence of small differences, we should judge that there is no particularly significant difference.

Consider the exquisite music composed by Beethoven and Mozart, the graceful Parthenon temple in Greece, and the beautiful balance of the golden proportion. All these achievements are appreciated by a multitude of people. From this fact, can't we say that humans have nearly identical qualia? What, then, is the difference between humans and bats? Although both of them are mammals, humans can never feel that they are "the same as bats." The basis of qualia may possibly have been the same for humans and bats at the outset because they were born and have evolved in the same environment of the Earth. We should judge, however, that our qualia are different from those of bats because humans and bats followed different paths of development thereafter.

The author understands that every living body possesses qualia from the time it is born. The author further believes that qualia are a phenomenon captured by the function of consciousness in the process of evolution or through developmental learning. The function of consciousness represents internal phenomena as it represents the phenomena of the external world. One of these internal phenomena is qualia.

The function of consciousness interacts with the external world and ascertains its existence by executing behaviors, or else by means of a scientific means to certify the existence of the external world. The difficulty with the qualia problem is the lack of a scientific means to ascertain the existence of qualia that are a phenomenon of the internal world, and as such, there is a possibility of agnosticism gaining power in the discussion of qualia.

Then, how about humans and machines?

Does a machine have qualia? Scientific arguments will go like this: Machines do not have the function to generate or feel qualia, and therefore machines cannot have qualia. Igor Alexander, a British researcher of machine consciousness, claims that MAGNUS, a conscious system he developed, has qualia (Aleksander, 1996). He asserts that machines have their own qualia. Although his argument has not been readily accepted, it can be one of the explanations deserving further investigation if the qualia problem is left for discussion at the mercy of Kant's agnosticism. Of course, agnosticism is a manifestation of a thought, and as such, there can be another thought to deny it. Discussions in this regard are not over. The author currently holds the same views as Alexander.

9.9 Problem of Free Will

The problem of free will was mentioned in Chapter 4. The topic is discussed in this chapter in detail while referring to the model of the consciousness system.

Research on human free will by Benjamin Libet, professor emeritus at the University of California, San Francisco (UCSF), and other researchers highlighted the theme of free will (Libet, 2004). These research studies doubt the common thinking that humans act based on their own free will. New scientific evidence found by the researchers is that the brain is activated about 0.55 sec before a human arbitrarily decides on and performs a behavior. Humans are unaware of the fact that activities regarding a behavior have already begun in the brain, and they become aware just when the behavior is about to be performed.

When humans think, they become aware of the content of the thought in some cases. For this reason, there are two types of cerebral activities by humans: those noticed and those not noticed. Cerebral activities occurring before one becomes aware may be an integrated process of what has been learned empirically all of one's life since one was born and the functions that one had already possessed when one was born. The latter refers to the innate functions acquired in the evolution process of life since life appeared on the Earth. Because these functions are built into and belong to individuals, the cerebral activities are performed with self-responsibility, irrespective of whether one is aware of them or not. This is the basic idea underlying the thought of free will.

Free will means not to receive instructions from others and that all decisions are made by oneself. On the other hand, when one becomes aware of the behavior selected by the cerebral activities, one is given the opportunity to scrutinize the proposed behavior and then perform the behavior. It can be said that one is acting with self-responsibility. We cannot be aware of the scrutiny activity of the brain, but we have the option not to do the behavior once it surfaces to our consciousness. This may be called free will. Wrist-flexing is not a very common social activity; so it may be difficult to use it to discuss the importance of suppressing behavior as it relates to free will. If wrist-flexing were replaced by the gesture of aiming a gun at a person, for example, the role of free will to suppress a behavior would be more keenly appreciated.

In later research, Libet found that an awareness of conscious will occurs about 0.15 to 0.2 sec prior to the physical action, and just at that moment, one can select between stopping the behavior or going ahead with it. Libet formulated a hypothesis that the possibility of this selection is the entity of human will. The brain is always ready to perform behaviors, but consciousness controls the execution of behaviors. In short, humans do have free will and are held responsible for their actions.

The author's model does not specifically use control signals for suppressing the execution of a behavior. However, the author's consciousness system can execute suppressed behaviors without jeopardizing the consistency of cognition and behavior (Suzuki *et al.*, 2005; Takeno *et al.*, 2005) Therefore, it is possible to suppress behaviors using signals to execute or not execute. A suppressed behavior means the act of imagining the behavior. A process seems to exist in the human brain in which the brain imagines a behavior just before the behavior is executed and determines the possibility of releasing the suppression by evaluating the condition of representations firing at that time.

It is possible for the consciousness system proposed by the author to have a similar function to the human mechanism of free will.

The problem here is the relationship between will and responsibility, i.e., who is responsible for the behaviors performed by humans? Generally, if a behavior is performed with awareness, the performer is responsible for the behavior. If, on the other hand, a behavior is performed without awareness or unconsciously, the performer may not be held responsible. And even if the unconscious performer is held responsible, the responsibility seems to be lighter than that assumed for a conscious performer. If a robot has the function of will, can the above concept be applied also to the robot? The problem is, who is responsible for the behavior of a robot? If we make the robot responsible, is there any practical meaning to stating that the robot accepts the blame? If a robot has a self, should it be responsible for its own behavior like a human? The author believes that although robots may have nearly the same functions as humans, the maker of the robot should be basically held responsible for the conduct of the robot, at least for the time being. This would be much like the case of an automobile, in which the owner or driver is held responsible, and various legal regulations should be established to ensure the safe use and operation of robots. Robot owners should also be subject to various legal controls to promote the safe use of robots. We should do our best to ensure the safety of robots so that conscious robots can work in society for the benefit of humans.

9.10 Summary and Observations

This chapter introduced the word's first self-aware robot capable of becoming conscious of itself reflected in a mirror. The scheme of consciousness is made up of recurrent neural networks. The basic idea is as follows: (1) The process for the robot to cognize information coming from sensors and the process to execute behaviors share a common neuron group K, and (2) the common neuron group K is represented by neural circuitry. Another feature of the author's robot is that behavior information is returned to the common neuron group K via somatosensory nerves. The robot thus cognizes information about the external world supplied via the sensors and information on somatic sensation, i.e., both external and internal information, as representations, simultaneously and at all times. This means that the robot distinguishes between the self and others. This scheme of consciousness has been demonstrated to be capable of learning and performing imitation behavior.

Using this scheme of consciousness, the author has constructed emotion, feelings, and association areas in the consciousness system of the robot. The structure, though simple at this stage of development, shows the possibility of the robot's consciousness interacting with emotion and feelings. Subconsciousness and explicit consciousness have been described. Lastly, the sense of a robot, i.e., the problem of qualia, free will, and problems of responsibility related to free will, have also been discussed.

Human consciousness has not yet been completely defined, but the author's scheme satisfies 90% of Husserl's 10 properties of consciousness. As presented in the online version of the Discovery Channel (www.discovery.com), the scheme of consciousness developed by the author is a solution to the problem of mirror image cognition, i.e., becoming aware of the self reflected in a mirror.

Chapter 10

Physical Demonstration of Successful Mirror Image Cognition by a Robot

Humans can be aware of their images reflected in a mirror, but the relevant mechanism — the reason why they can — has never been described physically and mathematically and still remains a mystery. The author has developed a computer program and conducted experiments to achieve awareness in a robot.

Three experiments have been conducted:

(1) A robot imitates the image of itself reflected in a mirror.
(2) A robot imitates another robot of the same type that is made to behave in the same way as the first robot.
(3) Two identical robots with the same functions were placed face to face to imitate each other.

The coincidence of the behavior between the self and the other is calculated in each experiment. The coincidence in Experiment 1 (mirror image test) is always higher than that observed in Experiment 2 (ordered behavior test). In Experiment 2, the other robot that is controlled by the self robot can be regarded as part of the self robot, like a limb in the case of a human. From these experiments, one can say that the image in the mirror is an existence

Self-Aware Robots: On the Path to Machine Consciousness
Junichi Takeno

ISBN 978-981-4877-90-9 (Hardcover), 978-1-003-26181-0 (eBook)
www.jennystanford.com

that is closer to the self than a part of the self is, i.e., it is none other than the self, and the self is controlling the mirror image. These were the first experiments to physically describe mirror image cognition, and we achieved a 100% rate of success in mirror image cognition with a robot.

This chapter contains detailed logical and physical descriptions of how this result was attained. It also includes observations and prospects derived from the experiments.

10.1 Introduction

Humans can easily identify their images reflected in a mirror despite the fact that since birth no one has ever seen his or her own face directly. This is strange. Humans say, "I know my face by looking at it in a mirror." However, this does not solve the problem. Humans are also said to become aware of their own images reflected in a mirror when they are about two years old (Amsterdam, 1972).

G. Gallup, Jr., proposed a mirror test to evaluate such awareness (Gallup, 1970). So far, chimpanzees, orangutans, dolphins, Indian elephants, and magpies have passed the mirror test.

Jacques Lacan, a French psychoanalyst and philosopher, announced his mirror stage hypothesis, in which he cited the mirror test as an essential milestone in human growth and development (Lacan, 1982).

Becoming aware of one's self image in a mirror is said to suggest the existence of self-consciousness because humans put on makeup and dress themselves in front of a mirror. An explanation of this becoming aware of one's self image in a mirror might possibly lead to a solution of the problem of consciousness.

When we refer to consciousness, we immediately run into a difficult problem. There are groups on two sides: those who do not acknowledge the existence of consciousness and those who do. The former group generally includes scientists on the engineering side, whereas the latter consists of scientists on the physical side. This split is due to a lack of a clear description of the phenomena, i.e., a definition, of consciousness.

The extremists in the former group believe that consciousness is a subjective phenomenon and cannot be described mathematically, and therefore its existence is not acknowledged. Other researchers in the same group also do not acknowledge the existence of consciousness but assume the existence of various functions of recognition, which are integrally combined to form something new, which they call "emergence." For them, consciousness is simply a phantom or an illusion.

The researchers belonging to the latter group acknowledge the existence of consciousness in humans and try to locate it in the structure of the human brain.

Given these circumstances, the effort to discover the scheme of recognizing one's self image in a mirror as a clue to solving the problem of consciousness may well bring on furious arguments between both sides.

The author decided to challenge this task in a way not attempted much by researchers in the past. I decided to define consciousness physically and mathematically by checking research studies on consciousness published in the past. The reason why such an approach had seldom been used in the past is a persistent spiritual resistance to any attempt to describe human consciousness physically and mathematically. This arises from the firm belief that humans are different from machines. Perhaps the remote possibility of actually attaining the aim may be another reason. It is generally believed that a definition of personal consciousness is far from a universal truth.

If the attempt to define consciousness concretely, both physically and mathematically, is postponed at all times, we cannot take even the first step toward our goal of understanding the mechanism of human consciousness. It may be one way of research to expect "emergent" phenomena resembling consciousness by combining various functions of recognition. In the field of brain science, a variety of knowledge may be obtained to prove the repetitive reactions between the brain and the body. Even in these cases, we will still be confronted, eventually, with the difficult problem of defining consciousness. Although we have not yet found a clear-cut and universally acceptable definition of consciousness, we should never give up answering the question, "What is consciousness?" I believe the process I am going to use is an important scientific technique for understanding human consciousness.

Why should we hasten in our efforts to understand human consciousness?

It is naturally great to enjoy the fruits of understanding human consciousness or the mechanism of thought or behavior. I would like to emphasize the following three merits in particular:

1. The contribution to the understanding and discovery of treatment methods for brain diseases, including schizophrenia
2. The development of a means of recovery after the loss of consciousness caused by accidents
3. The development of artificial limbs devised to feel like a person's original limbs for those who have lost limbs in accidents and due to other causes.

I mounted a computer program on an existing small robot. The program was written using architecture developed based on my subjective definition of consciousness. The architecture uses a top-down design based on my original ideas, but the program itself runs using a bottom-up approach. The program not only drives the robot to physically demonstrate mirror image cognition but also describes most of the items related to consciousness that are known today (Takeno *et al.*, 2005).

The author's robot imitates the behavior of another robot placed in front of it according to the built-in program and calculates the rate of coincidence of behaviors between the self and the other. The three important experiments conducted with the robot fully and physically describe the problem of mirror image cognition. This is the meaning of what was described as a 100% successful experiment. The experiment was successfully conducted on September 1, 2004 (Takeno, 2005). The experiment was introduced on the online version of the Discovery Channel 1 TV, the United States, on December 21, 2005 (see APPENDIX C, http://www.rs.cs.meiji.ac.jp/TakenoArchive/DiscoveryNewsAwareRobot211205.pdf). Now, after about 16 years have passed, I feel that the success of this experiment is becoming more and more important. In this chapter, I will check a large number of comments received during this period and re-evaluate the experiments and their effect.

I herewith declare the 100% success of the mirror image cognition test of my conscious robot. Lastly, some topics derived from the results of the experiments will be introduced.

10.2 What Is Mirror Image Cognition?

Mirror image cognition means that one is aware of the image of oneself in a mirror. I can tell my image in a mirror out of several co-occurring images (Fig. 10.1). I think other people can also identify their images in a mirror as I can do because they put on makeup and dress themselves in front of a mirror. Some say that this is a subjective experience and therefore is a typical theme that rejects a scientific approach. Scientific research is, however, necessary when we are asked to answer the question, "Why can I tell it's my own image in a mirror?" The author calls this problem the "mystery of mirror image cognition" and has been trying to unravel this mystery using robots or machine systems.

Figure 10.1 Where am I? (Starbucks in Washington, D.C.)

Unlike humans and animals, every part of a robot allows scientific investigation, and the process used is understandable for lay people.

I am now developing a robot to scientifically verify mirror image cognition. If a robot capable of scientifically proving mirror image cognition is available, we will be able to clarify the mechanism of mirror image cognition of humans by analyzing the robot in detail.

Gallup formulated mirror tests to evaluate the existence of high-level cognition ability. His tests are designed to check whether animals can tell their own images in a mirror. The mirror test was reportedly successful when conducted with chimpanzees, orangutans, dolphins, Indian elephants, and magpies. It is impossible, however, to scientifically research how these animals acquired the ability to cognize their mirror image. This is also true of humans.

Detailed scientific research is possible with robots. The author believes that demonstrations by robots will teach us their self-recognition process and assist us in scientifically proving the existence of consciousness in humans too. The author believes that the robots are mirrors showing us scientifically the existence of the self.

10.3 Robots for Mirror Image Cognition Experiments

Two methods are available: an engineering-based approach and a conscious system structure approach. The former attempts to achieve the goal through engineering without discussing human consciousness.

Gold and Haikonen have announced that their mirror tests “appeared” to be successful. It is very difficult, however, to describe human functions such as cognition and consciousness (Michel et al., 2004; Haikonen, 2007).

It is possible, for example, to produce robots capable of cognizing the members of a human family and displaying heartwarming facial expressions without using the functions of human cognition and consciousness. As this example shows, it is possible to achieve the functions of human cognition and consciousness using a set of recognition functions and drive programs totally irrelevant to consciousness and cognition. Nevertheless, the creation of a robot capable of mirror image cognition is very difficult when considering several points that will be described later. The author calls this procedure an "engineering-based approach." Obviously, a robot created using this approach cannot describe any little bit of the phenomena of human cognition and consciousness. This type of technique may be useful but generally does not touch the root of the research theme of the author.

With the conscious system structure approach, the subjective phenomenon occurring within the self called consciousness can be treated as a physical phenomenon.

Subjective functions can be built into a robot (Tani, 2002; Kawato, 2000). The purpose of Tani's and Kawato's studies is to clarify the truth of subjective phenomena objectively and physically through robot demonstrations. This technique belongs to scientific positivism. The author uses the conscious system structure approach.

Compared with the simple engineering-based approach, the conscious system structure approach has a higher chance of making a major breakthrough in the study of consciousness. This is because in the former method, part of the functions of a whole system is created individually, i.e., piece by piece, whereas in the latter method, the principle of the overall picture of consciousness can be described when successful. Specifically, with the former method, it is almost impossible to present futuristic and constructive hypotheses, whereas with the latter method, it is highly possible that a large number of hitherto unsolved problems may be collectively solved and many hypotheses full of interesting suggestions can be presented. If I were asked which method I would take as a scientist, I would select the latter because I acknowledge the scientific rationality of the heliocentric theory of Copernicus (Nicolaus, 1473–1543) compared with the geocentric theory.

10.4 On the Development of a Conscious Robot

The author attempts to build a conscious machine. Consciousness is a subjective issue, but we consider it to be a physical phenomenon and have developed artificial consciousness using a mechanical system.

Machine systems, including robots, allow us to perform objective and scientific studies and observations. They provide us a solid base for the scientific observation of subjective phenomena. We can understand the phenomena of consciousness as an objective reality using machine systems.

Steps of Study

(a1): Define the meaning of consciousness.
(a2): Define a concept model based on the definition of consciousness.
(a3): Replace the concept model with neural models.
(a4): Mount the neural models on a robot.
(a5): The robot achieves mirror image cognition.

Steps a1 through a4 were discussed in the previous chapter. Step a5 is described in this chapter.

There are four important points to consider when developing mirror image cognition robots. No human has ever seen his or her own face since they were born. Humans have little prior information about the image of themselves, their face in particular.

First, in consideration of the above fact, the robot to be used in the study of mirror image cognition should not be given perfect prior information about itself. Humans cannot cognize their own image in a mirror when they are born. They can do so at about the age of two.

Second, to solve the mystery of human consciousness, we need to describe the process of the development of cognition from the stage of inability to cognize the self image in a mirror through to the stage of being able to do so.

Third, consider that mirror reflex information is not always perfect. Reflectance and planarity are not always 100%. Even if the robot possesses information about itself, the mirror-reflected information may not theoretically agree with the original information.

Fourth, the functions of the robot arising from the built-in computer program must be able to describe facts that are generally known to have derived from the functioning of human consciousness.

These facts include, among others, self-recognition, multiplicity of consciousness, and consciousness of others.

10.5 Mirror Image Cognition Experiments with a Conscious Robot

We use Khepera II, a commercially available small robot, and mounted the neural network program on the robot. The program uses recurrent networks called MoNADs as the basic module (Fig. 9.11). Each network has a hierarchical structure consisting of three modules (Fig. 10.2).

The operating mechanism of a MoNAD is described as follows. A MoNAD performs neuro-computation to determine the current behavior and cognitive representation based on information about the external world and on cognitive representation and behavior

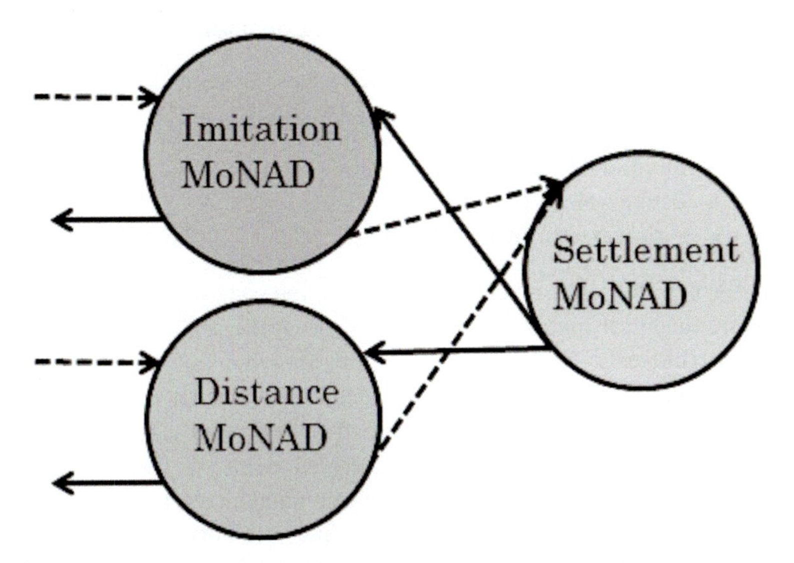

Figure 10.2 Three MoNADs are connected in layers in the experiments. The Imitation MoNAD cognizes and responds to the behavior of the other robot. The Distance MoNAD calculates the distance to the other robot. If the distance is small, the robot moves back. If it is too large, the drive unit (motor) is instructed to move the robot forward. The Settlement MoNAD restricts the behavior of the lower-level MoNADs.

information one step earlier. The derived information is used recurrently. The use of information one step earlier means that the current behavior is decided by taking into consideration past information (use of experience).

The robot imitates the behavior of the second partner robot in the mirror using MoNADs. The robot further cognizes the behavior of the self and the other simultaneously and calculates the success rate of behavior imitation (coincidence of behavior between the self and the other) using MoNADs. The coincidence success rate was about 70% in our experiments (Takeno et al., 2005). Although the success rate was not 100%, we came to the conclusion that the robot could physically discover its mirror image 100% of the time. Since the robot achieves mirror image cognition of the self, we call this robot incorporating the hierarchical MoNAD networks a conscious robot.

10.5.1 A Robot with an Embedded Conscious System

The author incorporated the consciousness system in the robot using three kinds of MoNADs (Fig. 10.2):

- Imitation MoNAD
- Distance MoNAD
- Settlement MoNAD

When the robot repeats an imitation, the consciousness system continually calculates cognitive behaviors of the self and the other simultaneously. The blue LED lights up when an imitation is successful as determined by the calculation (Fig. 10.3). Imitation coincidence rates are recorded. When the coincidence rate exceeds a certain value, the other entity is interpreted as the self. The imitation MoNAD interprets the behavior of the other and instructs the motors to perform the same behavior (a simple inference system). The distance MoNAD measures distance to the other. If the distance is small, the motor is instructed to back up and if it is large, the motor is told to move forward (a simple feelings system). The settlement MoNAD controls the behavior of the relevant lower-level MoNADs. I call this MoNAD formally the association MoNAD too. It is, however, not a central control tower such as a homunculus because

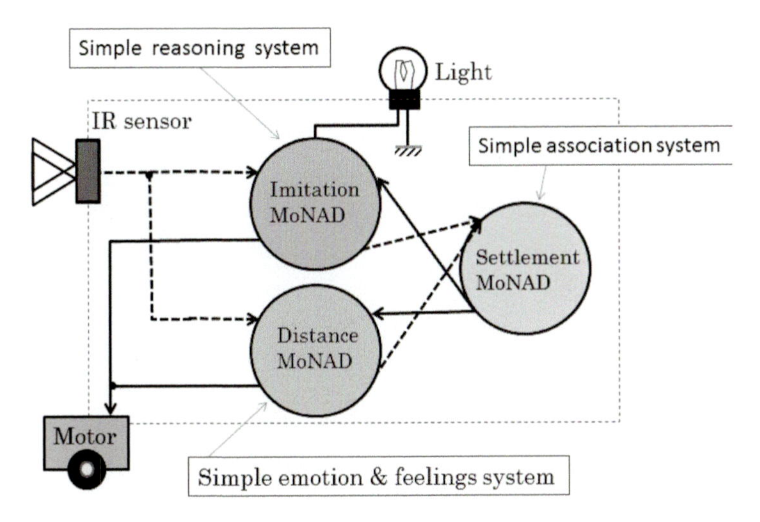

Figure 10.3 Structure of the mirror image cognition robot: The mirror image cognition robot continually calculates the behavior of the self and the other while repeating the imitation behavior. The blue LED lamp lights up when the imitation based on the calculation is successful.

its behavior is determined by the information supplied by lower level MoNADs (a simple association system). The LED controller compares the representations of the imitation MoNADs and fires the blue LED when the behaviors of the self and the other coincide with each other.

10.5.2 The Experiments

Three experiments were conducted.

Experiment 1: Robot Rs imitates the behavior of its mirror image Rm (Fig. 10.4). The IR reflectance of the mirror used in the experiment is 98%. (IR reflectance of ordinary mirrors used in daily life is typically 85%.)

(e1-1) Self robot Rs incorporates the consciousness system.

(e1-2) Self robot Rs compares and imitates the behavior of its mirror image Rm.

Experiment 2: Self robot Rs imitates the other robot Rc, which is completely controlled via cables by self robot Rs (Fig. 10.5).

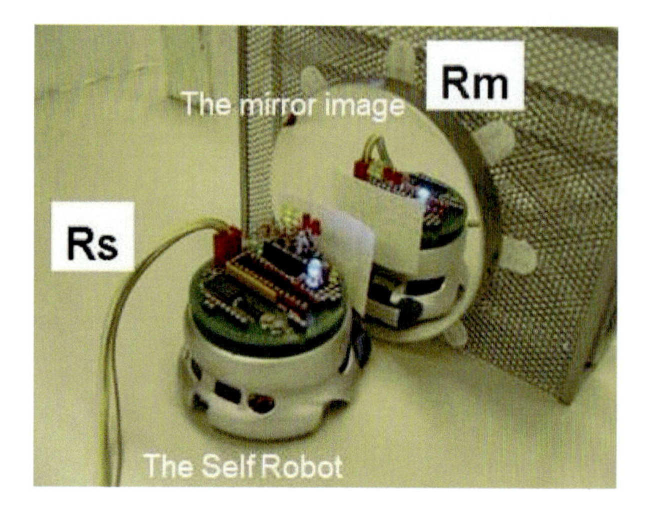

Figure 10.4 Experiment 1: The self robot Rs imitates the behavior of the mirror image Rm, which it views in the mirror in front of it. The mirror image Rm is used as the other robot.

Figure 10.5 Experiment 2: The self robot Rs imitates the behavior of the other robot Rc standing in front of it. The other robot Rc is connected to the self robot Rs via control cables. Using the cables, the self robot Rs instructs the other robot Rc to perform the same behavior as the self robot Rs (that is, Rs instructs Rc to imitate the imitation behavior being performed by Rs). Robots Rs and Rc look like they are imitating each other.

(e2-1) The other robot Rc is placed in front of self robot Rs (no mirror is used). Robot Rc is physically almost identical to robot Rs.

(e2-2) Both robots are connected by control cables. Commands are issued via cables to make the other robot behave in the same way as the self robot.

(e2-3) The other robot is programmed with a simple reflex system to execute the specified commands.

(e2-4) The self robot imitates the behavior of the other robot.

Experiment 3: Self robot Rs imitates the behavior of other robot Ro, which incorporates nearly the same hardware and software as self robot Rs. Both robots imitate each other repeatedly (Fig. 10.6).

(e3-1) Control cables, which were used to make the other robot behave in the same way as the self robot, are removed.

(e3-2) The reflex system software is removed from the other robot Rc and the same consciousness system as installed in self robot Rs is installed in robot Rc. Self robot Rs and the other robot Ro have nearly the same hardware and software.

(e3-3) The self robot and the other robot imitate each other.

10.5.3 Results of Experiments and Observation

The rate of coincidence of behavior of mirror image robot Rm, controlled robot Rc, and the other robot Ro are about 70%, 60%, and 50%, respectively. These values do not intersect with one another.

10.6 Why Are the Coincidence Rates Different?

All the robots used in the experiments have nearly the same physical features and functional specifications. The controlled robot Rc in Experiment 2 was also used in Experiment 3 as the other robot Ro. The different coincidence rates are due to different complexities of the robots in terms of physical property and functional specifications (Fig. 10.7).

Rs versus Rm: Self robot Rs and mirror image robot Rm have identical physical properties because robot Rm is a mirror image of robot Rs. Nevertheless, Rm is more complex than Rs because of the

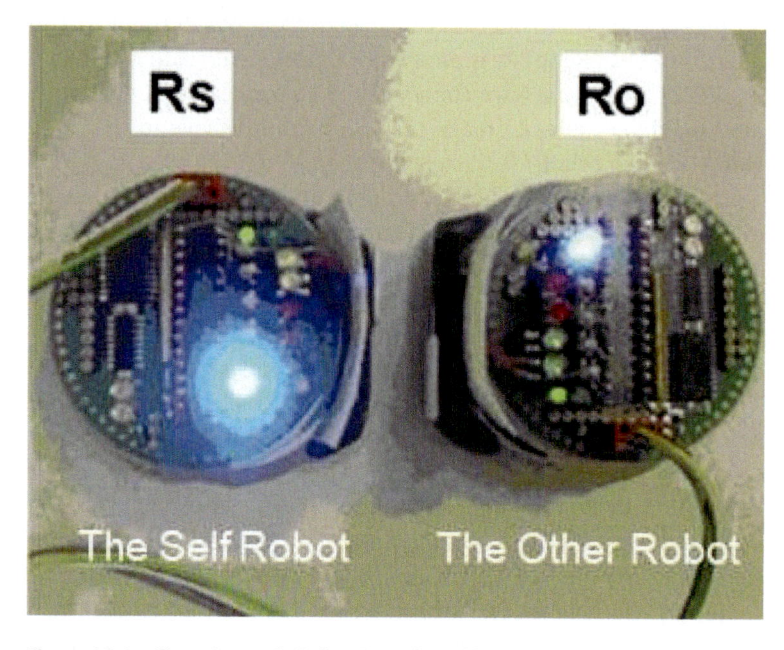

Figure 10.6 Experiment 3: Robot Rs is the self robot. Robot Ro has the same functions as robot Rs, that is, the robots are made from the same hardware and use the same software. Robots Rs and Ro perform the imitation behavior together.

effect of mirror reflectance (a), which will never reach 100%, and the interference of the external world with the IR sensors (b).

Rm versus Rc: Controlled robot Rc is physically more complex than mirror image robot Rm. Unlike robot Rm, robot Rc is free from the problem of mirror reflectance. Nevertheless, robot Rc is more complex than robot Rm because of friction when robot Rc runs on the floor (c); the individuality of the robots due to slightly different functions of motors and sensors (d); and the installation of the simple reflex system in robot Rc (e). For these reasons, robot Rc is, on the whole, physically more complex than robot Rm.

Rs versus Rc: Controlled robot Rc has basically identical physical functions to those of self robot Rs except that the functions of the motors (m1, m2) and sensors (s1, s2) are slightly different. The ground friction (f1, f2) of the drive wheels is also slightly different.

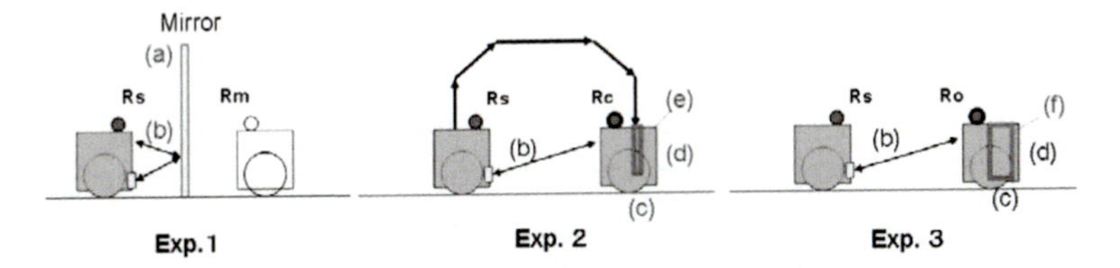

Figure 10.7 Discussion on the complexity of the other robots demonstrated in experiments: Here complexity refers to the difficulty of the forecasting behavior due to the effects of the hardware, software, and environment. Robot Rm is more complex than robot Rs. Robot Rc is more complex than robot Rm. Robot Ro is more complex than robot Rc.

This means that controlled robot Rc has its own individuality different from self robot Rs.

Rc versus Ro: Independent robot Ro is physically more complex than controlled robot Rc because it incorporates the consciousness system, whereas robot Ro has only a simple reflex system installed.

The difference between Experiments 1 and 2 is explained mostly by the increased physical complexity, whereas the difference between Experiments 2 and 3 arises from increased functional complexity, which is described in Fig. 10.8.

10.7 Summary and Consideration

The cable-controlled robot Rc is considered part of self robot Rs because it is connected to, and behaves according to the instructions of, self robot Rs. According to the results of our experiments and physical observations, the behavior coincidence rate of the mirror image robot Rm is always higher than that of the cable-controlled robot Rc. Based on these facts, the author concludes as follows.

Robot Rs determines that mirror image robot Rm is part of the self and is controlled by the self like robot Rc. According to our experiments, self robot Rs determines whether the other robot is the self or the other based on the coincidence rate (success rate) for behaviors. The threshold value is about 60% for self robot Rs. When the success rate is above 60%, the self robot judges that

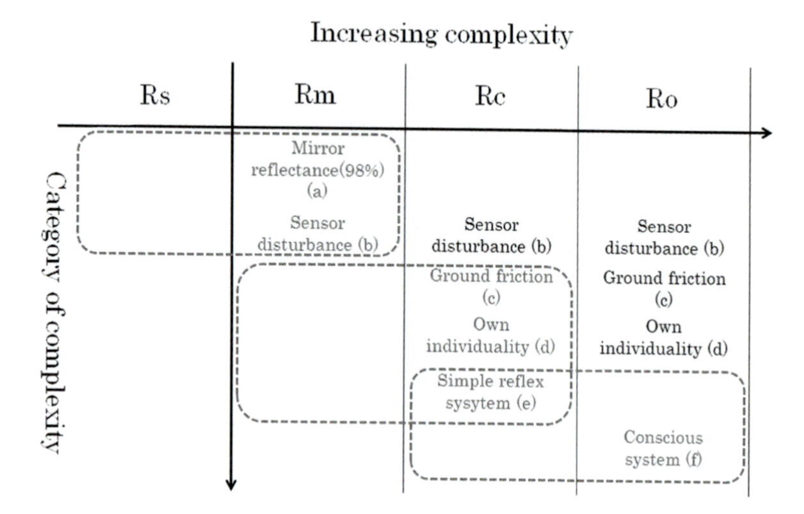

Figure 10.8 Summary of the complexity.

the other robot is actually the self. In other words, a robot judges whether the partner is actually the self or the other robot based on the coincidence rate of the behavior of being a part of the body of the self. For these reasons, we conclude that a 100% cognition rate has been achieved with regard to the condition that all robots used in our experiments possess the same normalized and functional specifications.

10.8 Investigations and Prospects

10.8.1 An Elucidation

Mirror image robot Rm is an existence closer to self robot Rs than robot Rc, which is part of the self (from the results of our experiments).

Humans become aware that mirror images are a part of the self (self-body theory). In other words, the author has discovered the physical meaning of humans being able to become aware that the image of the self in a mirror is the self. According to these experiments, the mirror image of the self is the other that is separated from the self. It is a special "other" and generates a sense of being part of the body of the robot. The LED lights up not because the coincidence rate of the behaviors between the mirror image and the self robot reaches 100%. The self robot recognizes that its image in the mirror is an existence that is closer to it than any part of it is. In other words, the robot's recognition that the mirror image is closer to the self than any part of self is has reached 100% (Takeno, 2008).

We have now solved the mystery: The robot recognizes the image ambiguously, and the mirror image is felt to be a part of the body of the robot.

10.8.2 Mirror Box Therapy

The human brain can feel the existence of lost limbs. After a person loses a hand or a leg in an accident, the person sometimes feels that the lost limb still exists on the body. This is called the phantom limb phenomenon. The phantom limb sometimes gives rise to phantom pain. Dr. V. S. Ramachandran, a US neurologist, succeeded in eliminating phantom pain from patients using his mirror box therapy (Fig. 10.9). The author's experiment-derived theory that the mirror image of the self is felt like a part of the body of the self provides physical support for the success of Ramachandran's mirror box therapy.

10.8.3 Mirror Stage

This observation is another physical demonstration of the mirror stage hypothesis (self-body theory) introduced by the French psychopathologist Jacques Lacan (1901–1981). According to the author's theory, the cognition of the self and the other is necessary before being able to cognize the mirror image of the self based on the results of the cognition of behaviors of the self and the other, i.e., to succeed in mirror image recognition. The mirror stage seems to refer to a stage where the behaviors of the self and the other are cognized separately and where the relationship between the self and the other (the meaning of the self and the other) can be cognized.

The mirror stage hypothesis assumes that infants establish the self as they cognize their mirror image and become aware of their

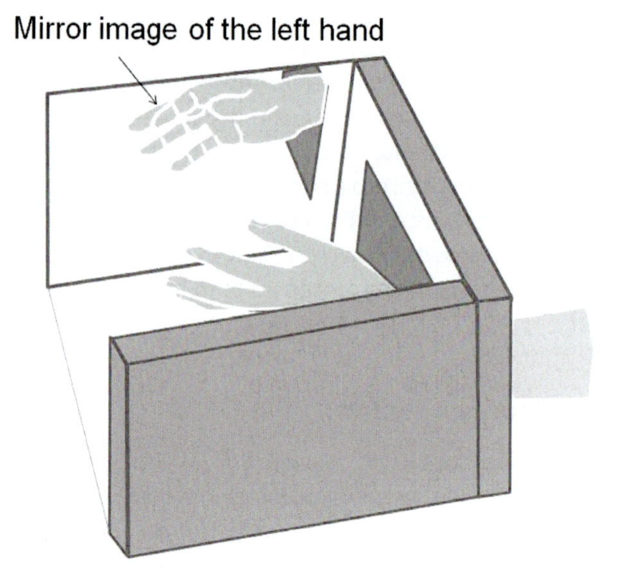

Figure 10.9 A scene of mirror box therapy.

integrated physical bodies in the early stage of development of their neural systems, which is called the mirror stage (Fig. 10.10).

10.8.4 Can Self Robot Discriminate Itself from Any Other Robots?

The answer is no. If the other robot possesses functions exceeding the self robot's capability (mobility and sensor capacity), the self robot will determine that the other robot is part of the self. Actually, an imaginable super robot cannot exceed the performance of its image in the mirror. This observation provides the grounds for believing that any artificial limbs that exceed the function of the self are a real part of the self even though they are not the real living limbs (Fig. 10.11). It can be judged, therefore, that artificial limbs are the real parts of the self (artificial limb hypothesis). This hypothesis is good news for those who must live with the discomfort of using artificial limbs after loss of their own limbs, because it provides a physical theory for artificial limbs to be accepted as if they were real limbs by the brain.

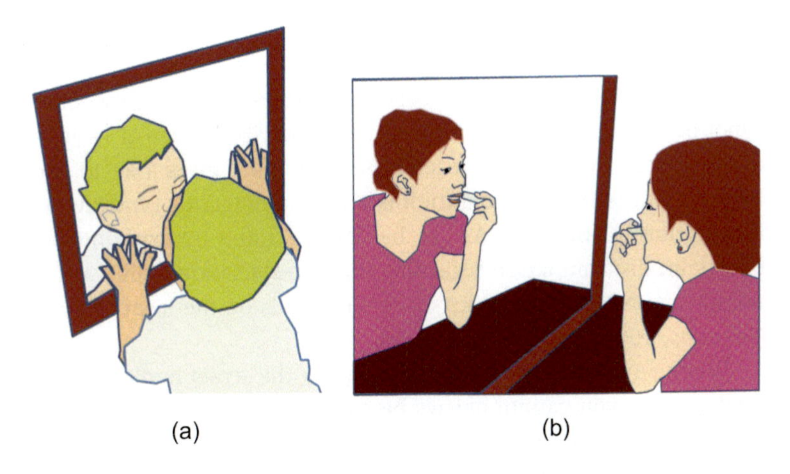

Figure 10.10 Jacques's mirror stage (a) and she knows her body (b).

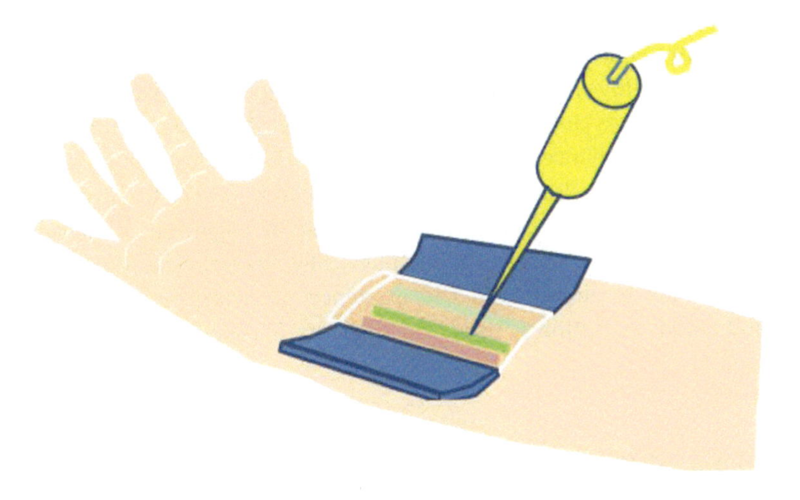

Figure 10.11 Artificial arm that feels like one's own arm.

The result of mirror image cognition experiments using robots supports Lacan's mirror stage hypothesis that humans become aware of the image of the self reflected in a mirror at infancy and thereafter further develop their cognitive functions.

10.8.5 Mysteries of Illusions of Reality

The result of Experiment 1 suggests that the coincidence rate of the robot behavior cannot theoretically reach 100% because of various disturbances existing in the environment. This gives rise to the hypothesis that human cognition is always ambiguous (ambiguous recognition hypothesis). As can be demonstrated by experiments, two-point discrimination regarding the skin sensation of humans is not always successful, but its success depends on the location (van Nes *et al.*, 2008).

When you look at people walking along the street and try to tell their gender, you cannot always be correct. Doctors cannot always correctly diagnose lesions hidden in the body.

The above hypothesis might have been proved true, but the following mystery still haunts: However ambiguous sensations may be, humans understand them as if they represented "sure existence." Examples include the phenomena of phantom limbs and phantom pain. These phenomena are called illusions of reality.

The cognitive function of humans is made up of "living machines." Since machines are involved in the cognitive function of humans, various disturbances in the external world (including physical disturbances generating from the human body) affect the process of cognition in the brain. And the result of cognition is always ambiguous both theoretically and physically. Nevertheless, the brain never fails to understand the existence of reality even though ambiguous cognition is used. This mystery is called the illusion of reality.

10.8.6 Study of Human Brain Using Mirror Image Cognition Robot

The MoNAD modules proposed by the author can describe many phenomena of human consciousness because the author's definition of consciousness is based on knowledge of the phenomena of human consciousness. Scientific knowledge tells us that the human brain consists of about 100 billion brain cells, with information entering and leaving these cells, i.e., being transmitted to other cells. In addition, some regular information is sent from certain areas in the brain to other areas or circulates between them. For example, information arrives from the body, passes through the spinal cord, and reaches the parietal lobe via the central part of the brain. Some other information is exchanged between the parietal lobe and the frontal lobe, and between the frontal lobe and the central part of the brain. Although there remains the dream that unknown substances are related to the function of human consciousness, it is necessary to try to identify the function of human consciousness on the basis of only currently available scientific knowledge. The author believes that human consciousness is formed not just by information circulating in the brain and that the information circulating between the brain and the body also plays a critical role. The author supposes the existence of MoNADs from these information circulations and judges that the MoNADs can describe human consciousness physically (Manner, 2011). Since the author's robot with built-in MoNADs has succeeded in achieving mirror image cognition, it is natural to believe that the MoNAD structure of the brain can be the first step toward describing human consciousness physically (brain MoNAD hypothesis).

Chapter 11

Further Developments

In this chapter, we present some concrete examples of the conscious system that uses the MoNAD.

As stated earlier, the MoNAD has the features of a conscious module.

It is based on my subjective observation that the source of human consciousness is "consistency between cognition and behavior." Because the term "consistency" is indeed a literary expression, and the terms "cognition" and "behavior" belong to different categories, it is necessary for me to unite them.

I have always thought about how to construct an electronic circuit that satisfies this requirement, and I happened to conceive such an electronic circuit of the neural network. That was MoNAD. In my idea, consistency means the state in which information is circulated in the electronic circuit and eventually a convergence value is stably established. Also, I simply considered the difference between categories as a variation of the information pattern.

As a result, this information system using MoNAD successfully carried out the experiment of mirror image cognition by a robot. In other words, this experiment provides scientific proof that the system can recognize the "self-image in a mirror" as a special

Self-Aware Robots: On the Path to Machine Consciousness
Junichi Takeno

ISBN 978-981-4877-90-9 (Hardcover), 978-1-003-26181-0 (eBook)
www.jennystanford.com

existence that reacts as if it were a part of one's own body even though that image is completely separate from the body.

It may be a scientific elucidation regarding the longstanding issue of how humans can be conscious of their "self-image in a mirror." I thought that the experimental results from MoNAD could be an important basis for explaining the process of human consciousness.

The MoNAD is an information circuit composed of two closed loops. The first loop acts as a sort of primary reflex system, and the second loop can be expressed as the secondary reflex system that reflects the reaction from the first system. Therefore, the MoNAD has the feature that it can basically evaluate the state of the primary reflex system by using the secondary reflex system. Such feature is called the "self-referential" function or "metalogical" function.

In the previous chapter, I described the experiment of the robot's mirror-image cognition through the connection of several MoNADs. In this experiment, the information circuit has three outstanding functional portions that have been developed to be mounted on the robot: the Reason System, the Emotion & Feelings System, and the Association System (Fig. 9.13).

Each of these consists of a combination of MoNADs. The first system reacts to an object distant from oneself (for example, another person). Visual and auditory senses belong to this system. The second system reacts to the status of one's own body. The third system realizes the communications between the Reason System and the Emotion & Feelings System. In other words, the Association System coordinates the integration or the contention between the Reason System and the Emotion & Feelings System.

This functional segmentation of three systems is derived from the knowledge of conventional philosophy, psychology, or brain science. In summary, a single MoNAD or a combination of several MoNADs enables the development of a more complicated information system, which I call the "conscious system." By configuring such conscious system by using the Reason, Emotion & Feelings, and Association systems, I thought that a more sophisticated system similar to human consciousness could be developed.

Then I attempted to find out, with my students, how useful the conscious system using MoNAD could be for understanding the human consciousness and its limitations. This chapter describes our activities.

11.1 A Robot Recognizes the Unknown

Conventional robots have been required to correctly perform work using their own knowledge in order to achieve the purpose. Without doubt, this is an important point of view. Robots are expected to play a role, not only in a defined working environment in a plant but also, sometime in the future, in an unlimited natural environment or at a disaster site that might be impacted by unexpected environmental change, or a dangerous site where access is physically impossible, or even in cosmic space or other unknown environments.

For example, in March 2013, a giant earthquake and tsunami caused cataclysmic catastrophe in the Tohoku district and led to the meltdown, a supposedly impermissible incident, in the Fukushima nuclear power plant. Obviously, people could not enter the building with the meltdown reactors, and robots were expected to work in the area. Due to earlier explosions, however, concrete blocks and architectural materials were scattered throughout the building and the dispatched robots soon exhausted their knowledge and stopped. It is not possible to make an assertion that a robot responsive to such incident can be developed in the future, but robot researchers recognized the importance of improving the robots' environmental adaptability.

The study to improve the environmental adaptability is well known as "machine learning." Also, the terms "Q-learning" and "reinforcement learning" are often used. These are prominent learning theories among robot researchers. For example, some four-legged robots shake their legs vainly at first and then gradually start moving in one direction. This is because the robot is equipped with a program that finds, as the result of initial trial and error, a motion element toward one direction, then gradually reinforces this motion. This research is targeted at enhancing the robot's ability to more efficiently learn the intended action.

I can appreciate the importance of these studies but consider that this learning theory gives no indication about human consciousness although the information cycle of cognition and behavior is provided. Rather, this theory seems to show the fundamental learning process before humans started evolving their consciousness.

I have considered the problem of learning from the viewpoint of the conscious system. As stated earlier, the conscious system is the information circuit that achieves consistency between cognition and behavior. A series of consistent information converges in a stable pattern within the MoNAD information circuit.

11.1.1 Detailed Considerations on MoNAD

This section explains the MoNAD by using Fig. 9.11 in Chapter 9. The information on external environment and the information from the lower-layer MoNAD are continuously input to neuron group A (inputL). Neuron group B (outputL) is the terminal used to output the behavioral representation of MoNAD. Such information becomes the behavior instructions for the motor or other actuators that can change the external environment or for the lower-layer MoNAD group.

In summary, each portion of the MoNAD carries out the neural calculation regarding the information given to terminal A and determines the value to be output from terminal B. When the information given to A has already been learnt by MoNAD, the value of terminal B becomes a convergence value, namely a fixed point of the neural network. With this explanation, readers might get the wrong idea that the MoNAD is a simple neural network in which the input information for A determines the output information for B.

The MoNAD is indubitably a sort of neural network, but it provides a special information structure, which uses two recursive intelligence cycles to determine output B using input A. One cycle is "A, K, B" and the other is "K, Z, Z′" (Fig. 9.11).

Readers might think that the first cycle A, K, B is enough to convert input value A into output B. As far as a simple neural calculation is concerned, it is true that the second cycle K, Z, Z′ is an unnecessary calculation. I added the second cycle, however, for a special purpose. The second intelligence cycle is a circuit called the "redundancy system."

The redundancy system has been added for several reasons:

1. The MoNAD repeatedly carries out spontaneous driving.
2. The MoNAD simultaneously carries out the processing for both behavior and cognition.
3. The MoNAD is an information processing unit of the self-reference type.
4. The calculation results are expected to be used outside the MoNAD.

These four reasons are derived from the purpose of representing the human conscious functions in the neural network. Therefore, I call the redundancy system the self-reflective system.

The following is a more detailed explanation:

For reason (1), the human consciousness is activated mainly by external stimulus (information), but some stimuli are internal. For example, in the former case, foraging behavior is triggered upon seeing a ripe apple and, in the latter case, hunger triggers the foraging behavior. Behaviorists may criticize the latter case by arguing that hunger does not exist.

However, it is not necessary to discuss this issue here, because it is less problematic whether human consciousness is activated by internal stimulus or not. In addition, the existence of internal stimuli in human consciousness is not questionable because the activation by internal stimuli is pertinent to the purpose of life extension. A stomach problem urges people to seek medical advice.

Besides, human consciousness usually never stops. Knowledge of brain science indicates that some sort of iterative process drives the human consciousness (Carter, 2002). For example, a movie continuously presents 24 individual frames of still images per second, but humans interpret them as a seamless motion picture. This indicates that the human conscious process is activated at least at intervals of approximately 42 milliseconds.

Regarding reason (2), the human consciousness can perceive the synchronization between behavior and cognition. For example, open your hand. This behavior can be observed by seeing the opened state of the hand (behavior & cognition) or by knowing the muscle conditions of the fingers and hand (behavior & cognition). Also, you can envision the action of opening your hand (behavior & cognition).

Again, for this point, there may be a rebuttal from behaviorists. However, most readers, I believe, cannot deny my argument based on the phenomena.

Thus, I noticed that human consciousness was relevant to an odd correlation between cognition and behavior. And it could be argued that the terms "cognition" and "behavior" were conceived as different notions by a human. Who other than a human could conceive these terms? Cognition and behavior imply different phenomena, but, under these phenomena, they may share some common "neural network activity." If so, it may not be necessary to make a distinction between cognition and behavior. While I devoted myself to this thought experiment, an Italian researcher showed that such neuron group existed in the brain of a monkey (Rizzolatti et al., 2006).

It is called a "mirror neuron," and is used for both cognition and behavior. I did not know about this finding when I devised the MoNAD. However, neuron group K installed within the MoNAD acts exactly as both cognition and behavior (Fig. 9.11). I gave the name "primary representation" to this K, which is the neuron group common to the aforementioned two recursive intelligence cycles. In other words, both cognition and behavior are concurrently processed in the K section of MoNAD.

For reason (3), as stated earlier, the MoNAD mediates two recursive intelligence cycles via neuron group K. When the intelligence cycle A, K, B is deemed as the lower cycle and K, Z, Z′ as the higher cycle, it can be said that the higher cycle supervises the lower cycle.

Logically speaking, the higher cycle is the meta-logics of the lower cycle. Here, the term "lower" simply means that it is closer to the external environment. I agree with a researcher who calls the MoNAD structure "Second-Order Cybernetics."

In terms of robot engineering, the MoNAD can be regarded as an information processing model of the self-reference type.

For reason (4), the higher intelligence cycle in the MoNAD has the input/output terminals for exchanging the information with other information processing units. They are inputH (Z′ terminal in Fig. 9.11) and outputH (Z terminal in Fig. 9.11). I call these "behavioral representation" and "cognitive representation," respectively. This is because, when the former has received a new piece of information from another information unit, the MoNAD tries to take on a new behavior. Also, the latter additionally conceptualizes (abstracts) neuron group K of the MoNAD. K stores the neural calculation results of the information received from inputL (X in Fig. 9.11), the information from outputL of the previous cognition/behavior cycle (Y′ in Fig. 9.11), and the value Z′. This means the acquisition of new information (X in Fig. 9.11) through the behavior instructions of MoNAD (Y in Fig. 9.11).

More specifically, the cognitive representation Z of MoNAD is the result derived by integrating and abstracting the new information (X) generated by behavior instruction (Y) and the values Y′ and Z′. In addition, under normal conditions, the MoNAD has a process that copies the results of cognitive representation (Z) onto the behavioral representation (Z′). This causes the MoNAD itself to function so as to maintain the current cognitive representation. However, as described above, when a new piece of information (inputH) is given to the behavioral representation Z′ from an external information unit, this copy process becomes ineffective.

The MoNAD functions to maintain the current status as in a normal case, but since the higher information unit has determined a new behavioral representation, the MoNAD has to follow the instructions.

The above explanation persuades the readers, I believe, to understand that the second intelligence cycle is not a useless processing operation.

Getting back to the main topic, as mentioned earlier, when the MoNAD generates a stable pattern, external and internal environmental information is combined and expressed as a representation. The convergence of MoNAD itself into a representation is an important point in this section. Because of the features of MoNAD, such converged representation is consistently associated with the external environment. The representation is also associated with the internal environment, but, in this section, for the purpose of convenience, the internal environment is fixed to "I stand still" for example.

An example situation is "I stand still and see a red color." The action of seeing is related to the action of "turning one's eye to something." However, in this section, I disregard this relationship in order to simplify the explanation. In the robot experiment, this situation can be translated into "red color input to the eyesight of the robot."

In the experiment, a conscious system recognizing three colors is preprogrammed in the robot. These are red, green, and blue, i.e., the three primary colors. The section aware of the three primary colors is called "prim-reason."

In the MoNAD, each color is associated with a behavior of the robot: stop, move forward, and move backward, respectively. For example, as long as the green color is input to the eyesight, the robot continues to move forward. Some robot scientists might say "such robot can be constructed with a conventional mechatronics model." I agree, but the reason why the conscious module is used will be understood later.

Assume that the color purple is input to the robot. Purple is a mixture of the abovementioned three primary colors: red ($R = 128$), green ($G = 0$) and blue ($B = 128$). Eventually, purple is analyzed as a mixture of red and blue. The robot cannot conclude that this is red or blue. Needless to say, green is not possible.

Next, we study a case where purple is input to the prim-reason. Does the prim-reason become conscious of the purple color and answer that this is purple? Or, does it cause the robot to take some kind of action? Without a doubt, the actual prim-reason cannot be conscious of the purple color, and regards it as red or blue because it has learnt only the consciousness of red, green, and blue. In other words, the knowledge of prim-reason is limited to three concepts (red, green, and blue). Therefore, it has no choice but to understand the purple as one of three primary colors.

In the experiment, it is conscious of red if the red constituent is richer among the three primary colors, and the robot stops. When the red and blue constituents have the same value, it vacillates between red and blue and the decision is deferred. However, in this case, the settings of the actual experiment environment are so difficult that probably one of the three primary colors will be selected due to noise.

Normally this process provides the same result as the three-primary-color recognition system using the neural network.

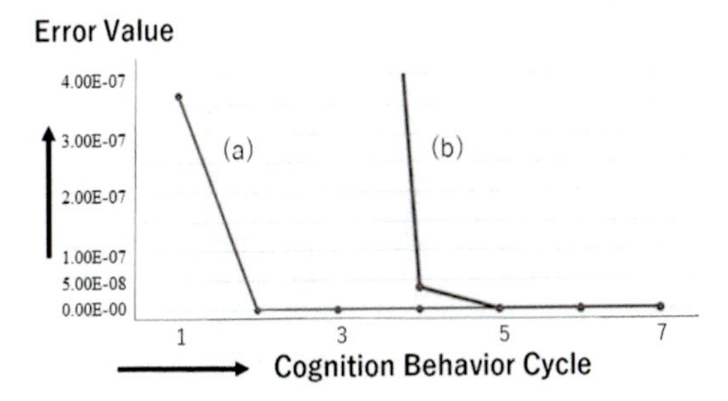

Figure 11.1 The convergence state in response.

It is certainly incorrect if it answers "red" when purple is input. So, I hope that it answers "I do not know" or "unknown."

The study on this problem is a main theme in this section.

If the robot can be aware that the input information is "unknown," the robot can change it to "known" information by learning such unknown information. This would be a breakthrough for a new type of robot.

Let us discuss what unknown information is.

As explained above, the prim-reason has learnt the red, green, and blue colors. This implies that the prim-reason, namely the conscious MoNAD, promptly provides the convergence state in response to three types of input information (red, green, and blue) (Fig. 11.1a). It outputs behavioral information which instructs the robot to behave based on the color information.

In summary, the prim-reason can quickly react to the already learnt input information to determine the output information. For the input information of purple, such a quick reaction is not possible and therefore the prim-reason defers the processing (Fig. 11.1b). This is because the prim-reason has no orientation in regard to the input information of purple. As explained above, this is a general property of the MoNAD which circulates the information in a duplex recurrent network. When the input information is known, the MoNAD converges to a certain value quickly, but when the input information is unknown, it defers the processing. By using this feature of MoNAD, it is possible to easily determine whether the input information is known or unknown.

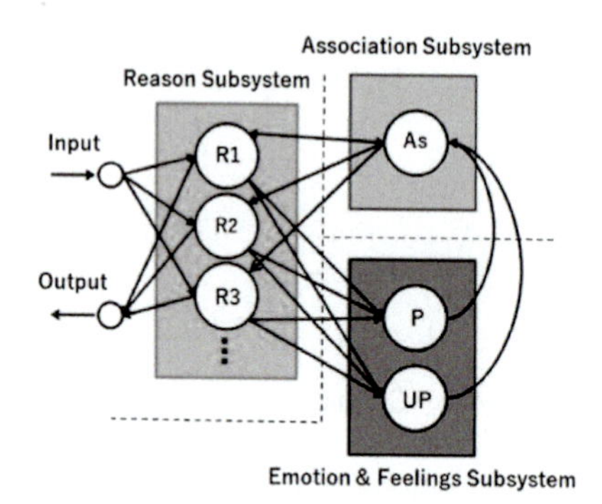

Figure 11.2 The basic conscious system for this experiment.

With the students, the author tried to configure the robot to recognize that the information is unknown.

The conscious system (Fig. 11.2) was developed according to the basic model shown in Chapter 9 (Fig. 9.13).

In this development, we converted the expression "system" into "subsystem." For example, "Reason system" was changed to "Reason subsystem." This is because we wanted to emphasize the meaning of "a part of the conscious system." In addition, "B" and "C," i.e., Emotion MoNAD in Fig. 9.13, were not used, because the experiment of this section did not use the "emotion" connected to body sensors. Instead, in this experiment, we prepared three sets for the "reason system A and G" in Fig. 9.13 and named them R1, R2, and R3, respectively. Here, "G" is the MoNAD unit which measures the state of "A." Also, in this experiment, "D" and "E" in Fig. 9.13 were the MoNADs named "P" and "UP." The Association MoNAD "F" was called "As" in this experiment. The MoNAD "P" represents "pleasant" and "UP" represents "unpleasant"; we will discuss what is "pleasant" and "unpleasant" later. In this experiment, "pleasant and unpleasant" correspond to "good and bad" of convergence in the Reason subsystem (Fig. 11.2).

R1 of the Reason subsystem is the prim-reason of this conscious system. Therefore, it recognizes the red, green, or blue color, and outputs the associated behavior (stop, move forward, and move backward) to the Output terminal (Fig. 11.2). R2 and R3 are the MoNAD which has neither orientation nor information at the initial stage of this experiment. We call them null MoNAD. When a color other than the three primary colors (red, green, blue) is presented to the Input terminal of the conscious system, the convergence state of prim-reason R1 becomes degraded. The conscious system represents "UP" MoNAD through the G unit in regard to this situation.

When the MoNAD represents a value or converges to a value with known information, the author expresses this situation as "the MoNAD *shoukies*." This is an analogy of the expression "the neuron fires." Using this expression, it can be simply said that MoNAD R1 *shoukies* when red, green, or blue is presented to the robot. When the purple color is presented to the Input, however, R1 does not *shouki* because of poor convergence. As described above, the state of poor convergence causes the MoNAD UP to *shouki*.

At this time, in order to add new knowledge to the null MoNAD R2 in the Reason subsystem, the MoNAD As in the Association Subsystem establishes the network for Input, Output, P, UP, and As (Fig. 11.2). Readers may wonder which subsystem is responsible for this network connection.

Although it is not indicated in the conscious system, actually a fundamental system exists to back up the conscious system. This system can be configured by MoNAD or simply by a neural network, but in this experiment, this has been developed by a simple program (C++ and others). I believe that the fundamental system which supports the conscious system does not have the function of being conscious.

However, obviously it can be assumed that when the fundamental system becomes active, the conscious system starts to work and becomes conscious of the activity of the fundamental system. Therefore, it is important to note that the conscious system can eventually be conscious of the activity of the fundamental system.

Let us return to the topic of the experiment.

Since R2 was given, the Reason subsystem can have two MoNADs (R1 and R2). In addition, the fundamental system prompts R2 to learn the information of "purple" presented to the Input terminal. However, at the same time, it is also necessary to learn the behavior in response to the purple color. R1 is associated with three behaviors (stop, move forward, and move backward). What behavior should R2 learn in response to "purple"? This is a major issue, because there has never been a discussion about the function with which the conscious system can determine or generate such behavior by itself. This issue is important because it is related to the "act of creating" by robots.

This issue is considered in detail below.

The MoNAD module has been designed in accordance with the definition that consciousness has consistency between cognition and behavior. Specifically, cognition is associated with behavior and is learnt by MoNAD. Therefore, this associated behavior is supposed to be generated internally within the conscious system which has evolved till now. This is the act of creating because the conscious system produces a new behavior by itself. However, since there are no clues to solving this problem, I would like to leave it as a future issue.

As a last resort, in this experiment, we decided that the behavior was to be instructed by a human being. For example, if a new color is the k-th to be learnt, then the robot turns [$45 \times k$] degrees to the left or right instantly. Since purple is the first color to be learnt, the robot turns 45° to the left or right instantly.

After this learning, whenever the purple color is presented, the robot carries out the left- or right-hand rotation of 45°. In addition, the robot is capable of stopping, moving forward and moving backward in response to red, green, and blue, respectively, as it used to be.

Hence, it can be said that the robot found an unknown color (purple), and learnt a new consciousness (R2). In a later experiment, the robot found another unknown color (yellow) and learnt another consciousness (R3).

11.1.2 Results of Experiment and Discussion

Figure 11.3 outlines the experiment.

The robot used in this experiment was the commercially available e-puck robot. To present a color to the robot, we used the display of a cell phone. In Fig. 11.3a, the robot stopped in response to red. In Fig. 11.3b, the robot moved backward in response to blue. In Fig. 11.3c, the robot moved forward in response to green. In all cases of Fig. 11.3a–c, the robot turned on the P lamp (green LED).

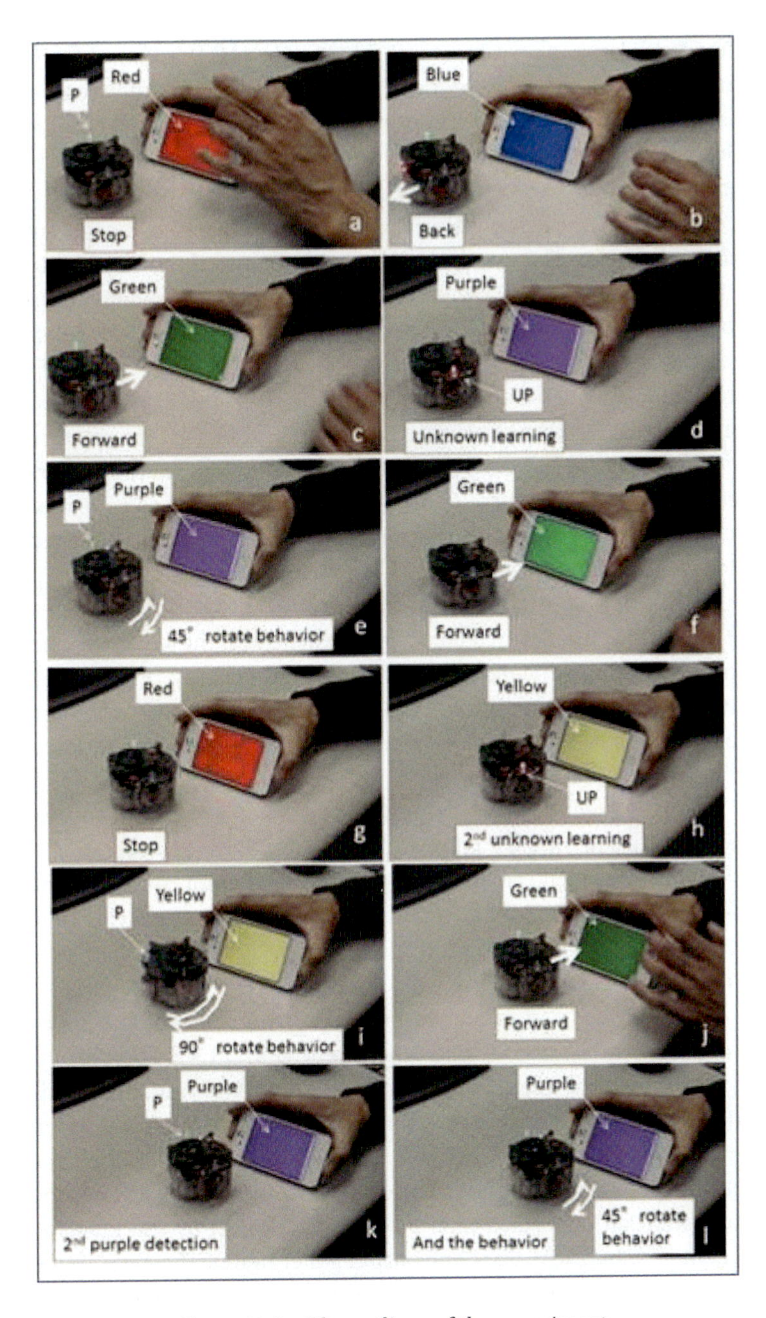

Figure 11.3 The outlines of the experiment.

When the P lamp is lit, it indicates that the MoNAD P is in the *shouki* state in the conscious system. P means "pleasant." In other words, this lamp indicates that the robot has learnt the red, green, and blue colors, and therefore it can be conscious of these colors. Then, the purple color was presented to the robot (Fig. 11.3d).

As a result, the robot recognized the purple color as an "unknown" color. This can be understood because the robot turned on the UP lamp (red LED), which means "unpleasant." Then the robot proceeded to the "unknown learning" process in which the robot had to learn an unknown color (in this case, purple) (Fig. 11.3d). After the learning, the robot turned off the UP lamp and turned on the P lamp. At the same time, the robot carried out the "45° rotation behavior" (Fig. 11.3e). At this stage, the robot has learnt four colors (red, green, blue, and purple). Accordingly, when the green color was presented again (Fig. 11.3f), the robot moved forward. When the red color was presented again (Fig. 11.3g), the robot stopped.

Next, a new color (yellow) was presented. The robot cognized the yellow as an unknown color again, turned on the UP lamp, and proceeded to the second unknown learning process (Fig. 11.3h). After the learning, the robot turned off the UP lamp and turned on the P lamp. At the same time, the robot carried out the "90° rotation behavior" (Fig. 11.3i).

Then, the green color was presented again. As a matter of course, the robot turned on the P lamp and moved forward (Fig. 11.3j).

Finally, the purple color was presented. The robot turned on the P lamp (Fig. 11.3k) and carried out the 45° rotation behavior (Fig. 11.3l) because it had already completed the learning of the color purple.

The next illustration shows how the inner state of the robot changed in the course of the experiment (Fig. 11.4). The vertical axis indicates the amount of change in error. The horizontal axis indicates the elapsed time by fixing the processing time unit based on the cognitive/behavioral cycle to "1." This cycle covers the time from the input of color information to the Reason subsystem, the representation of P (pleasant) or UP (unpleasant) by the Emotion & Feelings subsystem, up to the output generated by the Association subsystem (or the correction by the Association subsystem in regard to the behavior representation of R1, R2, and R3 of the Reason

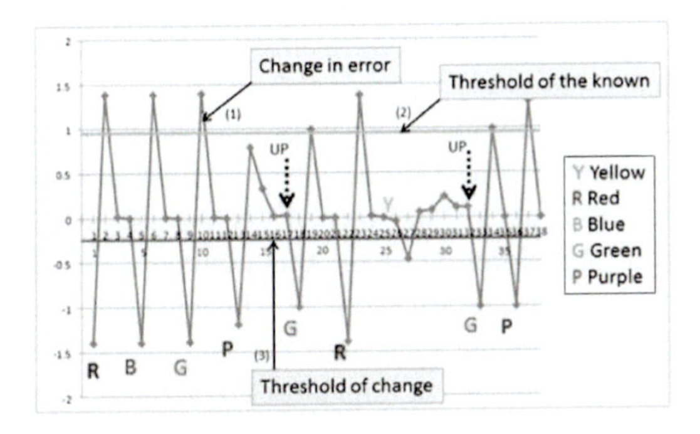

Figure 11.4 This illustration shows how the inner state of the robot changed in the course of the experiment.

subsystem (Fig. 9.11, d′) if unknown information is to be learnt). On the graph, the scale mark "1" is set to the time when the red color is presented for the first time in the experiment and the robot acknowledges the change in color. The variation of the line (Fig. 11.4(1)) indicates the amount of change in error actually calculated by the conscious system at each time. The horizontal line (2) indicates the threshold to determine whether this information is known or not. The horizontal line (3) indicates the threshold to determine whether the color has changed or not. R, B, G, P, and Y (Fig. 11.4) are the colors presented to the robot at that time. Two dash lines are the time when the UP lamp (red LED) of the robot goes ON to notify that it is in an unpleasant state. Accordingly, it recognized that the degree of unpleasantness increases and the presented information was unknown. In this experiment, it occurred at the time of 17 and 32. A more detailed explanation is as follows. Firstly, the process from Time 1 to Time 12 is examined. The known colors, i.e., red, blue, and green, were presented to the robot one after another. At each occasion of cognition, the amount of change in error was less than −0.24. At this point, one could judge that the color had changed (Times 1, 5, and 9). Then, at the next point in time, it was clear that the information could be judged as known because the amount of change exceeded 0.95 (Times 2, 6, and 10). The known information rapidly decreases the error value for convergence in the Reason subsystem MoNAD. At Times 1 to 12, the P (pleasant) lamp (green LED) of the robot was always ON. This means that the colors presented at Times 1, 5, and 9 were all known. However, the P lamp went OFF instantaneously at Times 1, 5, and 9.

Then, when the unknown color (purple) was presented to the robot, the degree of unpleasantness started increasing, exceeded a certain value at Time 17, and turned ON the UP lamp (red LED) indicating the unpleasant state. At this point in time, the amount of change in error did not exceed 0.95. This means that the Reason subsystem MoNAD did not show any rapid convergence. At the same time as lighting of the LED of unpleasantness, the learning process of unknown information was carried out. Times 18 to 21 in Fig. 11.4 clearly proved that the known colors were correctly recognized even after the learning of unknown color. The green color presented to the robot at Time 19 was judged as known information because the amount of change in error exceeded 0.95. In addition, when the purple color was presented again, the robot immediately swung left and right (45° rotation behavior), and when the green color was presented, the robot kept moving forward. This means that the purple color had been changed from unknown information to known information.

After that, the robot recognized the yellow color as unknown, and learnt it to change to known information. As a result, the robot became capable of carrying out the 90° rotation behavior in response to yellow (Time 25 to Time 32).

11.1.3 Conclusion

This section described, with more generality and in a specifically developed style, the conscious system (Fig. 10.3) for the mirror-image cognition experiment discussed in Chapter 10.

The section dealt with the subject of the robot recognizing unknown information and learning it as new information. With regard to learning by robots, machine learning technologies such as Q-learning are well-known, but this section proposed a completely new learning method. Conventional machine learning cannot cope with "unanticipated" unknown information. Conventional machine learning is based on the idea that, in order to attain a certain target, the rules for praise and blame are applied depending on whether or not the target has been achieved. A robot developed according to this idea can do nothing when it encounters an "unanticipated" situation and stops immediately. The experiment in this section tried to build up a system in which the robot could continue to work even in an "unanticipated" environment.

The key point is that the robot can "know an unknown." If this is possible, the robot can newly learn "unknown information," convert it to known information, then take a behavior in response to that information. Thus, this issue of "knowing an unknown" is a completely new research challenge for robot technology.

This challenge was readily resolved by the MoNAD of the conscious module. This is because the MoNAD facilitates judging whether input information is known or unknown. More specifically, the MoNAD circulates the input information in two recursive neural networks to converge to an oriented value. In this case, the phenomenon that the convergence occurs promptly if the information is known and the convergence is delayed if it is unknown has been logically proved. Our experiment utilized this phenomenon. In the experiment, the conscious system found the presented color as unknown information and learnt it; then the behavior in response to this color was successfully stored in the conscious system through human instruction. Also, this learning process added the newly generated conscious module MoNAD to the Reason subsystem, thus expanding and developing the network of the current conscious system. This experiment suggests a future image of robots which can evolve by themselves and continue to function even under "unanticipated" situations.

11.2 A Self-Aware Robot

In the first edition, the author discussed the phenomena of self-awareness. After describing the phenomena of self-awareness of human beings, the author then examined whether or not the robot having a conscious system under development could satisfy the fundamental requirements of these phenomena.

Exactly what is self-awareness of human beings? The discussion should start with this question. The literature on this subject mainly examines philosophical ideas, with no scientific explanation. Like consciousness, the phenomena of self-awareness are difficult to understand and very little is known about them.

The author's ideas on this subject are summarized below. The main requirements for the phenomena of self-awareness are

(1) to cognize one's own behaviors,
(2) to perform consistent cognition and behavior,
(3) to have a sense of cause and effect,
(4) to cognize the behaviors of others, and
(5) to have a sense of controlling others as one wishes.

In regard to requirement (1), the author's conscious module MoNAD provides feedback of one's own behavior information to the cognition process (Fig. 9.11, Y′). Therefore, one's own behavior can be cognized.

The conscious system has consistency between cognition and behavior in MoNAD. Accordingly, requirement (2) is also satisfied (Fig. 9.11, K).

When the behavior representation of the conscious module MoNAD (Fig. 9.11, Z′) decides to carry out the behavior, as a result, the cognition representation of MoNAD (Fig. 9.11, Z) will be determined. Therefore, the behavior has a causal correlation with the subsequent cognition. According to requirement (1), the behavior is cognized as one's own behavior (Fig. 9.11, Y′). It can thus be said that requirement (3) is also met.

In addition, from the cognition of one's own behavior, the behavior of others can be calculated (Fig. 9.11, Y′, X, and Z). Therefore, requirement (4) is also met.

Because the robot can deem the image in a mirror as a part of its own body (see Chapter 10), requirement (5) is also met in principle. Under this requirement, however, the description "as one wishes" needs to be further examined.

Concerning "sense" in requirements (3) and (5), the conclusion must be postponed if it means "qualia."

The author's robot is equipped with a conscious system, and as stated earlier, an experiment of mirror image cognition has been successfully completed. The author's experiment, which was published on the Internet, was reported as "Robot Demonstrates Self

Awareness" (Discovery News, December 21, 2005, see Appendix C). Mr. Michio Kaku, professor of physics at City College of The City University of New York, U.S.A., introduced my experiment in his book, stating, "This is the first time in history that a robot has been built specifically to have some sense of self-awareness" (Kaku, 2014).

Accordingly, after the first edition, the author and students carried out several new experiments on self-awareness, focusing on requirement (5). This requirement states that self-awareness must "have a sense of controlling others as one wishes." The word "sense" is used again in this requirement, but it can be temporarily replaced with the word "recognition."

As described in Chapter 10, the results of the author's experiment of mirror-image cognition by a robot indicated that it was difficult for the robot to distinguish between the behavior of the self-image in a mirror and that of its own body. Here, the word "behavior" means the action which the robot is continuously taking toward the observed object or, more specifically, moving forward, moving backward, and stopping via "imitation behavior." It is difficult to distinguish because, in this experiment, the behavior of the self-image in a mirror and one's own behavior are alike in every way. I am proud of "providing a scientific explanation through experiments using a robot" to the longstanding unsolved philosophical question: How can human beings easily cognize their self-image in a mirror? I think this was a historic event.

However, some readers might think of this as strange, because the robot's inability to distinguish between behaviors is perplexing. Actually, for the outside observer who watched the experiment, it was difficult to distinguish between behaviors, except that obviously one of them was just "an image of a robot in a mirror." Nevertheless, some image analyzers using highly sophisticated cameras may be able to distinguish between behaviors, and the author would agree with this. Such an external unit, located outside the experimental environment, observes both the robot and the robot's self-image in a mirror. Therefore, this external unit has "God's eyes" or a "third-person perspective." In this experiment, however, the robot's own judgment, namely the first-person perspective, is involved, because requirement (5) refers to the subjective judgment of the robot. Some readers may consider that the expression "robot's own subjective judgment" is odd. Put simply, the experiment observer, i.e. the human being located outside the experimental environment, examines the calculation results derived by the program built into the robot. More specifically, the calculation results of the robot (robot's subjective information) are examined by the human being (objective information on experimental results). My explanation above gives importance to the robot's subjectivity because the concept of "robot's consciousness" is merely one's understanding of the "robot's subjective evaluation" from an objective viewpoint. Still, some readers may question this.

Such issues are probably related to what happens if a highly sophisticated external unit is installed in a robot. Without discussing the details here, I would like to comment only that the judgment on the behavior of the robot will be more complicated, in other words, the conditions for judging the image in a mirror as a self-image will become stricter. Ultimately, the strictness of these conditions depends on the processing cycle (resolution) of the external unit. If the cycle is infinite, it will be difficult to judge the image in a mirror as a self-image, but such an external unit does not exist in the real world.

I have spent too much time on the preliminaries. Next, I would like to discuss the experiment on self-awareness.

Above, I referred to the successful experiment of mirror-image cognition by a robot (Chapter 10). This experiment was based on a process in which the robot calculates by itself the consistency of behavior between the moving robot Rs and the self-image of the robot Rm in a mirror (Fig. 10.4).

The author believes that research on "self-awareness of a robot" can be based on the experimental results of mirror-image cognition, because this mirror-image cognition experiment provides calculations for a very important process for self-awareness, i.e. the process in which the conscious system of a robot deems the image in a mirror to be its self-image or the image of another.

Based on this mirror-image cognition experiment, the author and students have continued to examine the problem for robots: Where does the boundary lie for distinguishing between oneself and others?

11.2.1 Discussions for Experiments

Judging from the results of the mirror-image cognition experiment, one can argue that the reaction time required for determining the coincidence of imitation behavior between oneself and others is an important parameter.

Therefore, we examined the events which disrupted this coincidence.

The most likely event is that the conscious system of self-robot Rs becomes unstable for some reason. Accordingly, the cognition-behavior cycle (Cbc) becomes unstable, namely shorter or longer. This is called the first cause (C-1). The second possible event is instability of communications between the self-robot and opposite-robot (Rc). This occurs when there is a problem in information flow from the conscious system to others. This is called the second cause (C-2). In addition, the opposite-robot has the problem of instability in operation. This is the third cause (C-3). In this experiment, however, we assumed that the problems C-1, C-2 and C-3 never occur for Rc. Such assumption is invalid in the case of human beings and other animals but, for the experiment using a robot, the people involved in the experiment can easily set up the robot and the environment.

Let us further examine C-1 and C-2. First, C-1 is caused by the instability of the conscious system. There may be several major sources of this instability but, in this paper, we firstly examine the case where the processing is delayed in the conscious system. This is because the Cbc of the system can be regarded as the most important parameter for the conscious system to calculate the coincidence of behavior. Also, a variety of sources are conceivable for C-2, but the communication delay on the path between self-robot and opposite-robot can be regarded as the most important. These two sources are called Dti and Dto, respectively.

With regard to the first source Dti, to deal with the parameter of information delay, it is necessary to study the problem: What is the subjective internal time for robots? If this parameter is not fixed as objective data, the objectivity of the experiment would disappear. Odd as it may sound, this experiment aims at "objectively examining the subjective data of a robot." Since a robot is generally operated by one or more computers, the instruction cycle (Ic) of the computer can be used as internal time. However, in this case, the experimental results are influenced only by the performance of each CPU used, which defeats the purpose of the experiment, which is to study the subjective phenomena of robots.

Here, the Cbc can be defined as the internal reaction time between input and output within the robot used in this experiment. As a matter of course, the Cbc can be objectively calculated from Ic and the program execution process. Further analyzing the execution process of the conscious system, it can be divided into two types. In one type, all the MoNADs constituting the conscious system operate in parallel mode; in the other type, they operate in pseudo-parallel mode. Both are available with current information technology but, as of now, the latter type seems to be more realistic. In both cases, the Cbc can be calculated from Ic. The experiment of Chapter 10 used the latter type.

The other delay Dto is related to the communication path and is determined by the external environment. Recall the experiment in Fig. 10.5 of Chapter 10. The self-robot Rs and the opposite-robot Rc were linked via a communication cable. In this experiment, Rs conducts the imitation behavior of Rc. The behavior of Rc is continuously instructed by Rs. In other words, Rc is the robot controlled by Rs. The instruction is given via the communication cable between Rs and Rc. Although the transmission delay Dto on the communication cable can be precisely calculated in a logical manner, the initial value Dto may be set to zero for the purpose of this experiment. This experiment examined the case where Dto was excessively generated. Generally, Rs creates the instruction of its own behavior and immediately outputs that information to Rc via communication cable (Dto = 0). However, in this experiment, the transmission delay is artificially created. In particular, Rs does not output that information to the transmission cable immediately, but creates a certain delay by inserting a simple loop program.

In summary, this experiment is intended to observe to what extent the robot Rs can cognize the opposite-robot Rc as a part of itself via imitation behavior by varying two delays Dti and Dto.

In terms of human beings, the former delay can be likened to cerebral hypofunction caused by aging, accident, or drugs (decrease in fundamental processing speed of the brain). The latter delay rarely occurs in ordinary life, and probably corresponds to the case where one watches one's own behavior via a surveillance camera and the behavior is displayed on a screen with a certain delay.

As stated earlier, the experiment presented in this section is similar to the mirror-image cognition experiment of Chapter 10, but several modifications have been added to the conscious system.

The internal design of the Reason subsystem and the Emotion & Feelings subsystem has been slightly changed. In the experiment of Chapter 10, a single imitation MoNAD carried out the imitation of the robot and its associated behavior. In the experiment of this section, however, the imitation MoNAD (Im) carries out the imitation and separately another MoNAD (Ac) conducts its associated behavior. This modification was made so that the experimenter can vary the Cbc for MoNAD Im during the experiment.

Also, this experiment provides the emotion-related MoNAD group (Pa, So) and the feeling-related MoNAD group (P, UP). In this respect, a major modification has been added, compared to the experiment of Chapter 10. In the experiment of Chapter 10, a distance MoNAD was installed whereas in this experiment, four MoNADs cover this function. The reason is described in Section 9.7.2.

At any rate, similar to the experiment of Chapter 10, the conscious system conducts the imitation of the object (the opposite-robot) located in front of the robot.

In addition, the robot used in the experiment has been changed from "Khepera" to the commercially available "e-puck" (Fig. 11.5). A touch sensor and IR sensor are installed on the front face of the robot. The touch sensor detects a collision of the robot against an object, and the IR sensor calculates the distance to the object (access sensor). The touch sensor detects the collision status against the mirror or opposite-robot while the robot conducts the imitation behavior. A collision causes the touch sensor to generate a signal, then the signal (In2) activates the MoNAD Pa related to emotion and leads to the representation of pain within the conscious system (Fig. 11.6). Furthermore, this representation causes the robot to activate the MoNAD UP and to represent unpleasantness. When the robot has created the representation of pain while moving forward, the robot is designed to create the instruction to move

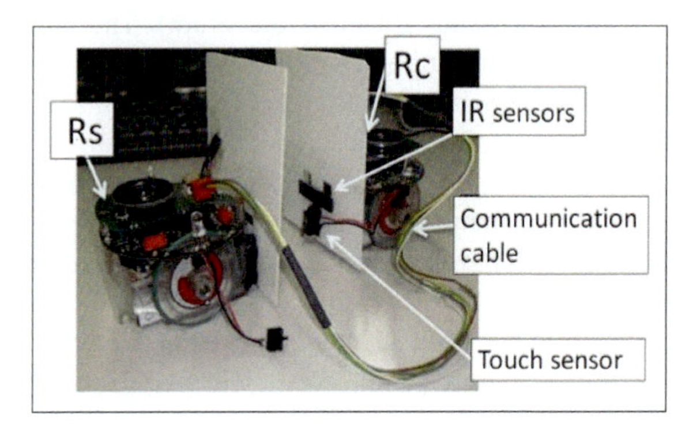

Figure 11.5 Two robots in the experiment.

backward. In the experiment of Chapter 10, this instruction was created by the distance MoNAD (i.e. emotion & feelings subsystem). However, this experiment was changed so that the instruction is to be created by the Reason subsystem. In Chapter 10, the robot was designed to perform reflective behavior in response to pain. However, the modification was added to the present experiment for the purpose of simplicity, and all the behaviors of the robot are carried out by the Reason subsystem (Fig. 11.6). The information of the IR sensor activates two MoNADs (MoNAD Im and MoNAD So) via input terminals In1 and In3. The MoNAD Im conducts the imitation behavior. Therefore, when the object moves forward, the self-robot is meant to also move forward. This portion is the same process as in the previous experiment of Chapter 10. The role of MoNAD So is to cope with the abnormal circumstance where the information of the IR sensor cannot be detected while the imitation behavior is being conducted. When the input information from In3 connected with the IR sensor is missing, the MoNAD So related to emotion represents solitude, and the MoNAD UP related to feeling represents unpleasantness. Subsequently, the system is designed to transmit this representation from the Association subsystem to the Reason subsystem in order to create the move-forward instruction. The move-forward instruction is created because unpleasantness was represented while the robot was moving backward. This portion is also the same process as in the experiment of Chapter 10.

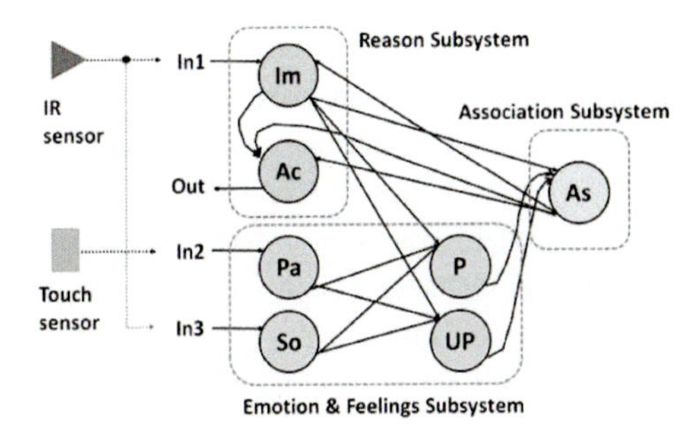

Figure 11.6 The conscious system for the experiments.

11.2.2 Experiments

Figure 11.5 shows the circumstances of the experiment. The robot on the left is the self-robot Rs. The robot Rc is the opposite-robot and conducts the same behavior as Rs according to the instruction from Rs via the communication cable. When Rs moves forward, Rc is instructed to move forward. During the experiment, Rs continuously imitates the behavior of Rc and calculates the coincidence of behavior between itself and the other. As mentioned above, the author believes that the evaluation of coincidence is an important parameter for the cognition, i.e. self-awareness. The experiment of Chapter 10 proved that the self-image in a mirror showed much higher coincidence with the original behavior than the opposite-robot Rc given a behavior instruction via a communication cable. On the basis of this result, the author concluded that it was difficult to distinguish between the behavioral reaction of self-image in a mirror and the reaction to one's own behavior. The author argues that this experimental result is a scientific basis for the phenomena that human beings are capable of easily cognizing their self-image in a mirror, or it is an example showing the scientific reason for the "mirror stage hypothesis;" why can two-year-old children cognize their self-image in a mirror?

Readers may have other questions about the relationship between this experimental result and self-awareness. As already discussed in Chapter 10, this experiment provides strong suggestions for the distinction between oneself and others, which is one of the most important functions of human consciousness. The experiment of Chapter 10 cannot be directly connected to human functions, but it was the first-ever case where a machine system, i.e. a robot, cognized a self-image in a mirror as "a part of itself." This experimental result indicates a mathematical model in which one can cognize a self-image in a mirror as "a part of oneself separated from others." If this is true, it would be possible for such a mathematical model to contain "the process for the machine system to cognize a part of itself separately from others" and the process yields a valuable clue for "the process by which human beings cognize a part of themselves separately from others." From these considerations, the author believes that this process may provide a scientific basis which contains important clues for "self-awareness," which is an important function of humans. This is because self-awareness literally means "to become aware of the self," which begins with "cognizing a part of oneself separately from others."

This experiment evaluates whether or not one can distinguish between oneself and others by evaluating the coincidence of behaviors between oneself and others. The "evaluation of coincidence" is, in simple terms, the calculation of to what extent the behaviors match between oneself and others on the basis of internal time determined within the robot. In the experiment of Chapter 10, the coincidence rate of imitation behavior against "self-image in a mirror Rm" was approximately 70% while the coincidence rate against "opposite-robot controlled via communication cable Rc" was approximately 60%. In the latter case, Rc was connected to the self-robot via a communication cable, and via this cable the behavior instruction was given from the self-robot. Accordingly, Rc is a part of the self-robot. Namely, the reference point to determine whether or not the self-robot can distinguish between itself and others could be slightly higher than 60%. Because the self-image in a mirror provides a coincidence rate of 70%, the robot can naturally judge it as others, not a part of itself. Conversely, when the coincidence rate is lower than 60%, the robot does not judge it as others, not a part of itself. Readers may consider that well-known elements other than coincidence rate are involved in the function to distinguish between oneself and others. For example, the stimulus of emotion and feeling within the body enhances the clarity of one's own existence. Obviously, this element could become an important element to distinguish between oneself and others. This element was excluded from the present experiment, however, because it is less involved in the function to cognize the self-image in a mirror as a part of itself. I leave this matter for future discussion.

In this experiment, the judgment may be deemed as a subjective evaluation of the self-robot but, in reality, the cognition-behavior cycle Cbc of the self-robot is an important parameter as the basis of the judgment. In addition, such value can be determined by the instruction cycle Ic of the computer. In other words, the subjective judgment of the robot becomes the study object which can be objectively verified by human beings.

Let's return to the main subject.

This experiment sets up Rc similarly to the experiment of Chapter 10. Rc uses the same hardware as Rs. As software, however, it does not use the conscious system of Rs but uses the program to simply receive the behavior instruction via the communication cable and then conduct the specified behavior. Therefore, the software of Rc is simpler than that of Rs. Since this program has fewer steps and structural loops, the reaction time from input to output can be regarded as constant "0" in comparison to the Cbc of Rs. Note that the behavior of Rc is always determined by Rs. It should also be noted that Rs imitates the behavior of Rc and therefore the behavior of Rs is determined by that of Rc. From the standpoint of the experimenter (third party), Rs and Rc imitate each other although they are connected by the communication cable. Accordingly, it is not difficult to differentiate this situation from the experiment of Chapter 10 in which Rs imitates the self-image Rm in a mirror.

During this study, the author decided to investigate the influence of instability on the behavior coincidence rate due to C-1 and C-2. In general, when the instability increases, the coincidence rate should decrease, because this rate corresponds to the count of behavioral coincidence within the reference time.

Next, consider the instability of each cause.

With regard to C-1, there are several causes of instability of the conscious system. In principle, the conscious system determines the next behavior according to internal and external environments. The result of behavior is assumed to change the structure of one's own conscious system. In this experiment, however, no such mechanism which could cause instability is conceived in the conscious system. Therefore, the structure of the internal program will not change during the behavior. The conscious system used in this experiment runs based on its own common cognition-behavior cycle Cbc. As described above, Cbc can be calculated on the basis of the robot's basic instruction cycle (Ic), the complexity of programs for the conscious system (Cp), and a problem in the robot's hardware (Hp). The volumes of Ic and Cp are already known, and Hp is a constant value with no instability. As a result, Cbc can be assumed to be constant. In summary, this experiment has no risk of instability of the conscious system caused by C-1.

Next, we discuss cause C-2.

This is the delay of information transmission over the communication cable (Dto). In this experiment, the information transmission is unidirectional, from Rs to Rc, and this transmission may have a variety of unstable elements. One major problem for Rs is the case where the information is ready for transmission on Rs but Rs enters a temporary waiting status, and so the information has to be sent later. As described above, this case is referred to as delayed presentation of information in Rs (C-2). There is no advance transmission in this experiment (a negative value of Dto is not considered). This is because no internal change of conscious system is supposed in this experiment. Next, with regard to the information delay on the transmission path, as already established on a logic basis, it is evaluated as almost zero (similarly to an optical communication system) in this experiment.

The next case involves the information delay of Rc itself. In this case, although the input information for Rc has already been established, Rc itself does not obtain that information. No delay is assumed in this experiment in order to avoid the complication of Rc for the purpose of the experiment, i.e. subjective self-awareness of self-robot Rs. Also, such a case is unlikely to occur in general, because it is not reasonable that the robot postpones by itself the acquisition of external environment information. From the above, in principle it can be considered that there is no instability of transmission information due to C-2.

Thus, to intentionally apply instability to the coincidence rate, two sources are conceivable: The first source is to make Cbc variable (C-1) and the second source is the instability caused by C-2.

The instability due to Cbc can be expressed with a decreased cognition-behavior cycle, as stated above. Briefly speaking, it implies that the time interval required from input to the conscious system to the determination of output information (one cognition-behavior cycle time) becomes longer. In general, this situation corresponds to a decrease in the resolution ability of the conscious system. Taking a video camera as an analogy, it is equivalent to filming video at a lower frame rate. Actually, this change is realized by forcing the Reason-subsystem MoNAD Im to conduct the same calculation for the conscious system two times successively, though one calculation is sufficient in normal operation (Fig. 11.6).

This experiment uses two types of conscious system: Cs1 and Cs2. For each conscious system, three fundamental experiments (①, ②, ③) were conducted. These fundamental experiments correspond to the three types used for mirror-image cognition in Chapter 10, namely the experiment where the self-robot Rs imitates the self-image in a mirror (①), the experiment where the self-robot Rs and opposite-robot Ro imitate each other (②), and the experiment where Rs imitates the opposite-robot Rc connected via the cable (③). Then, five types of transmission delay experiment (④, ⑤, ⑥, ⑦, ⑧) are added to experiment ③. Experiment ④ includes a delay of 0.5d, ⑤ 1d, ⑥ 2d, ⑦ 3d, and ⑧ 4d. Here, the variable "d" is the basic delay time. In the experiment, after the self-robot Rs has generated its own behavior instruction, it runs the redundant program to determine the delay time, then it sends the signal of the behavior instruction to the communication cable. The program count corresponds to adding integer "1" 100,000 times. As the transmission delay Dto is gradually increased from experiment ④ to ⑧, the observer checks how the evaluation changes in regard to the rate of behavioral coincidence between oneself and others. As for the difference between Cs1 and Cs2, the Cbc of Cs2 is approximately half of the Cbc of Cs1. Cs2's resolution ability of consciousness degrades by approximately 50% just like a slow shutter of a camera. The observer checks how the rate of behavioral coincidence between self and others changes.

No.	State	Delay	Rate	Rate′
①	Rm	0	87.80%	95.75%
②	Ro	0	53.40%	56.40%
③	Rc	0	73.80%	67%
④	Rc(0.5)	0.5	55%	73.80%
⑤	Rc(1)	1	61.60%	67%
⑥	Rc(2)	2	60.20%	72.40%
⑦	Rc(4)	4	43%	58%
⑧	Rc(6)	6	39.60%	57.60%

Figure 11.7 Experimental results.

11.2.3 Results and Considerations

The variations of coincidence rate for Cs1 and Cs2 are shown as Rate and Rate′, respectively in Fig. 11.7. Also, the variations in response to transmission delay are shown in Figs. 11.8 and 11.9.

First, Fig. 11.8 is examined.

In the fundamental experiments ①, ②, and ③, the coincidence rates were 88%, 53% and 74%, respectively. In the experiments of Chapter 10, they were 70%, 50% and 60%. In the experiments in this section, all the values are increased in comparison with those of Chapter 10. One can argue that the values of ①, ②, and ③ show almost the same trend as those of Chapter 10.

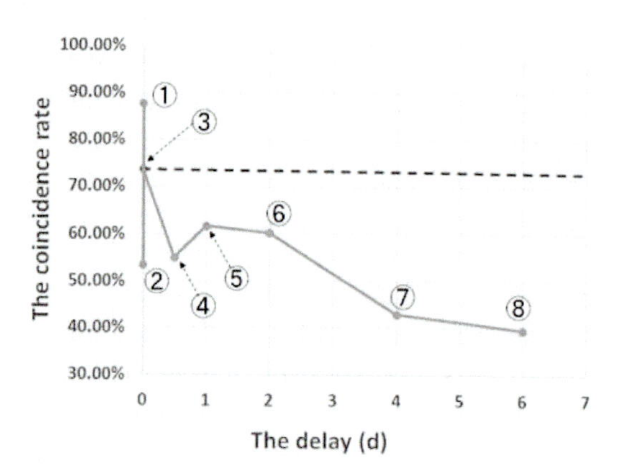

Figure 11.8 The variations of coincidence rate for Cs1 and the variations in response to transmission delay.

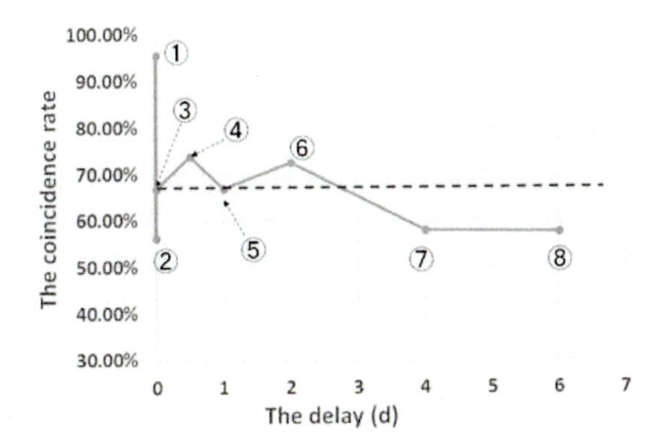

Figure 11.9 The variations of coincidence rate for Cs2 and the variations in response to transmission delay.

It is likely that the difference between the two experiments is attributable to the performance of the robots. In Chapter 10, Khepera II was used as the robot, and in this section e-puck was used.

In this experiment, the experimental value (③) of the cable-controlled opposite-robot Rc is used as the criterion to evaluate the distinction between oneself and others. The coincidence rate of judging as others was 60% or lower in the previous experiment but it changed to 74% or lower in this experiment. For the mirror-image cognition experiment, 70% changed to 88% (①). Therefore, in both experiments, the robot can judge the self-image in a mirror as a part of itself. Concerning the cognition of opposite-robot Ro, both coincidence rates are almost 50% and obviously the opposite-robot is cognizable. Therefore, in both experiments, the self-robot Rs can clearly distinguish the opposite-robot Ro.

Next, Fig. 11.8 and Fig. 11.9 are compared. The following is a comparison between experiments ①, ② and ③.

The coincidence rate ① is 88% in Fig. 11.8 but increases to 96% in Fig. 11.9. The rate ② increases from 53% to 56% (Fig. 11.7) while ③ decreases from 74% to 67%. Although rate ③ decreases, the cross relationship between ①, ② and ③ is unchanged; value ① is the highest, ③ the second, and ② the third. Therefore, it can be argued that both the self-robot and opposite-robot are distinguishable. Value ③ falls between ① and

②, but value ③ in Fig. 11.9 is closer to value ② than in Fig. 11.8. Accordingly, the criterion for distinction between self-robot and opposite-robot is decreased. This is because, as mentioned above, the conscious system becomes unstable or the cognition-behavior cycle Cbc changes to 1/2. In other words, the resolution ability for distinction between oneself and others degrades to half. This degradation also decreases the resolution ability to cognize one's own body.

With regard to experiments ④, ⑤, ⑥, ⑦ and ⑧, a certain delay of information transmission is added to experiment ③. The higher the experiment number is, the greater the transmission delay becomes. The results of both experiments indicate that the distinction between oneself and others degrades as the experiment number increases. Some reverse phenomena can be observed in experiments ④ and ⑤, but this can be regarded as almost a parallel shift.

The dotted lines in Fig. 11.8 and Fig. 11.9 indicate the value of coincidence rate in experiment ③. This is the criterion to determine whether or not the opposite-robot Rc is a part of self-robot Rs. When the coincidence rate Rate or Rate′ is below the dotted line, it can be considered that the robot concerned is likely to be a stranger. When the line is above the dotted line, the robot may be a part of itself.

From the above, only ① (Rm) and ③ (Rc) are regarded as a part of itself in Fig. 11.8, while ① (Rm), ③ (Rc), ④ (Rc (0.5)), ⑤ (Rc (1)) and ⑥ (Rc (2)) are regarded as a part of itself in Fig. 11.9.

These experimental results suggest that the instability of the conscious system (reduced Cbc) in the self-robot Rs decreases the criterion for the robot to distinguish between itself and others. In addition, the delay of information presentation (information transmission) from self-robot Rs to others decreases the criterion for the robot to distinguish between itself and others. Note here that the variation of Cbc and the delay of information transmission can be objectively measured by means of robot-internal time.

11.2.4 Conclusion

This section describes the considerations on the problem of a robot's self-awareness. First, we focused on the phenomena of human self-awareness and then tried to reproduce the phenomena on the robot. The experiment of Chapter 10 dealt with the problem of cognizing the self-image in a mirror, and eventually created the "Self-Body Theory" that the self-image in a mirror is a special stranger (another person but a part of self). The self-image in a mirror is not only another person but also an external object which can be freely manipulated by self, just like the hands and arms controlled by a human being. This is precisely the "issue of self-awareness" because the self-image in a mirror is cognized as a part of oneself and is freely manipulated.

To clarify this issue further, the following points were examined by using the robot and conscious system used in the mirror-image cognition experiment (Chapter 10). The first task was to find the parameter which affects the distinction of the robot while the conscious system is distinguishing between itself and others. Two parameters which can cause variations of distinction were found in terms of experimental mechanism. The first parameter is the instability of the conscious system, and the second is the delay of information presentation. In the actual experiment, the first parameter was realized by reducing the behavior-cognition cycle Cbc, and the second by introducing a delay into the transmission of information to others. Thus, the experimenter observed how the coincidence rate of behavior between oneself and others was affected.

As a result, it was observed that the first parameter had changed the criterion for the distinction between oneself and others and the second parameter had also changed the criterion. More specifically, after these parameters had been added, the self-robot Rs mistakenly cognized as a part of itself the object that had previously been judged as others. Hence, it was found that both parameters caused a decrease in the criterion for distinction between oneself and others. Furthermore, the criterion for distinction was proved to be eventually derivable from the robot's internal processing time Ic.

The requirement (5) for self-awareness, i.e. "to have a sense of controlling others as one wishes," was described. This experiment showed the method of determining the criterion for distinction through the robot's process of distinguishing between oneself and others. The condition defined for the self-robot Rs to judge as a part of self was that "the coincidence rate for that portion must be equal to or more than the criterion value." The experiment revealed that the portion was "subjectively and freely controlled as if it were the body of self-robot Rs."

Here, the question arises as to whether or not the self-robot Rs and the self are identical. In my opinion, if the concept of "self" exists as represented by the conscious system and all the activities of self-robot Rs are associated with this "self," then Rs and the self are related. I consider that the representation "self" is ultimately associated with the activities of hands, legs, and most other corporeal faculties. Actually, this consideration provided the first-ever explanation of the issue of a robot's self-awareness, and may serve as the principle of self-awareness "to control as one wishes."

Also, the experimental result that the distinction between oneself and others is affected by the instability of the cognition-behavior cycle in the conscious system could shed light on the research of pathogenesis for high-level brain dysfunction attributable to the problem of distinguishing between oneself and others or of unstable self-awareness.

11.3 A Robot with Episodic Memory

In conventional robot development, it appears that the functions of human memory have not been adequately studied. The reason for this is that the purpose of robot development has been primarily for engineering applications.

Regarding the study of human memory, there are disciplines such as psychology and brain science, but in a simple classification, human memory is said to be based upon short-term memory, long-term memory, as well as procedural memory and episodic memory. There are also terms that are derived from memory such as forecast and expectation.

The author considered incorporating a mechanism to approximate the functions of human memory in a conscious system installed in a robot. In order to enable the robot to continue working in an external environment, I gave top priority to the safe behavior of the robot based on its experience. And in order to enable the robot to work safely, it is important for the robot to expect a dangerous situation based on the previous work it has experienced so far, and it is important to enable the robot to avoid such a dangerous situation. To achieve these functions, it is necessary to first incorporate a mechanism of episodic memory on the robot. The meaning of the word "episode" can be explained simply as "a somewhat interesting story." In terms of words that relate to a robot, an episode may be considered a "chain of events that resulted in an emotion." To restate this a little more clearly, "When an emotion such as pain occurs in the body of the robot while the robot is acting, an episode is the event that caused the emotion and the chain of several events leading up to that event." Thus, episodic memory is a chain of events leading to an emotion. This episodic memory is a combination of chain of events and emotional information, so for the robot, it can be expressed as "a single experience." If the robot has already had this experience, it traces this memory and observes the current chain of events, so that it is possible for the robot to select other actions in expectation of emotions that might occur in the near future. For example, if the robot can recognize that the current chain of events is a process that will cause it pain, the robot can stop or change its action. In this section, the author only considers the emotions of pain and solitude.

The conscious system developed by the author and his students is shown in Fig. 11.10. In its standard configuration, the conscious system comprises a reason subsystem, an emotion & feelings subsystem, and an association subsystem. The episodic memory subsystem (Ep) was added to the conscious system shown in the previous section (Fig. 11.6). Ep is not a MoNAD consciousness module; rather it is constructed as a combination of conventional neural networks (NN). In our experiments, Ep consists of three NN layers that comprise the (a) part which is the basic part, and it is additionally connected to one NN layer which is the (b) part. When the robot is acting and represents an emotion, the representation of various MoNADs firing is learned as teacher data at the (a) part of Ep.

As an example, assume that an emotion of Pa or So is represented at time $(k + 1)$, (where k is an integer). Due to this representation, the UP MoNAD has an unpleasant representation. At that time, in

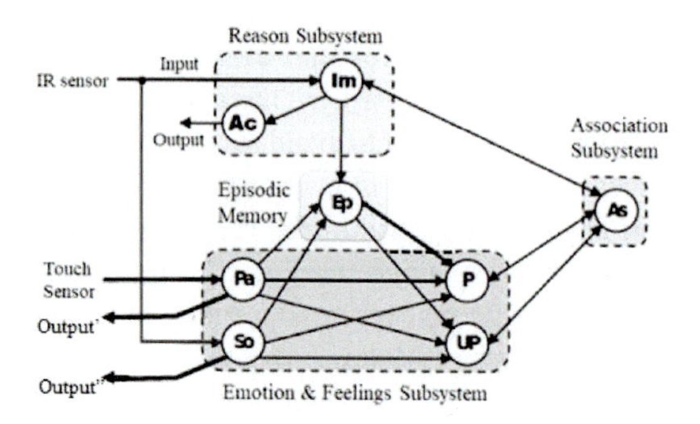

Figure 11.10 Conscious system that expects emotions.

the (a) part of Ep, the behavior representation BL $(k+1)$ of the Im MoNAD at time $(k+1)$ and the cognitive representation RL $(k+1)$ become Ep input and output information and are learned as teacher data. Simultaneously, the behavior representation BL $(k+1)$ and cognitive representation RL $(k+1)$ of Pa and So are also learned as teacher data.

11.3.1 Model of Episodic Memory and Expectation

A diagram of the model is shown in Fig. 11.11.

In the (a) part, when an unpleasant representation occurs at time $k+1$, the behavior representation BLi $(k+1)$ of the Im MoNAD, the behavioral representation BLp $(k+1)$ of the Pa MoNAD, and the behavior representation BLs $(k+1)$ of the So MoNAD are used as input information, and also the cognitive representations RLi $(k+1)$, RLp $(k+1)$ and RLs $(k+1)$ of the Im MoNAD are used as output information, and these are used for supervised learning. Furthermore, using that output information from one NN layer (Fig. 11.11b), supervised learning of the unpleasant or pleasant information performed. (Only unpleasant information is used in these experiments, and that currently uses 1 bit.)

This mechanism can be understood by knowing the information processing method of the MoNAD, which is a consciousness module developed by the author. The cognitive representation RL (k) of the MoNAD generated at time k is usually copied directly to the behavioral representation BL $(k+1)$ of the MoNAD at time $k+1$.

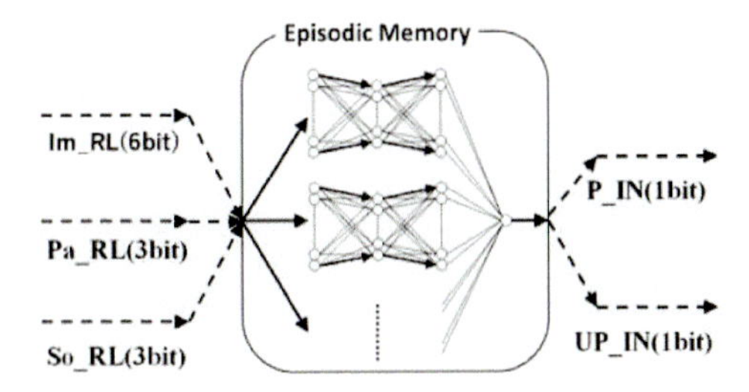

Figure 11.11 Diagram of episodic memory and expectation model.

That is, because the input information of the episodic memory at time $(k + 1)$ is automatically determined by RL (k), the information in Ep at time $(k + 1)$ can be calculated by the information at time k. In the episodic memory, learning has already been done at time $(k + 1)$ in conjunction with the emotion information (experience), so even in situations where emotion has not yet been generated as a result, by using episodic memory it is possible to expect the occurrence of emotion at time $(k + 1)$. However, the representation BL $(k + 1)$ may be given an external representation from outside the robot, but of course, expectation becomes possible if there is already episodic memory for the BL information. If it does not exist, a new episodic memory can be created.

In addition, the reason that the cognitive representation RL at time $(k + 1)$ of each MoNAD is used as teacher information for the output information of the episode memory is because, by making that information learned recursively at the input part of the episode memory, the author and the students want to enable the episodic memory to respond dynamically to time series changes in the future. However, this is not used in this section.

11.3.2 Experiment Using a Robot

An e-puck is used for the robot. Like the robot in Section 11.2, a white panel is used for reflecting the infrared light, and an IR sensor and a touch sensor are arranged on the front of the robot. A conscious system equipped with episodic memory Ep is installed on the robot. The conscious system performs imitation behavior by using external information from the infrared sensor

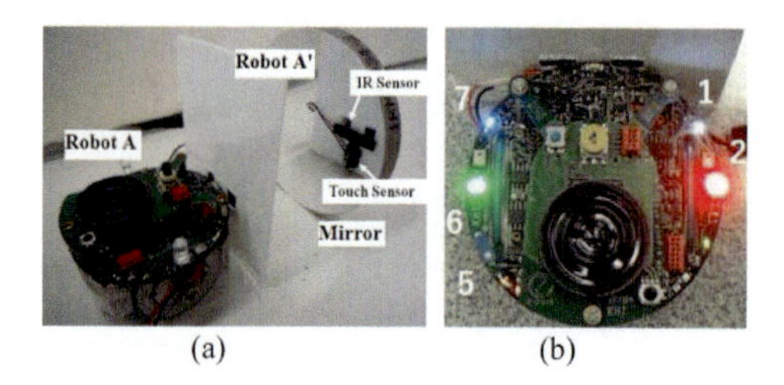

(a) (b)

Figure 11.12 (a) Robot A and the mirror used in the experiment. The sensors used on the e-puck robot can be seen in the mirror. (b) Arrangement and meanings of the LEDs on the top of the robot: LEDs 1, 2, 5, 6, and 7 indicate solitude, unpleasant, pain, pleasant, and imitation, respectively.

and information from the robot as it recognizes its own movement. In this experiment, as in the previous section, the robot performs actions to imitate its self-image in the mirror. In this imitation behavior experiment, the robot first executes a forward movement by imitating the movement of its image reflected in the mirror. If robot continues to imitate the forward movement following the image in the mirror, the robot eventually collides with the mirror. At that time, the distance between the robot and the obstacle (in this case the mirror) becomes almost 0 in the distance as measured by the IR sensor. Upon collision, the touch sensor switches on. From this information, the robot stops and then starts moving backward. Then, since the robot that is reflected in the mirror is moving backward, Robot A (see Fig. 11.12a) continues to move backward by similarly imitating this action. However, this backward movement eventually makes it difficult to measure the distance using the IR sensor. That is, the distance becomes the maximum value at that time. Then, Robot A stops and starts moving forward once again, performing the imitation behavior in a forward movement.

In this experiment, note the fact that Robot A collides with the mirror while moving forward when performing the imitation behavior. Robot A is colliding with the mirror. While performing the imitation behavior in the experiment, Robot A cannot avoid always colliding with the mirror, and when the collision turns on the touch sensor, pain is represented, and unpleasant is represented.

However, since Robot A in this section is structured with an expectation function using episodic memory, even if the robot comes

Figure 11.13 Collision with the mirror when the robot is performing imitation behavior. Because the touch sensor is activated, LED 5 (pain) and 2 (unpleasant) are lit up.

close to colliding with the mirror, the representation of pain does not occur and the information (experience) previously stored in the episodic memory is drawn out and unpleasant is represented. As a result, this enables Robot A to stop in expectation of a collision with the mirror. That is, the touch sensor does not turn on at that time. Thus, Robot A memorizes (learns) the episode of pain representation due to a collision that it experienced before, so that a similar behavior will not cause the same painful episode twice. This is the expectation function achieved by using episodic memory.

Robot A collided with the mirror when performing the imitation behavior (Fig. 11.13). When observing the lighting of the LEDs on the top of the robot in this figure, we can see that LED 6 is lit indicating a pleasant representation, and while performing the imitation behavior LED 7 is lit. When a collision with the mirror occurs, pain is represented and LED 5 lights up, and at the same time the LED 2 lamp lights up indicating the unpleasant representation.

At this time, when Robot A collides with the mirror, it learns the episode and stores it in the Ep of the memory layer.

At the end of learning in the experiment, the colliding mirror is subsequently reinstalled away from the robot at a certain distance (about 10 cm). Then, Robot A starts moving forward again and imitates its advancing image in the mirror. This imitation behavior would result in Robot A colliding with the mirror, but the robot stops

Figure 11.14 In expectation of the occurrence of pain representation, Robot A stopped in advance of a collision using episodic memory. The robot does not collide with the mirror and it stopped the unpleasant representation.

because unpleasant is represented by Ep immediately before the collision would occur (Fig. 11.14). In other words, Robot A stopped in expectation of a collision from the experience of a previous collision (an episode of pain generation). It can be said that Robot A was able to avoid the occurrence of pain by expectation.

11.3.3 Conclusion

This section described the achievement of episodic memory, which is a kind of memory or storage system, for the conscious robot. If the robot can perceive emotions such as pain and can automatically learn several representations in a time series, it is possible for the robot to expect in advance the emotions that it will have in the future. If it can do this, the robot can avoid future emotions in advance, and this makes it possible to further enhance the safety of the robot's behavior.

Together with a student group, the author loaded episodic memory in the conscious system of a robot. The basic part of the episodic memory Ep used a normal 3-layer feed forward type of NN. For the supervised learning of Ep, the input information gives behavioral representations of the Im, Pa, and So MoNADs that fire when detecting emotions, and the output information gives each cognitive representation of those MoNADs that occurred at that time. In addition, one NN layer, whose output information is used as input information, is concatenated to calculate the pleasant and unpleasant information. Learning like this, Ep can expect the emotion that would occur the next time by using a similar episode that the robot would encounter in the future. This expectation enables the robot to avoid in advance emotions that are disadvantageous to it. The author and the students implemented a conscious system with an episodic memory on a small robot and conducted experiments that demonstrated that the robot can expect the occurrence of a disadvantageous emotion and avoid that emotion.

11.4 A Pavlov Robot

Readers may be familiar with the name of Ivan Pavlov (1849–1936), a Russian physiologist who won the Nobel Prize (1904) for his physiological experiments on digestion. Pavlov is famous for his experiments on the salivary glands using dogs. When given food, dogs physiologically (automatically) secreted saliva in their mouths. This is generally referred to as an unconditional reflex. However, if a bell was rung whenever the food was served, the dogs gradually learned to salivate to the sound of the bell alone. This is referred to as a conditioned reflex.

Although little was known about human brain functions when Pavlov was active, his observations that innate phenomena such as salivation could be controlled merely by the sound of a bell was met with great surprise by people of that time. Probably they thought that some change was made to the dogs' brains as a result of an acoustic stimulus alone.

No doubt I am not alone in thinking that Pavlov's experiments opened the door to human brain science. His work influenced American psychologist John B. Watson (1878–1958), who started and extensively developed behavioral psychology, focusing on changes in external behavior caused by external stimuli. Pavlov's influence also spread to Europe, where Austrian neurologist Sigmund Freud (1856–1939) and French psychologist Pierre Janet (1859–1947) pioneered psychiatric medicine, focusing on changes inside the brain caused by external stimuli.

Pavlov's experiments were indeed a great milestone in brain science.

This section describes an experiment conducted by the author and his group of students, focusing on unconditioned and conditioned responses in robots as a theme of their project.

11.4.1 Configuration of the Conscious System

To set up the system used in this experiment, some changes were made to the conscious system described in the previous sections.

First, MoNAD Br was added to the reason subsystem. This MoNAD can represent whether or not a white light was irradiated on the robot. Also, this subsystem included MoNAD Re and MoNAD Ac. MoNAD Re gives an instruction for the robot to move forward, and MoNAD Ac conveys the instruction to the drive motor. MoNAD So was removed from the emotion & feelings subsystem because it would not be used in this experiment. However, MoNAD Pa, which represents pain, and MoNADs P and UP, which represent pleasant and unpleasant feelings, respectively, were retained in the subsystem. Furthermore, Ep, an episodic memory described in the preceding section, was incorporated into the newly developed memory subsystem, where MoNAD Li, a subsystem that integrates various representations, is connected with Me, which is a neural network (NN) that stores event memories. However, Ep is basically used as is for Me. Finally, the association subsystem, As, is retained in the system.

11.4.2 Unconditioned and Conditioned Responses in a Robot

First, let us define the unconditioned and conditioned responses in terms of psychology and biology. Imagine that you subconsciously touch a boiling kettle. Instantly you feel an intense pain in your hand when touching the kettle and simultaneously experience your hand being reflexively pulled back from the kettle. This phenomenon is referred to as a subconscious behavior. The word "subconscious" here refers more or less to the state in which you are initially unaware of the behavior you are engaging in. Shortly after that, however, you become aware of the burning pain in your hand and of the fact that your hand has pulled back from the kettle. The process is not well elucidated but generally explained by saying that it takes a little time for the pain and the hand movement to be recognized (Bloom et al., 1985). In other words, you are definitely aware of (conscious of) a change in your physical condition but with a temporal delay in the process of awareness, and if you are not aware of this temporal delay, that is because the delay is something that cannot be recognized subjectively.

There is another aspect of the unconditioned response—a response that is innate. An imitation response would be an example of an innate response (Meltzoff and Moore, 1977). All infants exhibit this kind of response, although with a few exceptions. A conditioned response is categorized as a learned response. Although a learned response is generally described as something that occurs also subconsciously (Bloom et al., 1985), this description seems too simplistic to the author because, if thinking is regarded as a kind of behavior, it seems undeniable a learned behavior is recognizable.

For example, a sort of trauma immediately causes an emotion, and the affected person is overwhelmed by (or aware of) the emotion. Given this discussion, it seems that, whether it is conditioned or unconditioned, a response arises in the subconscious state, which is immediately replaced by the conscious state. According to Libet's experiment, a direct stimulus to the brain is recognized with a temporal delay of the maximum 0.5 sec. (Libet, 2004), which agrees with our discussion.

So, can a robot have innate functions? The body of a robot may have innate properties. The function of a robot of moving on wheels or two legs is undoubtedly an innate function. The various programs originally installed in a robot can be referred to as innate functions. Given this thinking, it seems more practically important to discuss what kinds of functions should be configured in a robot as innate functions than to define the meaning of an innate state for the robot. The innate functions thus determined are subconsciously activated and sometime later become represented in the consciousness of the robot. All functions originally configured in the robot can be referred to as innate functions. When a robot learns something new, all responses caused by the learning can be referred to as conditioned responses.

Through this discussion, the author and a group of students attempted to equip a robot with a function to learn a conditioned response in a way that mimics Pavlov's experiments.

The function to learn a conditioned response was incorporated into the artificial consciousness as an advanced version of the episodic memory described in the preceding section. When an emotion such as pain occurs in the body, the representations from the various MoNADs in the conscious system are learned as events by Ep, which is a forward feedback neural network (NN) (see Section 11.3). The generation of an emotion is considered to be an unconditioned response. The emotion & feelings subsystem in the conscious system is considered to be an innate function. The evolutionary development of this is very intriguing but is not discussed here. In this experiment, we used a white light as a stimulus to elicit a conditioned response. The generation of pain was regarded as a substitute for the dogs' salivation in Pavlov's experiments, and the irradiation of a white light as a substitute for the sound of the bell. In humans, the direct exposure of the eyes to a white light causes an emotional response such as a dazzled feeling. In the robot used in this experiment, however, the irradiation of the white light can serve as a neutral, conditioned stimulus unless the conscious system is connected to the emotion & feelings subsystem.

The generation of pain means that pain is represented by MoNAD Pa in the conscious system when the switch on the front of the robot is turned on. The representation of pain means that an unpleasant (UP) feeling is represented automatically by the feelings subsystem, thereby causing the robot to perform an action—for example, to stop.

The important point of this experiment is that when the robot represents pain, it simultaneously receives a white light stimulus via an optical sensor. The two events—the generation of pain and the irradiation of the white light—are then integrated as a set of information by Li and stored as a memory in Me. Using this information stored in Me that associates pain with the white light, the robot can represent pain in response to the white light stimulus given at a later time—even if pain is not generated by a collision of the robot—and thus can stop the movement of the robot by representing an unpleasant feeling.

11.4.3 Construction of an Artificial Conscious System

The MoNAD [n1, n2, n3, n4] configuration for the light detection MoNAD Br was MND [3, 3, 5, 7], where n1 was the number of bits for the input information IN (inputL), n2 was the number of bits for the behavioral representation BL (inputH) and cognitive representation RL (outputH), n3 was the number of bits for the somatic sensing unit (SSU) and output information OUT (outputL), and n4 was the number of bits for the primary representation (K).

The configuration for the action MoNAD Ac was MND [3, 3, 1, 6].

The configuration for the pain MoNAD Pa in the emotion subsystem was MND [3, 3, 5, 7].

The configuration for the pleasant and unpleasant feeling MoNADs P and UP in the feelings subsystem was MND [3, 3, 0, 12], where n3 consisted of 0 bits, meaning that there was no terminal for the somatic sensing unit (SSU) or output information OUT (outputL) for use in the experiment.

The configuration for the information integration MoNAD Li in the memory subsystem was MND [8, 8, 5, 16].

The storage unit Me in the memory subsystem was not a MoNAD but consisted of conventional three-tier neural networks (Fig. 11.11).

The configuration for the MoNAD As in the association subsystem was MND [4, 4, 1, 8].

The configuration for the MoNAD Re in the reason subsystem was MND [1, 2, 1, 3].

The entire design of the conscious system used in this experiment is shown below (Fig. 11.15).

For the connections between the MoNADs, however, especially for those between the subsystems, the number of bits for connection may be reduced due to limitations in the robot's specifications as long as it does not compromise the rationality of the configuration.

First, we explain the light detection MoNAD Br (Fig. 11.16).

Input IN for MoNAD Br uses three-bit data to express the intensity of light in four grades (000, 001, 011, and 1111) from weak to strong. RL and BL for MoNAD Br use the same information (pattern). Output OUT for MoNAD Br uses five-bit data to express the intensity of light in eight grades (00000, 00001, 00101, 00100,

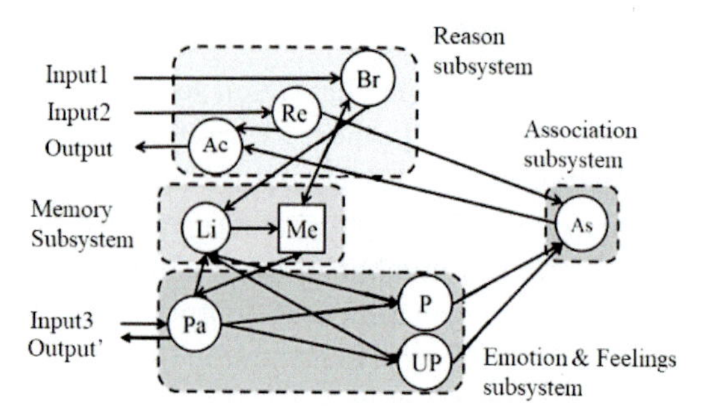

Figure 11.15 Structure of the conscious system used in the experiment.

01110, 01100, 11110, and 11111) from weak to strong (Fig. 11.16). The graded expression of the light intensity includes the change in light intensity that occurs when the light is turned on or off. For example, the strongest light (not the instant the light is turned on or off) is expressed as 11111, but the light intensity at the instant when the light is turned on is expressed as 11110.

Next, we explain MoNAD Pa.

The data used for MoNAD Br are used for MoNAD Pa. However, the graded expression includes the change that occurs when the

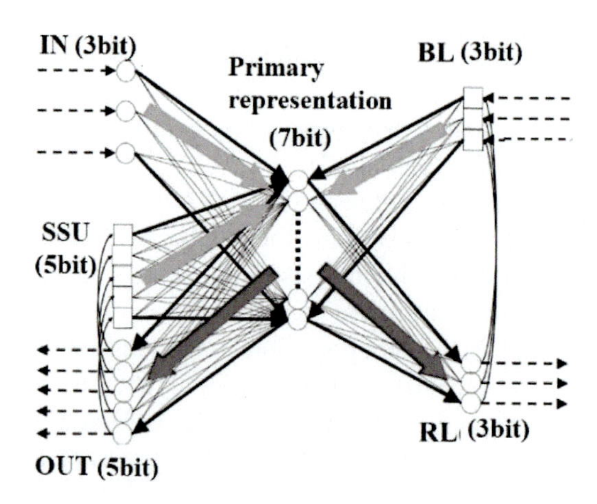

Figure 11.16 Circuit diagram for the light detection MoNAD Br (same for MoNAD Pa).

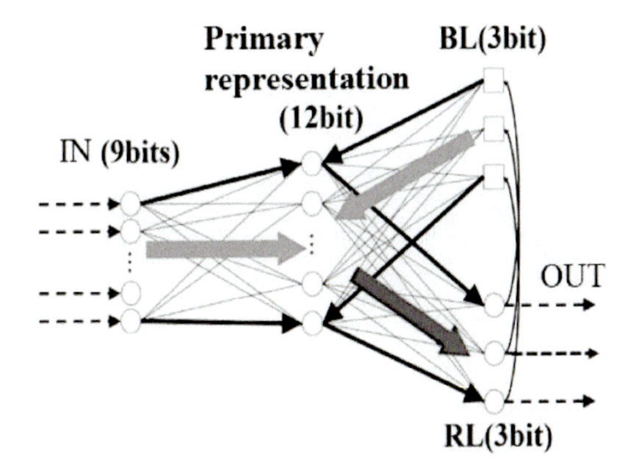

Figure 11.17 Circuit diagram for the pleasant feeling MoNAD P (same for the unpleasant feeling MoNAD UP). SSU and OUT were not used.

touch sensor is turned on or off. The status of the touch sensor is expressed in four grades (000, 001, 011, and 111) from off to on.

MoNAD Re in the reason subsystem constantly sends an instruction to OUT for MoNAD Re for the robot to move forward. We omit those details here.

Next are the MoNADs P and UP (Fig. 11.17).

MoNAD UP is explained as follows.

When IN for MoNAD UP receives information that the touch sensor is on (e.g., 111), RL for MoNAD UP represents three-bit information 111, and MoNAD P represents 000. Conversely, when IN for MoNAD P receives information that the touch sensor is off, MoNAD P represents 111, and MoNAD UP represents 000. At this time, the representation of a pleasant or unpleasant feeling is expressed in four grades (000, 001, 011, and 111) from weak to strong.

Next, we explain the role of the information integration MoNAD Li (Fig. 11.18a).

When MoNAD Pa represents an unpleasant feeling (the touch sensor is on) to IN for MoNAD Li, and if MoNAD Br represents the exposure to a strong light continuously or simultaneously, the information (11011) is output at OUT from MoNAD Li. This bit pattern instructs the information storage Me to store the current event as a memory (Fig. 11.15). The event that causes an unpleasant

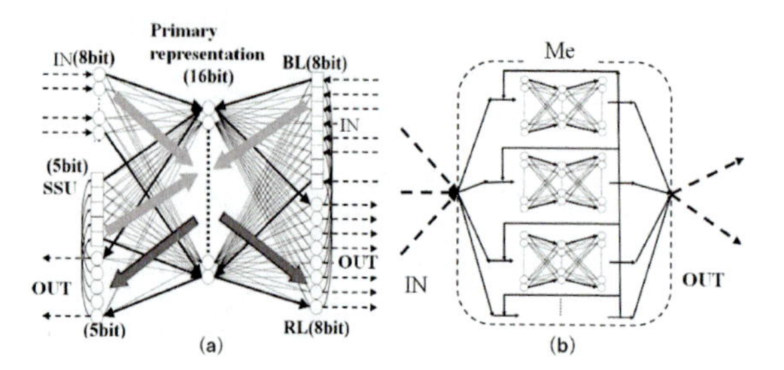

Figure 11.18 (a) Circuit diagram for the information connection MoNAD Li. (b) Circuit diagram for the information storage neural network Me.

feeling at time $(k + 1)$ means information that consists of the behavioral representation (BL $(k + 1)$) of MoNAD Br and the cognitive representation (RL $(k + 1)$) of MoNAD Pa. The circuit of Me is basically the same as the circuit of Ep described in Section 12.3 (Fig. 11.11); however, Me in this experiment uses paired three-tier neural networks (Ep elements) (Fig. 11.18b). Input information IN for MoNAD Li consists of eight-bit information calculated from the representation of pain Pa (using three bits) resulting from the turning on of the touch sensor, the representation of light detection Br (using three out of the five bits for simplicity), and the representation of a pleasant feeling P (using one bit) or an unpleasant feeling UP (using one bit). For example, if the IN information for MoNAD Li is 11000101, it means that pain is represented, a strong light is not represented, and an unpleasant feeling is represented in the conscious system. If the information is 11011001, it means that pain, a strong light, and an unpleasant feeling is represented in the conscious system. At this time, MoNAD Li sends an instruction for Me to store the information (11011). Me at time $(k + 1)$ has a behavioral representation BL $(k + 1)$ and a cognitive representation RL $(k + 1)$ currently coming from MoNADs Br and Pa which are learned as input and output information for the paired neural networks (in the robot experiment, however, information for either BL or RL is expressed in one bit to simplify the circuit). The pairing of the neural networks allows Me to output the cognitive representation of a strong light or a severe pain simultaneously to Br and Pa (11.3) when only either pain is represented by Pa or a strong light is represented by Br. If one

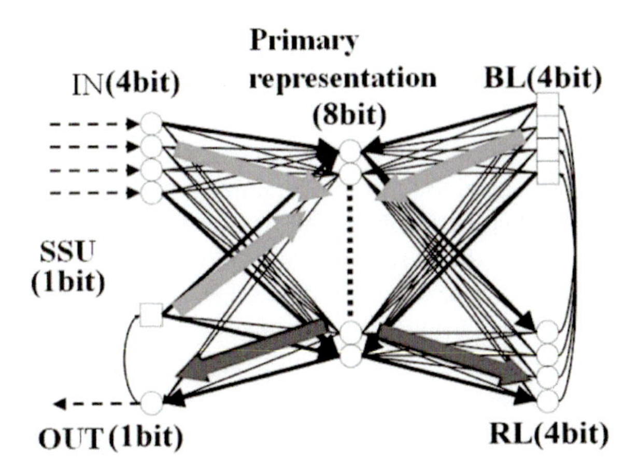

Figure 11.19 Circuit diagram for the association subsystem MoNAD As.

of the paired neural networks allows Br to recognize a strong light at time k (RL (k)), the cognitive representation of a severe pain by Pa can be simultaneously calculated (RL ($k+1$)), and the information of the unpleasant feeling generated at this time ends up terminating the instruction to move forward that had been sent out by the action MoNAD Ac from the reason subsystem Re through the association subsystem As—i.e., the robot stops. If the other neural network allows Pa to recognize a severe pain (RL (k)), the cognitive representation of a strong light by Br can be calculated (RL ($k+1$)).

Finally, we explain the association subsystem MoNAD As (Fig. 11.19).

MoNAD As sends out an instruction to the action MoNAD Ac based on the information from MoNAD Re in the reason subsystem and MoNADs P/UP in the feelings subsystem. MoNAD As instructs MoNAD Ac to stop the action if MoNAD UP represents an unpleasant feeling at that time regardless of what is represented by MoNAD Re.

11.4.4 Experiment with a Robot

The robot described in the preceding section (11.3) was used in this experiment. A white board and touch sensor were placed on the front (forward-moving direction) of the robot. The LED lights mounted on the upper surface of the robot are intended to show the internal state of the robot. The meanings of the LED lights when they are turned on are different from those in Fig. 11.12. LED No. 7 was not used. LED No. 1 was modified for the detection of light (white). The other LEDs were the same as shown. LEDs No. 2, 5, and 6, were designed to represent an unpleasant feeling (red), pain (orange), and pleasant feeling (green), respectively.

The robot was moving forward when the experiment was started. Since the touch sensor was turned off when the robot was moving forward, the green LED light on the robot was turned on to indicate the representation of a pleasant feeling (Fig. 11.20a).

Next, the robot collided with an obstacle (block). At this time, the touch sensor on the front of the robot turned on, and MoNAD Pa represented pain (Fig. 11.20b). The information that pain was represented caused the unpleasant feeling MoNAD UP to *shouki*. The term "*shouki*" here is used as an analogy of the firing of neurons to describe a MoNAD representing a certain pattern (information), and this word may be used later. At this time, the robot turned on the red LED light on its front because MoNAD UP represented a certain pattern (an unpleasant feeling).

Next, while the robot was moving forward, a strong white light was irradiated on the robot from its left using a flashlight. In this experiment, the irradiation of the white light did not cause MoNAD UP to *shouki*, but the light itself was detected (LED No. 2 turned on) (Fig. 11.20c).

Then the robot kept moving forward and encountered a Pavlov state. A Pavlov state refers to a condition similar to the Pavlov's dog experiments, where a bell was rung (neutral stimulus) when dogs salivated in the presence of food (unconditioned stimulus). In this robot experiment, the collision that caused the representation of pain and an unpleasant feeling served as an unconditioned stimulus, and the exposure to the white light as a neutral stimulus. Although humans may experience an unpleasant feeling when exposed to a strong white light, we should emphasize that in this experiment, the white light was not designed to cause an emotional response (Fig. 11.15).

At this time, the robot was in the Pavlov state—i.e., the robot collided with an obstacle and represented an unpleasant feeling while simultaneously being exposed to a white light. In the robot,

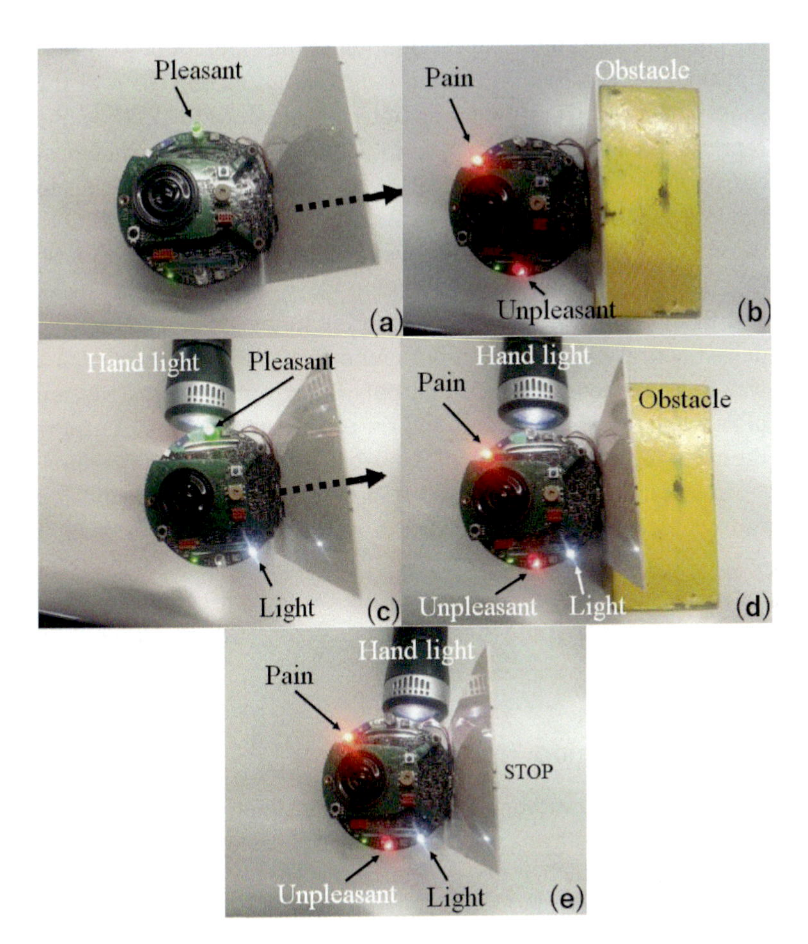

Figure 11.20 Sequence of events in the experiment. (a) The robot is moving forward. The LED for the pleasant feeling is on. (b) The robot collides with an obstacle. The LEDs for pain and an unpleasant feeling are on. (c) The robot is exposed to a white light. The LED for light detection is on, and the LED for a pleasant feeling is on. (d) The robot is in the Pavlov state. Collision with an obstacle and exposure to a white light occur simultaneously. The robot detects pain, an unpleasant feeling, and the white light. (e) Exposure to the white light alone causes the conscious system to represent pain and an unpleasant feeling and to make the robot stop moving.

the collision caused the representation of the emotion of pain (Pa) and thus the representation of the unpleasant feeling (UP). At the same time, the white light was detected by MoNAD Br. Then the represented information in Pa, UP, and Br caused the *shouki* of the information integration MoNAD Li and sent an instruction to the information storage unit Me. As described earlier, according to the memory instruction, Me receives a behavioral representation BL $(k + 1)$ and a cognitive representation RL $(k + 1)$ from MoNADs Pa and Br at time $(k + 1)$. These representations are learned by the paired neural networks as input and output information in such a way that one neural network outputs the cognitive representation of Pa when Br is represented, and the other neural network outputs the cognitive representation of Br even when only Pa is represented.

In summary, in the robot used in this experiment, the collision caused MoNAD Pa to produce the representation of white light detection and the representation of an unpleasant feeling simultaneously, or only the exposure to the white light could cause the emotion of pain (Pa) and the unpleasant feeling (UP) simultaneously (Fig. 11.20e).

11.4.5 Conclusion

Using a robot, the author and his group of students have reproduced the famous Pavlov's dog experiments by Nobel laureate Ivan Pavlov. Readers may know that when given food dogs automatically secrete saliva in their mouths. The salivation, which occurs in any dog, is generally referred to as an unconditioned reflex. Pavlov himself called this phenomenon a psychological response because it was considered an innate function of dogs. However, if a bell was rung when food was served to the dogs, after several attempts they learned to salivate to the sound of the bell alone—even in the absence of food. Since the sound of the bell was a stimulus irrelevant to salivation before the experiment, this response was called a conditioned response.

The author and his group of students installed a conscious system under development on a robot. The conscious system consisted of three MoNADs for the reason subsystem, three MoNADs for the emotion & feelings subsystem, and one MoNAD for the association subsystem. In addition, a memory subsystem was added to this experiment. The memory subsystem consisted of one MoNAD and multiple neural networks.

The robot was moving forward when the experiment was started. The robot collided with an obstacle, and the conscious system represented pain and then an unpleasant feeling. When the obstacle was removed, the robot resumed its movement forward. While the robot was moving forward, a white light was irradiated from the left side of the robot relative to its direction of movement using a flashlight. The robot represented the detection of light and a pleasant feeling. When the irradiation of the light was stopped, the robot resumed its movement forward. While moving forward, the robot collided with an obstacle and simultaneously was exposed to a white light irradiated from the left side of the robot relative to its direction of movement. That is, the robot was exposed to a white light at the time that the collision caused the representation of pain and an unpleasant feeling. The robot stopped moving as a result of representing pain and an unpleasant feeling due to the collision, and the exposure to the white light and the generation of pain were learned as a Pavlov state by Me, a unit consisting of the information integration MoNAD Li and neural networks in the memory subsystem. When the obstacle or white light was removed, the robot resumed its movement forward again. While moving forward, the robot was only exposed to the white light irradiated from the left side using a flashlight. As a result, the robot used the information stored in Me of the memory subsystem and represented the emotion of pain and the unpleasant feeling and stopped moving.

This experiment may be regarded as a reproduction of the Pavlov's dog phenomenon in a robot, and, as such, we refer to this robot installed with our conscious system as a Pavlov robot.

11.5 Color Vision Capabilities of a Conscious Robot

At this point, the author and his group of students were interested in the extent of the consciousness capabilities of the conscious system. In short, it was a question of to what degree one conscious system could have the capability of consciously distinguishing things in the natural environment. For example, if we could perform an evaluation test of how much color vision capability was possible with the current conscious system, it would also serve as one sort of capability evaluation of the conscious system. This question and a proposal arose when we were already conducting the research in Section 11.1 of Chapter 11, but at that time we were not capable of addressing the question. Now we had a chance to tackle this question again. The research in Section 11.1 was an important theme: "A Robot Recognizes the Unknown."

So here we considered one capability issue. This is a test of how many colors could be consciously recognized by having the conscious system learn various colors. Readers are likely to wonder how a conscious capability differs from a "cognitive capability." As previously defined by the author, our conscious system is a "system based on the consistency of cognition and behavior." We focused on the odd relationship between "behavior and cognition" in human consciousness. Put in simple terms, we think that the phenomenon at the core of human consciousness is "understanding (cognition) what one is doing (behavior)," and that "what one is doing (behavior) is the same as what one is thinking (cognition)." This is a review of our definition of consciousness. The author discovered that a recursive neural network called a MoNAD had functions very similar to this consciousness definition, and as such, a conscious system is what we call a hierarchically structured group of these MoNAD consciousness units.

The conscious system used in this experiment has already been shown. (Fig. 11.2).

At this time, the conscious system consists of three subsystems: reason subsystem, association subsystem, and emotion & feelings subsystem. This is the standard configuration. However, there are parts in the reason subsystem that are marked R1, R2, R3, ... These are MoNAD consciousness modules that form a recursive neural network with a special structure. Once again, a MoNAD is a self-referencing secondary cybernetics structure that possesses most of the properties of human consciousness. I believe that such a MoNAD structure exists at the core of human self-conscious capabilities.

Another researcher reports that he has found a MoNAD structure in the human brain (Manner and Takeno, 2011).

Here, an explanation of the MoNAD group in the reason subsystem is necessary.

The author and his group of students are attempting a new challenge in this experiment. This is a verification of the process by which the conscious system continues to learn. This process has already been mentioned in Section 11.1, but in this section we discuss the limitations of the process. The basic idea is that a MoNAD is never made to relearn when learning new knowledge. That is, the method of learning involves connecting an additional new MoNAD to the reason subsystem of the conscious system. Increasing new knowledge not only increases the volume of information of the conscious system, but also increases the network connections of the conscious system. This method is one way that does not require a relearning process for the network. We have already proposed this method in Section 11.1, but we had not yet examined the capability. In this section, as in Section 11.1, we already have the system learn to recognize (to be conscious of) the three primary colors of red, green and blue in MoNAD R1. However, in Section 11.1, MoNAD R1 was made to learn so that the robot moves forward if the color seen by the robot is green, stops if the color is red, and moves backward if the color is blue. That is, if a green color code (to be described later) is given to the Input of the conscious system, MoNAD R1 will issue an instruction code to Output that causes the robot to move forward. And R1 has finished learning that if the color is red, it outputs an instruction code to stop, and if the color is blue, it outputs a code to move backward. However, in this experiment, we did not have the conscious system learn to move the robot, we only displayed the name of the presented color. For example, the color green is presented. At that time MoNAD R1 reacts with "*shouki*" (the term described earlier that we use as an analogy of the firing of neurons when a MoNAD represents certain information), and displays "green" on the screen according to the Output information. The same occurs for the colors "red" and "blue." Thereafter, let us suppose that yellow is presented in the experiment. At that time, the conscious system will start the "mechanism to distinguish between the known and unknown" that we introduced in Section 11.1. The essential part of this mechanism is to evaluate the state of convergence of the MoNAD. We evaluate the error at k-time of the MoNAD's cognitive representation RL (k) and behavior representation BL ($k - 1$). (This is called an RB error.) Since BL ($k - 1$) is equivalent to RL ($k - 1$) due to the nature of the MoNAD, we can calculate the change in the RL error (i.e., the difference between RL (k) and RL ($k - 1$)). In this experiment, if the error difference exceeds a certain threshold for all MoNADs in the reason subsystem, then the information presented at the Input of the conscious system is judged to be unknown. In other words, the presented yellow color does not cause R1 to *shouki* at this point, but R1 receives this result and causes MoNAD UP to *shouki* and represent the unpleasantness of the emotion and feelings subsystem. Subsequently, this unpleasant representation activates a process that sends a signal to the association MoNAD As and learns the unknown information to make it known information. Although this process is not shown graphically in the conscious system already described, it is positioned as a foundation system (Fs) that supports the conscious system. The Fs program can be described using MoNADs, but in this experiment it is written as a regular program (e.g., in the C language). The role of Fs is to make the new "empty MoNAD" (R2 in this case) in the conscious system prepared to learn the "yellow" color information now being presented. At this time, since the new MoNAD is a consciousness unit, behavior information is required for the cognitive information (yellow in this case). However, since we have not yet found a way to automatically make a judgment on this information at present, the experimenter currently enters the name of the color manually from a keyboard. The means that automatically making this judgment is likely to be deeply related to "creativity," and the author is currently researching this issue further. Now, once R2 finishes learning, the Fs program will continue and connect R2 to the conscious system. This means that R2 is connected to Input, Output, UP, P and As. In this way, the conscious system can now represent not only green, red, and blue, but also yellow. For example, R1 will *shouki* if red is the information presented at Input, and R2 will *shouki* if yellow is presented. If a new unknown color is presented, a new MoNAD R3 can learn it using the procedure described above.

The learning method described in this research is different from the conventional learning method, and readers should be aware that we are presenting a new method that adds newly learned information to the existing conscious system connectively without relearning.

11.5.1 Conscious System Used in This Experiment

The conscious system uses part of the system described in Section 11.1.

MoNAD R1 is shown in Fig. 11.21. The consciousness structure is MND [9, 2, 12, 5] (Fig. 11.21). In the first round of the experiment R2, R3, ... are empty MoNADs. Empty MoNADs have the consciousness structure of MND [9, 1, 10, 1] as shown in R2 (Fig. 11.22). The connection weight and threshold value of a neuron in a neural network that positions the functions of a MoNAD are given as random numbers. That is, an empty MoNAD is a MoNAD with no directionality (it cannot *shouki*). The output of R2 is reduced to 1 bit. This is because the purpose of R1 is to be conscious of three kinds of color vision. The purpose of R2 and subsequent MoNADs is to be conscious of monochrome color vision. For the same reason,

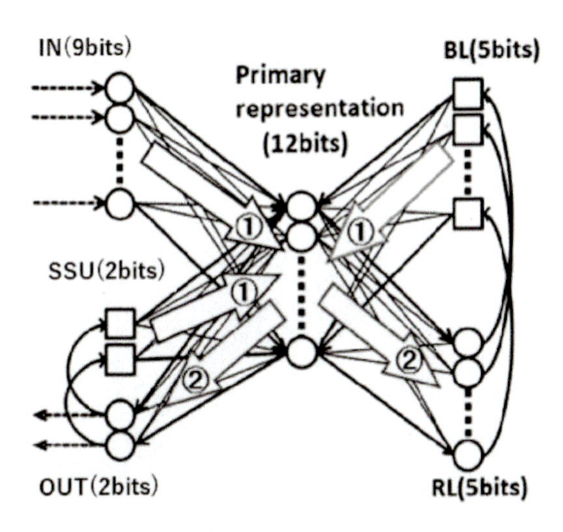

Figure 11.21 Neural network circuit of MoNAD R1.

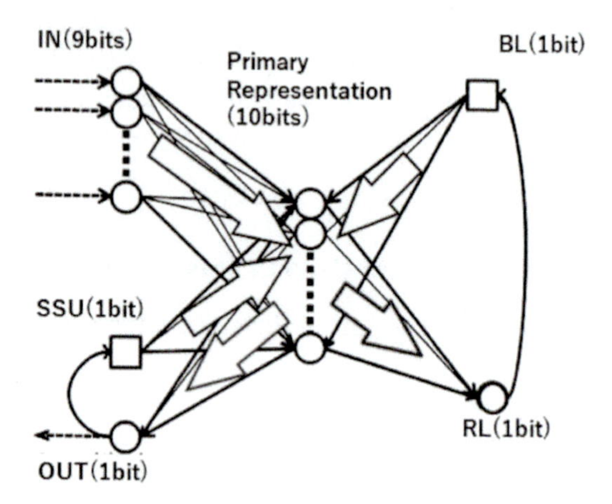

Figure 11.22 Neural network circuit of MoNAD R2 and subsequent MoNADs.

the primary representation is reduced from 12 bits to 10 bits. The MoNADs U and UP of the feelings subsystem are shown in Fig. 11.23. The consciousness structure is MND [2, 3, 0, 7]. The Output of the feelings subsystem is omitted as it is not used in the experiment. Also, the emotion subsystem is omitted as it is not needed in the conscious system of this experiment. Furthermore, the association subsystem MoNAD As uses MND [4, 1, 2, 6].

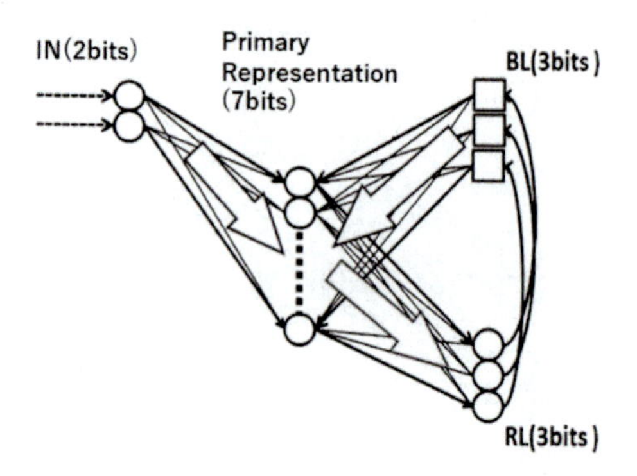

Figure 11.23 Neural network circuit of U or UP of the feelings subsystem.

11.5.2 Color Vision Capability Experiment and Results

The important point about this experiment is the problem of what are the standards of human color vision. Color vision is a subjective phenomenon in which electromagnetic waves emitted from the environment react with the cone cells and rod cells that are photoreceptors in the human retina. It is said that mainly cone cells are related to color vision, and rod cells are related to brightness. Also, color vision seems to have individual quality particular to each person.

The author and his group of students referred to the color test proposed by Prof. Diana Derval as a human color vision standard. Although Derval proposes 39 colors in total, only 20 colors were used in this experiment (Fig. 11.24). The colors selected for use in the experiment were clearly distinguished manually, and we assumed that the pigments had prominent names.

The 20 selected colors are as follows. Each color is shown ordered by its id number and pigment name.

① maroon, ② brown, ③ indianred, ④ lightsalmon, ⑤ burlywood, ⑥ khaki, ⑦ palegoledenrod, ⑧ yellow, ⑨ greenyellow, ⑩ lawngreen, ⑪ aquamarine, ⑫ lightseagreen, ⑬ mediumturquoise, ⑭ cornflowerblue, ⑮ steelblue, ⑯ royalblue, ⑰ mediumslateblue, ⑱ plum, ⑲ mediumorchid, ⑳ indigo

These 20 kinds of pigments are shown in Fig. 11.24 as a three-dimensional display with the three primary colors (RGB) as the axes. This figure shows that each color has its own separate RGB value, which indicates that it is appropriately distributed in the three-dimensional RGB color space.

When entering colors into the conscious system, the RGB565 color format was used. In other words, the RGB values of each color are converted into a total of 16 bits with 5 bits used for red, 6 bits for green, and 5 bits for blue. For example, the RGB value of indianred is 205 for red, 92 for green and 92 for blue. When converting these RGB values to 5, 6 and 5 bits, respectively, red is 25, green is 23, and blue is 11. After that, each of the values converted to RGB565 is converted into 4 levels of depth (dark [111], normal [011], light [001], none [000]) using 3 bits. (This is called color normalization.)

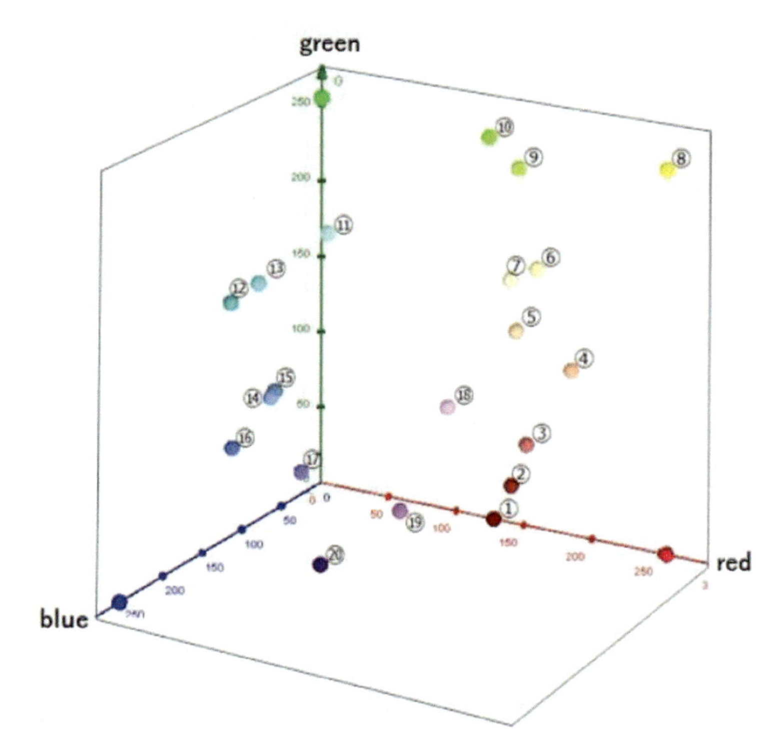

Figure 11.24 RGB three-dimensional display of the 20 colors used in the experiment.

That is, if indianred (③) is represented by a bit pattern of 9 bits in total, it is RGB (111,000,000). Thus, in principle, the type of color to be learned is 64 (= 4 × 4 × 4) through this normalization.

Now, we will describe the experiment. First, mediumorchid (⑲) is entered into the conscious system (Input, Fig 11.2). The order of entering the colors in this experiment was decided arbitrarily. At this time, there is only the R1 MoNAD in the reason subsystem of conscious system. R1 will *shouki* only with the primary colors of red, green, and blue, but with mediumorchid it cannot *shouki*. At this time, the conscious system considers mediumorchid to be unknown information (an unknown color), makes an empty MoNAD R2 learn the unknown color, and adds it to the reason subsystem. After learning is finished, the experimenter gives the name of the newly learned color to the conscious system by entering it from the keyboard.

At this stage, the conscious system will proceed to learn to recognize (to be conscious of) the four colors of red, green, blue and mediumorchid. Subsequently, brown (②) is learned at R3, and khaki (⑥) is learned at R4 in the conscious system. Then, palegoldenrod (⑦) is learned at R5, yellow (⑧) at R6, lawngreen (⑩) at R7, aquamarine (⑪) at R8, mediumturquoise (⑬) at R9, and indigo at R10. The next input was indianred (③), but R1 reacted with *shouki* and recognized it as a known color. Here, the human experimenter judges indianred to be an unknown color, and forces the learning (pressed learning, or simply called PL) of indianred as a new color in the conscious system. We have not explained this process, but we incorporated this mode into the program for the experiments to enable the learning of more neutral colors. Therefore, indianred could be learned at R11. At this time, both R1 (as red) and R11 will *shouki* in the conscious system. The system should give priority to the new learning result at this time and use the result of R11, but in this case R1 is used. The reason is that R1 is basic data in this learning process. That is, indianred was learned as a new color, but it was absorbed in red. Next, when lightsalmon (④) was input, it was learned as R12. Continuing this learning procedure, R13 is burlywood (⑤), R14 is greenyellow (⑨), R15 is lightseagreen (⑫), R16 is cornflowerblue (⑭), R17 is steelblue (⑮), R18 is royalblue (⑯), R19 is mediumslateblue (⑰), and Plum (⑱) is learned as R20. However, PL was performed from R13 and subsequent MoNADs. However, due to the PL of steelblue, the lightseagreen of R15 was excluded from the consciousness targets. And due to the PL of mediumslateblue, cornflowerblue of R16 was also excluded from the consciousness targets. In addition, due to the PL of plum, mediumorchid of R2 was excluded from the consciousness targets. These auto-exclusions depend on the previously mentioned rule in which, since a MoNAD acquired by new learning and a previously learned MoNAD in the conscious system will *shouki* at the same time, the newly learned MoNAD is adopted as a consciousness target.

The resulting color vision capability achieved in this experiment is shown in Fig. 11.25. The colors with white X marks are those

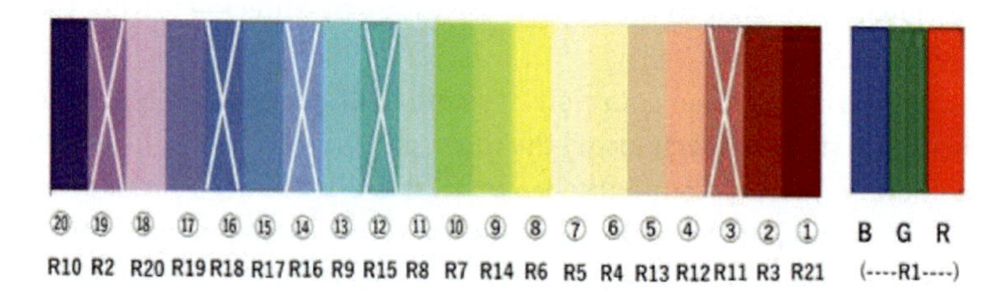

Figure 11.25 Color bar of colors learned by the conscious system.

that were learnable but could not be recognized as a consciousness capability due to the PL (i.e., these colors were excluded). Indianred (③) is the same as the R1 Red, lightseagreen (⑫) is considered to be the same as the R17 steelblue (⑮), cornflowerblue (⑭) is considered to be the same as the mediumslateblue (⑰) of R19, and royalblue (⑯) was judged to be the same as the Blue of R1.

These color vision results are reddish, greenish and bluish colors, respectively, and it was felt that it was difficult even for the experimenter to practically distinguish between those colors.

As a result, in addition to the three primary colors of blue(B), green(G), and red(R), the color vision of the system learned 15 new colors, thus through the learning in this experiment, the system became capable of consciously recognizing 18 colors in total.

The color id numbers and B, G, R for the three primary colors are shown in the middle row.

The bottom row shows the number of the MoNAD that learned the color above it, and this number indicates the sequence in which that color was learned.

The white X marks on the colors in the color bar at the top indicate colors that were not consciously recognized by the system due to the PL learning process.

11.5.3 Conclusion

The subject of this section is an investigation into the capabilities of an artificial conscious system composed of MoNADs.

The object of this capability evaluation is the color vision consciousness of the system. And this should be taken as an evaluation of the system's consciousness capabilities. In short, we sought to answer the question of how many colors this system was capable of recognizing. In conclusion, the system was conscious of 18 colors. This result is somewhat lower than that of human color vision ability (no more than 30 colors (Derefeldt, 1995)).

In addition, the point to focus on in this section is the issue of MoNAD relearning. Since a MoNAD is composed of a combination of conventional neural networks, MoNAD relearning is basically neural network relearning. Considering that human consciousness has such qualities as its lack of momentary disruptions and its persistence, momentary disruptions of a MoNAD should be avoided. If the momentary disruption of a MoNAD occurs, the overall balance of the conscious system would collapse and the outcome would be unpredictable. That is, a MoNAD is set as an independent consciousness unit that does not stop.

We already addressed this question in Section 11.1. In other words, the conscious system first judges whether the input information is known or unknown, and if it is unknown, the system has a new MoNAD learn the information and add it to the conscious system. Through this method, the conscious system can change unknown information into known information. In other words, the conscious system will continue to steadily learn any information that the system itself has determined to be unknown. As such, the conscious system can steadily increase its knowledge.

However, although there are limitations, this section offers one suggestion to this problem.

Certainly, the amount of memory that the conscious system can use is one of its limitations. However, network technology continues to push against these boundaries, and such limitations are being eliminated. The point suggested by the experiment in this section is that the capability of a MoNAD is limited by the number of bits that make up the MoNAD. However, today's deep learning technology offers a way to solve this problem, and as described in this section, our method of having a MoNAD (neural network) learn new knowledge, and additionally connecting and increasing its known information is an approach toward lessening the problem. And as such, our approach may be called a neuromorphic method.

11.6 Absolute Pleasantness

In this section, we will describe our recent research developments. First, I will talk about the principle of a robot's representation of pleasantness and unpleasantness.

In Chapter 9, Section 9.7.2, the author has already explained the states of unpleasantness and pleasantness of the conscious system. That is, we explained that in the MoNAD unit, the emotion was pleasant if the behavior representation and cognitive representation were consistent, and it was unpleasant if they were not consistent. Let us explain this in more detail. If we provide unlearned information to the INPUT of the MoNAD unit, the convergence speed of the MoNAD decreases, and this phenomenon has already been explained in Chapter 11, Section 11.1. Here we introduce a study that uses this phenomenon to determine whether the information input to the MoNAD unit is known or unknown. In other words, the author thought that an evaluation of the convergence speed of the MoNAD could be defined as a representation of the emotion of the MoNAD unit itself, such as pleasantness or unpleasantness. In other words, if the convergence speed decreases, the emotion will be unpleasant, and pleasant if the speed becomes faster. In this way, a MoNAD can express the state of its emotion on its own. When explaining this phenomenon further, it seems that, regarding explanations that there is no way that anything can express emotions on its own, the explanations up till now do not hold at all, no matter whether it is a mechanical system or an electronic system. It is as though we are just saying, "We cannot explain this, so it is not possible." A MoNAD is expressed by combining neural networks (NN), but since it has a structure that enables super logic, that is, second-order or secondary cybernetics, such an expression of emotion is possible. As super logic, since secondary cybernetics can explain things like emotions, we can say, "We can explain this, so it is possible."

Now, regarding pleasant and unpleasant emotions, the work of Freud is historically famous (Freud, 1900). He said that living organisms have a "pleasure principle" and they live on that principle. He also said that living organisms respond to stimuli originating in their external environment as well as internally. And if a stimulus is viewed as harmful by the living organism, it tries to get it outside the organism as quickly as possible. Of course, living organisms can take action to avoid external stimuli. In addition, the internal stimulus corresponds to the stimulus arising from instinctive urges or desires (Freud, 1920). This is basically an unavoidable stimulus in living organisms. He said that both libido and ego are present in these desires. The former serves to preserve the species, while the latter preserves the individual. He explained that living organisms cannot directly see the path of avoidance behavior by themselves in both of these desires. Later he arranges them as two types of urges, the instinct to live (Eros) and the instinct to die (Thanatos).

And, according to Freud's explanation, this means that living organisms that are not stimulated both internally and externally are in a pleasant state.

The explanation of pleasantness and unpleasantness of the author seems to be very similar to that of Freud. If we strongly consider the differences, the explanation about pleasantness and unpleasantness by the author is based on the reaction of the consciousness unit described in the neural network called a MoNAD as incorporated in a living organism. We think this is related to the interval of time from when the information is input to the MoNAD to when it is accepted in that MoNAD. The more quickly the stimulus is accepted, the shorter the (convergence) time is. And if the time is prolonged, pleasantness changes to unpleasantness. Expressing this in a slightly different way, if the MoNAD has already learned some information, it quickly processes (converges) it to input that information. However, when unlearned information is encountered, the convergence is slow. The former corresponds to pleasantness, and the latter corresponds to unpleasantness.

Although the extent of unpleasantness at this time can be evaluated by the decreased convergence speed of the MoNAD unit, how is pleasantness defined?

If we take up Freud's explanation, then would "a living organism not being stimulated internally or externally" be akin to "a MoNAD that has learned not being stimulated internally or externally?" If so, it can be said that a MoNAD that has learned is in a pleasant state without being stimulated internally and externally. Perhaps an explanation that replaces a living organism with a MoNAD that has learned might seem somewhat rude, but we would like to ask the reader to indulge us in this mere thought experiment. The author calls this a state of absolute pleasantness. Here, the term absolute means that the state was defined as an absolute standard for a living organism. Therefore, if a MoNAD could temporarily learn a convergence speed that exceeds this absolute standard in the environment, note that this MoNAD could evaluate pleasantness that exceeds absolute standards. If so, the absolute standard of pleasantness may be considered to be capable of transitioning to a higher standard, and that could be brought about by new learning from internal and external stimuli. In other words, from the developmental myths of living organisms, one explanation is made possible. That is, developmental myths have explanations that justify the evolution and development of living organisms from the viewpoint of the rationality of the living organisms. However, the explanations so far do not yet explain such rationality. For example, in conventional explanations, human evolution and development have enabled humans to reach a better state by themselves, and as a result, humans have reached the peak of existence (Darwin, 1859). Furthermore, there is an explanation that the internal motivation arising from the foundation of the gods of humans has enabled evolution and development. However, such explanations have been pointed out to be certainly arrogant in modern times. Of course, it is difficult to prove “the peak of existence” or “the foundation of gods,” but the point here is that the fact that a living organism can achieve a better state by itself can be explained by the “absolute pleasantness theory” of the author. The reason is because, by setting the standard of “absolute pleasantness” and determining the evaluation of pleasantness and unpleasantness according to that standard, it is possible to place the process that can more quickly handle stimuli received from both the internal and external environments at a position of higher pleasantness for the living organism. And this theory can be assumed to be a case in which stagnation always occurs with adverse effects arising from too high pleasantness standards or where the pleasantness standards are at a low level. Such thinking would seem to be a model approach toward the understanding of human brain disorders.

11.7 Conflict of Concepts and Rubin's Vase Model

The reader is already familiar with Rubin's vase (Fig. 4.15). Rubin's vase is an important topic in research on human consciousness. When a human viewer focuses on the center of the image, the shape of the vase is recognized, and when the viewer's line of sight is directed to the left or right sides of the image, the recognition of the vase disappears, and a figure like a human face can be recognized there. The term used here is "recognized," but the viewer can also be said to be "conscious." Here as well, readers are likely to ask how recognition is different from consciousness. Using the term "consciousness" emphasizes the aspect of "human subjective perception." To be certain, the "vase" and "the faces of the people" can be recognized in different parts in the picture of Rubin's vase." But if you use the term "consciousness," you can say, "Now I am looking at a vase," or "Now I am looking at the faces of the people," and the human subjective process becomes an important issue. A characteristic of this consciousness phenomenon is that you cannot experience "consciousness of the vase" and "consciousness of the faces" at the same time. If you change your gaze and look at the vase, you become conscious of the vase, and if you change your gaze away from the vase, you become conscious of figures that look like human faces. Moreover, it is difficult to remain conscious of only the faces or vase at a time, as the process of repetition is automatic, and you are continually conscious of the vase and then conscious of the faces repeatedly. It seems as though human consciousness is always working (driving the gaze) in search of a target to be conscious of. And, finally leaving the image called Rubin's vase, we will search for new subjects. Again, while it is possible to replace the term consciousness with recognition, when using the term recognition it is difficult to connect directly to behavior. However, the term consciousness can directly link to behavior. The reason for this is because consciousness is a phenomenon that has orientation. And a MoNAD can substantially explain the functions of consciousness (Section 9.1.4). Of course, a MoNAD also has orientation.

Furthermore, there is one more point we would like to emphasize.

Regarding the MoNAD consciousness module that was also mentioned in the previous section, the point is that one MoNAD can express one concept. Of course, it can learn more than one concept.

In the previous section, we used one MoNAD to express the concept of one neutral color. For example, the color yellow (⑧) was learned as MoNAD R6 (Section 11.5). This makes it possible to express the concept of yellow with R6. And that concept is learned as the word "yellow" with the MoNAD. That is, R6 reacts with *shouki*, the term we use as an analogy of the firing of neurons to describe a MoNAD representing information, when yellow color information is presented to it and the consciousness unit replies that the color presented as input information is "yellow." R6 can be called a "yellow conscious" unit as a place of "consistency in cognition and behavior," and it can also be said that R6 is a MoNAD that expresses the concept of yellow.

Let us return and speak again about Rubin's vase. Continuing the discussion in this vein, it would seem possible that a conscious system comprised of MoNADs can express the phenomena of Rubin's vase. In other words, the conscious system can be conscious of the vase inside the presented image, and it can also be conscious of the faces.

Although the image presented is a single one (Fig. 4.15), if the line of sight of the visual device of the conscious system is oriented toward the center of the image, the system can be conscious of the vase; and if the line of sight is directed to the periphery of the image, the system can be conscious of the faces. If the machine system merely recognizes the vase and faces, it is only necessary to end with the recognition confirmed. But in the human subjective phenomena for this given image, consciousness of the vase and faces occurs periodically. There is still a mystery about how this repetitive phenomenon is caused. However, we can infer some causes from the studies of preceding researchers.

The first thing to be said is that the input image is always the same, so the cause of the repetition phenomenon can be sought inside the system. In short, it is an intrinsic factor.

If that is the case, is it not possible to seek the cause as being just an intrinsic cause in the function of consciousness itself?

The author and his group of students think that this "repetitive phenomenon" can be modeled as a "phenomenon of conflict of concepts" that constitutes consciousness. We will explain it a little more simply. When a human views the image of Rubin's vase, phenomena occur in which the person is conscious of the vase and conscious of the faces in turns. And both of these phenomena are repeating. It would seem as though there is a competition and reaction towards a subjectively conscious phenomenon in which the concept of the vase and the concept of the faces are being subjectively perceived. In other words, we can imagine a state in which a "conflict of concepts" is occurring within the human consciousness system in which the concept of the vase and the concept of the faces are respectively competing for consciousness.

Let us explain that model.

First, the conscious system is directed to the "vase" by MoNAD ActE (Fig. 11.26). And when the system is conscious of the "vase," MoNAD Va will *shouki*. And the system displays the word "vase" on the screen by means of MoNAD ActL. At this time, the system will cause the pleasantness MoNAD P to *shouki*. In this state, the system is temporarily stabilized.

In other words, the input information to the conscious system is a single image called "Rubin's vase," and as output the system simply displays the word "vase." Because the information internally and externally does not change, the system is stable. However, the concept of "Rubin's vase" (MoNAD Ru) has already been learned in the system (Fig. 11.26). The concept of Ru, the concept of the vase (Va) and the concept of the faces (Fa) are connected as associative memory (Fig. 11.26a,b). In other words, Ru expresses the knowledge of "Rubin's vase" as a MoNAD on the system. The role of Ru is to continue to make Fa *shouki* when the *shouki* of Va occurs (gives expectation information to BL of MoNAD Fa). Conversely, Ru will try to get Va to *shouki* through the *shouki* of Fa. Now that the *shouki* of Va is occurring, the system gives expectation information to Fa

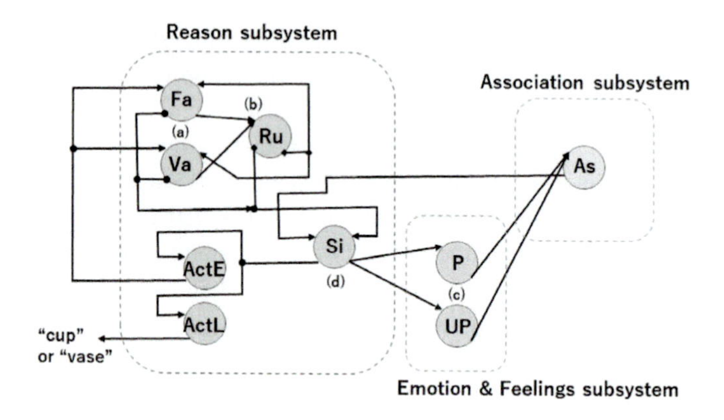

Figure 11.26 The conscious system for the experiment of Rubin's Vase.

through Ru. However, as the line of sight continues to be directed at the "vase," the input information of the system does not correspond to the expectation information of Fa, and the system will *shouki* the unpleasant MoNAD UP (here the pleasantness is also decreased, Fig. 11.26c). Then, due to this change in the emotion and feelings subsystem, the association MoNAD As signals MoNAD Si (to change of line of sight, Fig. 11.26d) and changes the line of sight of ActE. As a result, MoNAD Fa will *shouki* by giving "faces" information to the input of the system.

Now we will summarize the overall flow of the conscious system. First of all, when the system is conscious of the "vase," due to its knowledge (of the concept) of "Rubin's vase," it will have to expect the "faces". However, since the system still has "vase" information entered in it, the expectation of "faces" will not succeed. Therefore, while the system was conscious of the "vase," it was representing a pleasant emotion, but due to the failure of expectation, it will represent an unpleasant emotion. The internal change of this emotion leads to the change in the line of sight of the system and to the successful expectation of the "faces" by changing the input of the system to the information on "faces." In other words, the system was conscious of the "vase" before, but it changed to be conscious of the "faces" next, by making use of its knowledge of "Rubin's vase." And, we think that the process of changing the target of the consciousness can be understood to change next from the "faces" to the "vase" and this will be repeated until the system stops.

This study is an example of the modeling of human consciousness phenomena caused by a mysterious image (an ambiguous picture) simply called "Rubin's vase." However, this section suggests that the conscious system expressed using MoNADs makes it possible to further develop the modeling of human consciousness phenomena by means of a new expression called the "conflict of concepts."

11.8 Toward an Elucidation of Self-Development and Advanced Mental Illness

This section considers the question of what was responsible for triggering the development of the human ego and advanced mental illness in humans with relevance to the conscious system described in this book.

To begin with, the existence of the human ego must have been a very important theme in the last century and the century before that. However, it was the successful experiments with dogs by the Russian Ivan Pavlov that had a great impact on the world in the early part of the last century. The reason was that while dogs were known to salivate when fed even immediately after they are born, Pavlov rang a bell in his experiments that stimulated the dogs at the same time they were being given food. Gradually, the dogs would be stimulated to salivate only upon hearing the bell in the absence of any actual food. Am I the only one who thinks that the results of Pavlov's experiments enabled great strides forward to materialistic scientism? In Pavlov's time, humans were sentient beings with bodies and brains just like we are now, but those days were far from our present time when the structure and functions of the brain are already well known.

In Pavlov's time, it was not yet understood that many nerve cells are connected to the brain. With the extent of the knowledge of that time, when a dog was fed, saliva would be secreted into the dog's mouth as if it were an automatic reaction, and this phenomenon was also seen in small puppies. I think it was unavoidable that an argument would also be made that the secretion of saliva was due to the dog's psychogenesis, as the phenomenon had been observed in many dogs (Pavlov, 1927) (Popplestone and McPherson, 1994).

Today, I am rather reluctant to call a dog's secretion of saliva a psychogenic effect. This is because, although salivation will also occur in humans when we put something to eat in our own mouths, I think that this phenomenon is an innate conditioned reflex activity that humans acquired during our evolution rather than something to be classified as psychogenesis. Of course, it is possible to classify an innate conditioned reflex activity as a sort of psychogenic activity, but depending on how psychogenesis is defined, I feel somewhat reluctant to do so in these modern times. Nevertheless, I think such thinking was unavoidable in a time when research on the brain and body had not yet developed.

We can imagine that the results of Pavlov's research were met with wonder in such an age, because the sound of the bell alone directly caused the dog's psychogenic activity. And even in modern times, we should not underestimate the influence of those astonishing results. I have already mentioned in Section 4.6 that the astonishment at his experiments contributed to the creation of behaviorism in the United States, and its development continues to have a major impact today. Its importance was also emphasized in Section 11.4.

Behaviorist psychology in the United States quickly had a tremendous impact on psychologists all around the world, because the results of Pavlov's experiments offered significant implications. The experiments showed that a dog's psychogenic activity could be controlled by external stimuli such as the mere sound of a bell. There is no doubt that this brought about the thinking that a similar phenomenon might occur with psychogenic effects in humans. As an example, an attempt was made to improve the habits of heavy smokers. Although it appeared to produce great results, the behavioristic method was judged to be one that reinforced a conditioned reflex rather than reason, and improvement as psychology was seen necessary. B. F. Skinner (1904–1990) is famous both for the success and the criticism of his Skinner Box experiments using lab rats and pigeons. As a result, cognitive behavioral psychology began as a psychology that was based on a better understanding of human cognitive functions. Thereafter, cognitive psychology began with a focus on the study of the human brain.

Also, the dialectical materialism of Karl Marx (1818–1883) can be considered to have had a strong influence during Pavlov's time. Dialectical materialism is the idea that matter exists as the basis of human behavior, that is, social behavior and economic behavior.

Georg Wilhelm Friedrich Hegel (1770–1831) was thought to have originally formulated the concept, but it dates back to the times of ancient Greece. Marx's dialectical materialism was a very fascinating idea at the time, and it appears that people in those days felt that it could be used to explain everything in the world. The author agrees with the idea that matter is at the basis of human mental activity and, to put it frankly, matter is at the basis of the phenomena of human consciousness and the mind as well. However, how human consciousness and the mind are related to matter remains an unsolved and important issue. To be sure, no rational explanation for this issue existed even in Pavlov's time. Some religionists have claimed that the human heart and mind have nothing to do with matter, but even today there is no sign that this claim will be retracted. Also, some scientists have argued that the human heart and mind must be explained by matter. With the development of dialectical materialism, this claim was increasingly capturing the attention of scientists. In other words, I feel that many scientists at the time judged that the idea of dialectical materialism was "a more rational way of thinking." Rational thinking is the idea that a possible scientific explanation exists for any phenomenon.

Stated in a different way, this kind of thinking seems to have created an attitude of denying the existence of any phenomena that cannot be explained scientifically. The author basically supports this attitude, but I think one point needs our attention. This is about the attitude that if the existence of a phenomenon cannot be scientifically explained, one can claim that "the phenomenon does not exist." Looking back on history, I think that scientific thinking has been making progress every day. Newton's laws of motion were later superseded by Einstein's theory of relativity, and the science of quantum mechanics may have paved the way to new sciences. In other words, the claim that "unexplained phenomena do not exist," which may have seemed rational at the time was, in a sense, putting the brakes on the development of new science and the progress of science.

Also, I feel that the fascinating claim of dialectical materialism that "every phenomenon originates from matter" had leaned toward the dogmatic claim that "phenomena that cannot be explained by matter do not exist." Given the state of science at the time, only philosophers and religionists were able to resist this dogmatic claim. For example, if you were a philosopher, to begin with, from the idea of Kant's criticism of pure reason, when the idea that "human thinking is finite" changed from the former claim to the latter claim, you must have frowned at the harshness of it. And, too, if you were a religionist, you would feel like rejecting the former claim, because religionists would not accept the claim that God's existence is at least material. For example, is the "spirit" of religionists something with substance?

Thinking about it in this way, dialectical materialism appears to have been one of the conclusions of modern scientism. However, about 300 years ago, in his mind-body dualism thesis Rene Descartes (1596–1650) held that the mind and body had an existence distinct from each other. Descartes claimed that the two should be considered separately.

However, he did not state how the mind and body are integrated in humans. I think this idea was an important first step in the development of dialectical materialism. As mentioned previously, Descartes' mind-body dualism made it possible for humans to clearly separate the machine and the mind when thinking about the two. Perhaps I might be overstating this somewhat, but I think the Industrial Revolution in Britain was supported by this mind-body dualism, because people had become convinced that a human-made machine had no mind based on Descartes' claim. In other words, a steam locomotive had no mind and could be easily made and controlled by humans. And notwithstanding this example, one might say that Descartes' mind-body dualism spawned and promoted the first scientific principle that brought about the modern era.

Given these historical trends, Marx's dialectical materialism can be said to have promoted the second scientific principle that brought the modern age into being. In other words, the idea that "every phenomenon originates from matter" has served to advance a great deal of scientific thinking. I, too, also accept this idea. However, the idea that "there are no phenomena in the world that cannot be explained by matter" seems at this time to have been an overly ambitious claim. I think that the fact that there are many scientists who state that human consciousness and the mind are merely illusions, or that they do not exist, is related to the excessive claims of this time.

I think that one consequence of scientists who had an extreme interpretation of the dialectical materialism was the termination of the Berlin Wall in 1989. I feel that these scientists had a strong desire to develop dialectical materialism into an even more progressive form of scientism, and as a result, they had upset the balance of human understanding. I think it was unfortunate that despite the efforts of the Viennese psychologist Sigmund Freud to uncover the mysteries of the existence of the human mind, this shift in direction was not recognized at all, and it resulted in a major disruption of society on 1989 (Takeno, 2003).

At this time, humans once again experienced their beliefs and dogmatism leading to a horrendous calamity.

The results of Pavlov's experiments, together with dialectical materialism, showed that a dog's mental functions could be controlled only by external stimuli. And expanding this idea to human mental functions, I feel that there was a time when we thought that human mental functions could also be controlled by external stimuli. Certainly, this expanded interpretation has become a concrete way of scientifically verifying human behavior. On the contrary, it also proved that the mental functions of humans were never just such simple reactions as Freud's research had showed.

One more point: I would like to think about how we can avoid such beliefs and dogmatism.

I think the only way to approach this is to take a closer look at one's own intuitive reactions and the reactions of others. The author believes that humans are constantly striving to make sure that their own thoughts and the thoughts of others are reasonably consistent. I believe that the cause of these tragedies was that the dogmatism of those scientists came about when they denied human subjective reactions due to their extreme interpretation of dialectical materialism, and they also denied their own subjective reactions. Recall their self-contradictory statements: "My consciousness does not exist."

I am revisiting Pavlov's experiment in this section because it addresses the issue of the self in humans. The term "the self" is a topic taken up in both philosophy and psychology, like the terms consciousness and the mind, and is thought to be a difficult subject to deal with scientifically. I do not want to bring up Husserl's concepts of phenomenology here again. Rather, I wanted to consider this issue, as I have discussed the self up to now, as though we were treating consciousness and the mind as an information program. In short, what I want to try to do is to express "the self" as a conscious program. Various experiments (Sections 11.1–11.5) using a conscious system have been presented so far, such as A Robot Cognizes the Unknown, A Self-Aware Robot, Robots with Episodic Memory, A Pavlov Robot, Color Vision Capabilities of a Conscious Robot, Principle of Pleasant and Unpleasant Feelings in a Robot, About Absolute Pleasantness, and Conflict of Concepts and Rubin's vase Model. I wonder where the term "the self" would fit in these experiments. There is no suitable place anywhere in the conscious system where the term "the self" fits.

So, the author gave a suggestion to the students in a research group.

At the time, a head robot project that could reproduce the movements of a human head was active.

The timing was right, and this head robot was capable of almost the same movement as a human head. The robot was made of aluminum alloy and had almost the same shape as a human skull and cervical vertebrae. All parts of the robot were handmade by the students and the author (Fig. 11.27). We made a wax model of the parts, and then a lost-wax process was used to cast the aluminum alloy robot. We outsourced the lost-wax process to an expert in aluminum casting. The robot neck (Fig. 11.27, ①) was made thicker than a human neck and it was about five times thicker. A human neck has thick muscles that support the weight of the skull, but this robot head uses steel cables in place of the muscles. For this reason, the neck itself is thicker to support the weight of the skull. The structure of the neck also imitates human morphology.

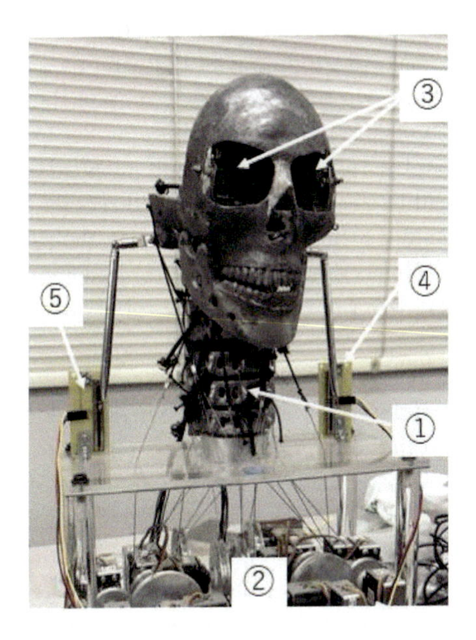

Figure 11.27 The Robohead. ① Robot neck, ② servomotors and cables, ③ two video cameras, ④, ⑤ inclinometers and rods.

In addition, the motion of the skull is driven by pulling wires connected to each part using a large number of servomotors installed below the stand on which the head (skull and neck) is mounted (Nakaguchi and Takeno, 2007). Two video cameras are installed in the eye sockets of the head robot where human eyes would be (③) to enable the robot to have forward vision. And additionally, inclinometers ((④, ⑤) using slide resistors) and rods that are used in the experiment described in this section are installed on both sides of the head. The inclinometers and rods connect the head of the head robot and the stand it is mounted on. The inclinometers measure the degree of inclination of the robot head. Hereafter, we will simply refer to this head robot as Robohead.

The basic part of the conscious system of this robot is used to perform imitation behavior. The conscious system incorporated in the robot implements the imitation behavior used in the mirror cognition experiments described in Chapter 10. And certainly some changes had been made to the system. The robot described in

Chapter 10 imitated the movement of its self-image reflected in a mirror (moving forward, stopping, and moving backward). For the robot in this section, we tried to have it imitate the left and right movement of a subject (a person) in front of it. For example, when I bring my face near the robot's head and tilt my head to the right, the robot tilts its head to the left, which is the direction of its own line of sight. Conversely, if I tilt my head to the left, the robot will tilt its head to the right. This is the so-called mirror behavior. Simply put, if I shake my head in front of the robot, the robot also imitates me by shaking its head.

Now I would like to discuss the experiment. I tilted my head to the right in front of the robot. The robot then imitates me and tilts its head to the left. At that time, I continue to tilt my head and gradually reach the limit at which I am not able to further tilt it to the right. My neck does not hurt when I reach the limit, but it feels cramped when being pressed. Of course, if I tilt my head more, I would begin feeling pain. In that state, if you look at the posture of the robot, the robot will be tilting its head while looking at me. If I stop tilting my head, the robot does the same, because the robot is imitating me.

However, let us suppose that in this state the robot is slightly less able to tilt its head than I can tilt mine. At this time, the robot's behavior is inadequate for imitating the inclination of my head. Put another way, the robot's head reached its inclination limit before my head reached its limit.

The conscious system installed in the Robohead at the beginning of the experiment uses four MoNAD consciousness modules: IB, ACT, Pa, and As (Fig. 11.28). The As′ MoNAD is automatically regenerated during the experiment.

The IB consciousness unit detects the movement of the subject's head as seen by the robot's vision system and imitates it. Information on the movement of the subject's head is sent from Input1 to IB. The vision system first detects the skin-colored area of the subject's face in front of the robot that can be observed through its cameras, and calculates the behavior of the subject using information on the movement of the center of gravity of that area. For example, if the vision system detects that the subject in front of the robot has tilted its head to the left with respect to the direction of the robot's line of sight, it sends that information from Input1 to

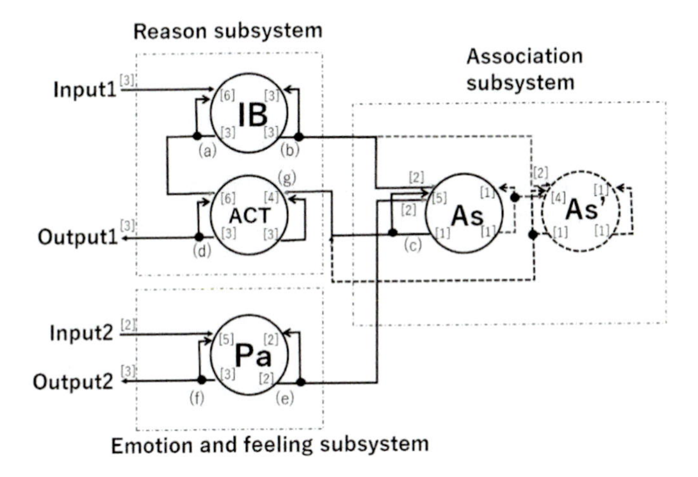

Figure 11.28 A conscious system of the Robohead.

IB. Then IB immediately calculates the imitation behavior and sends information to the ACT consciousness unit via the output terminal (a) of IB to tilt the robot's head to the right at that time. ACT is the MoNAD that actually commands the servomotors to perform the imitation behavior. At the same time, IB sends its cognitive representation to the As consciousness unit via the terminal (b). In other words, As receives information from IB that "the subject is tilting its head to the left in the robot's line of sight, so the robot is imitating this behavior." And normally, As uses that information and sends a signal from its output terminal (c) to the most significant bit of the ACT behavioral representation. (In the experiment, the actual value of the signal is 0.5 or less, that is, the logical value is 0.) ACT then combines the signal with its cognitive representation (3 bits) to determine the behavioral representation of ACT (for a total of 4 bits). As a result, ACT outputs a command from Output1 at the output terminal (d) to cause the robot to tilt its head to the left. This command causes the various servomotors that drive the robot's head to rotate, which pulls on the cables and makes the robot's head tilt to the left. Certainly, if the subject stops tilting its head, the robot's head will also stop tilting. Also, if the subject changes the position of its head from the left side and tilts it to the right side in the robot's line of sight, the robot's head also changes and tilts to that side. In addition, there may be a state in which information for ACT "to not perform the imitation behavior" is output from the output terminal (c) of as (when the behavior by ACT is suppressed). This is an essential part of this experiment and will be described in detail later.

The operator conducting the experiment can judge how well the robot is usually performing the imitation behavior during the experiment. This judgment can be made by visually inspecting the robot's behaviors as the experiment progresses as well as by using objective experimental data on the reaction of each of the consciousness units in the conscious system installed in the robot.

Now, let us discuss the new state that was mentioned earlier.

The new state occurs when the robot is not able to imitate the tilting of the subject's head. For instance, when the head of the robot performing the imitation behavior is blocked from moving by a third party, or when the imitation behavior is blocked by the physical constraints of the robot itself, and the third party is a problem with the conscious system itself. The case of a second party is considered in this experiment. The first party will be a subject to be taken up in future research. We will consider the third party later in this section.

The physical constraint of the robot is the limit to the range (called the degree of freedom) in which the robot can freely move its head on the body. A human, for example, is not capable of fully rotating the head (360°) around the spinal cord.

For instance, in this experiment, there is a limit on the inclination angle of the robot's head on its own body while performing the imitation behavior. The head robot used in this experiment simulates the human skeleton, and as such, its movement is similar to that of humans with the same physical constraints that humans have.

Now, we can see that in that new state the robot's imitation behavior has stopped in response to the subject's behavior (tilting to the left when viewed from the robot's line of sight). However, looking closely we can see that the robot's head has not stopped but is vibrating a little. High-speed vibrations appear to have stopped completely.

The cause of this vibration is behavior involving the Pa consciousness unit, which is part of the conscious system. Pa is the consciousness unit that represents the "pain" of the robot. There has been considerable debate about the "representation of pain" in robots. The limit point of the inclinometer installed on the body is regarded as the pain point in this experiment, and when the tilting of the robot's head reaches the limit point of the inclinometer, the robot expresses the pain as a cognitive representation of Pa. I can imagine the criticism that this explanation is too simplistic is likely to appear without delay. However, if we think of pain points as information integrated from multiple end sensors, I think this simplification is possible. Also, the author has already described one hypothesis for the explanation of mental pain (see Section 10.8.2, "The Mirror Box Therapy"). Simply put, that hypothesis positions the disruption of information in neural pathways (e.g., a chaotic state) as being associated with the development of "pain."

Pa continues to represent pain in the experiment while the robot's head is vibrating slightly (Fig. 11.28e). What is the cause of this vibration? It depends on the functioning of the Pa consciousness module. Pa obtains information from the inclinometer of the head upon reaching the limit point (as an anomaly), and at the same time as Pa is representing pain (e), an escape signal is output from Output2 of Pa (Fig. 11.28f) to the motors driving the robot head. This escape signal is an avoidance reaction by the so-called conditioned reflex. At the same time, Pa also sends a representation of pain to the As consciousness unit. Detection of an inclinometer anomaly here means that the inclinometer has reached the physical limit of the tilting of the robot's head. In other words, if the robot's head were to tilt any further, the robot's own neck would be physically damaged. Here, the escape signal sent to the drive motors means that, when the inclinometer (Fig. 11.27, ④) on the left side of the head (from the robot's line of sight) detects an anomaly, a motor command (escape behavior) signal is sent from Output2 to slightly back off and return the tilting of the head in the direction away from the limit point of the inclinometer. Since Pa is not merely a sensor but a consciousness module that represents pain, it has already learned to cause the above reaction upon a signal from Input2.

At this time, some readers may be thinking that something here is strange. This is because Robohead was performing an imitation behavior using the IB consciousness unit, but it has now stopped. In short, the IB unit performed an imitation behavior at first, but due to the physical constraints of the robot itself, the imitation behavior could not be continued.

"Ah, while the robot was imitating, it stopped when it couldn't continue the imitation behavior."

Yes, that is correct. However, although the head of the robot does appear to have stopped, if we look at it closely, we can see that it is actually vibrating slightly back and forth. So then readers might say something like, "Well, the robot is just programmed to do that." And that, too, would be correct.

I would like to explain in somewhat more detail how this repetitive behavior of the head occurs.

All in all, the explanation comes to this: Two consciousness units (IB, Pa) are functioning alternately in this state.

The IB unit is performing the behavior to "imitate the subject," and the Pa unit is performing the behavior to "escape from the pain." The question is, why is the repetition of these behaviors occurring?

When the escape behavior is performed upon the output from Output2 of the Pa unit, the robot's head can back away from the physical constraint, and the Pa unit will no longer represent pain. And then the imitation behavior that has not yet been performed can continue again. This happens because Pa is no longer representing pain as a result of the escape behavior. The IB unit can then perform the imitation behavior once again.

However, the subject in front of the robot keeps tilting their head further to the left in the robot's line of sight, and although the robot's head has backed away from its physical constraint, since the head can be tilted to the left by the IB unit again, the robot immediately reaches its physical constraint, the limit of its inclination, once more.

At that time, the Pa unit will once again represent pain and perform an escape behavior.

This is the cause of the small back-and-forth vibrations of the robot's head when reaching the physical constraint.

If we set the Pa unit to turn on a red LED when it represents pain, we will be able to observe the red LED turning on and off continuously for small periods.

Now, I would like add a slightly more detailed explanation of the As unit for better understanding.

The As unit has been explained many times, but allow me to briefly explain it here again.

The As unit is a consciousness unit in what we call the Association subsystem. This subsystem settles—it coordinates—between the Reason subsystem and the Emotion & Feelings subsystem.

In an usual conscious system, when a pain signal is sent from Pa to As (Fig. 11.28e) as a basic response, that signal causes other Reason subsystem processes, and as a result, various behavior commands are sent directly to the motors from the output terminal of the As unit (Fig. 11.28c). In this experiment, various behavior commands to the motors can be executed by sending some kind of signal to the ACT consciousness unit from the output terminal of IB (Fig. 11.28a). In short, the behavior commands from the Reason subsystem are performed via the ACT unit in this experiment.

The purpose of this is to make it easier to change the method of behavior by As through various behavior commands of the Reason subsystem via ACT.

I would like readers to understand that this behavior is different from the escape behavior mentioned earlier. In other words, the escape behavior is a behavior that does not involve the Reason subsystem, such as a reflex behavior.

However, there are also behaviors that involve the Reason subsystem using the representation of pain that are sent to the As unit in this way.

If we were to describe this function in terms of the behavior of the robot, the IB unit normally performs the imitation behavior, but when the Pa unit of the Emotion & Feelings subsystem functions, the robot will follow the Pa behavior. This means that this conscious system is programmed to prioritize the processing of the Emotion & Feelings subsystem behaviors over the Reason subsystem behaviors. That is how human programmers created the program. We might be criticized for programming it arbitrarily as we desired, but basically, that is the way it was made. There is not much evidence of this priority as of yet, but it seems that living organisms attempt to prioritize their own existence. Sigmund Freud also stated that the self-preservation instinct is preferentially acquired in the process of evolution (Freud, 1915). To begin with, I think that the basic design should give priority to the body of the self. And, certainly, humans also have self-sacrificing behaviors, so I think this is rather difficult to explain easily. I think that self-sacrificing behavior is a very interesting conscious behavior, but I would like to take it up as a theme for future research. I suppose that self-sacrifice can possibly be explained as a high-level conscious function related to morals that was learned very recently rather than during human evolution.

Now, let us continue to consider these vibrations that appear as though the imitation behavior has stopped.

Remember that when the robot's head is vibrating, the conscious system is continuing to represent pain repeatedly. The red LED is blinking repeatedly. And these vibrations never stop, and they would repeat infinitely. In the real world, the subject who is tilting their head in front of the robot may notice the continuous blinking of the robot's red LED and stop tilting their head. When the subject does that, the robot's head will stop vibrating, and the robot will be able to imitate the tilting of the subject's head again. In that way, the robot is in a good state to imitate the subject, and if there were a green LED on the head that indicated a comfortable state, the green LED would light up continuously.

Now, let us continue on to our main subject from here.

That is to say, the problem that we have is what to do when the head stops, the red LED is blinking, and the robot is not able to imitate.

Here the author and his student research group decided to tackle a new challenge.

We decided to try to find a way to have the robot resolve this situation by itself.

The reason for doing this is that the robot's conscious system has a mechanism for perceiving the representation of pain by itself, and we thought that this robot could devise a means to resolve its own difficult state by using that representation. However, since the problem involved the robot itself having a creative function, we thought it best to first focus in on a way into the robot's creative function.

An idea came to me at this time.

That idea was that, if a state of difficulty occurs inside the robot, the conscious system of the robot should act to generate a new consciousness unit by itself. In fact, we have already adopted this idea in previous studies. For instance, see Section 11.1, "A Robot Cognizes the Unknown," and Section 11.5, "Color Vision Capabilities of a Conscious Robot." In the former study, we demonstrated that when new information that had never existed in the robot before was input to a consciousness unit in the conscious system that had already learned some information, the consciousness unit was able to detect that that input information was unknown information. And, in the latter study, we demonstrated that the unknown information could be added to the conscious system as a consciousness unit and as new information. In short, robots can learn new information that they do not yet know and can use that information as their own knowledge.

Now, let us return to the topic where the robot seems to stop while performing the imitation behavior.

At that time, the conscious system of the robot detects the representation of pain Pa (e) and temporarily avoids it by using the Pa's own escape behavior (a reflex behavior). However, this avoidance allows the IB consciousness unit to perform the imitation behavior again. Then, the conscious system represents pain once again when pain is caused by that imitation behavior. And yet again, the escape behavior occurs, and this situation repeats without ending. Although the tilting behavior of the robot head appears to have stopped, in reality, the representation of pain inside the robot does not stop. The red LED blinks continuously. And, in principle, the robot cannot stop the continuous representation of pain in this state with the conscious system as it is. Put another way, this state can be said to be a type of conflict phenomenon between the imitation behavior of the IB consciousness unit and the pain of the Pa consciousness unit.

The As consciousness unit of the Association subsystem was designed so that the behaviors of the IB and Pa units would not conflict, but it was not able to cope with the unavoidable state of continuous pain.

Here, the problem is that the current difficult state of the robot is the robot's own subjective problem, with continuous representations of pain and its head making slight vibrations (a pseudo stop). The conscious system of the robot can subjectively infer this state from its own signal flow. When observing the appearance of the robot, an experimenter can only determine that the head has artificially stopped. However, since the experimenter has sufficient knowledge about the internal workings of the robot, by using electronic devices to observe the signal flow, the state of the sensors, and the state of the motor drives inside the robot that are operating in the conscious system, the experimenter can subjectively understand the difficult state of the robot.

Of course, the experimenter can also make modifications to the robot to resolve this problem. In addition, if the subject standing in front of the robot stopped the imitation experiment and moved the robot head upright (repositioned it so that the left and right eyes are horizontally aligned), the difficult state of the robot can be temporarily resolved.

However, here the conscious system of the robot is in a state where it can perceive the internal state of the robot on its own, so in a sense, this robot may be able to resolve the problem on its own. In other words, this robot has the potential to be self-aware and transition from its difficult state to a better state. In a manner of speaking, this method can be called self-aware intelligence (SAInt).

In other words, if we consider the current difficult state of the robot using SAInt, since the As unit is still functioning inadequately, it would be a good idea to subjectively generate a new consciousness unit in the Association subsystem.

Since experiments on self-development have been conducted with this conscious system, if this principle is used, the use of an internal signal indicating the difficult state of the robot can be considered as a method for generating a new As′ consciousness unit inside the Association subsystem (Fig. 11.28).

To do this, first of all, the As unit needs to be changed so that it can learn dynamically. Looking back, the As unit has always been a prelearned unit up to now.

The As unit is in the Association subsystem and it acts as a mediator coordinating between the Reason subsystem and the Emotion & Feelings subsystem. In this experiment, As basically outputs a signal (actual value is 0.5 or less) to cause the imitation behavior to be performed by IB (Fig. 11.28c). The logical value of the output signal is 0. In the experiment, pain is represented by

Pa. This representation of pain is sent to As as one input value (Fig. 11.28e). Then, at the same time that Pa is representing pain, an escape behavior is performed caused by the output from Pa. At this time, the signal that allows the imitation behavior to be performed by IB is still being sent from the output (c) of As, but the escape behavior performed by Pa in the Emotion & Feelings subsystem is given priority. As mentioned earlier, the behaviors of the Emotion & Feelings subsystem have a higher priority than the behaviors of the Reason subsystem. And, the escape behavior of Pa causes the representation of pain by Pa to disappear. As a result, As will once again be able to enable the imitation behavior to be performed by IB. I think our readers can understand that this imitation behavior causes the representation of pain by Pa to reappear. This repetition of the escape and imitation behaviors causes the continuing occurrence of pain as mentioned above.

If As is not capable of accepting a dynamic stimulus at this time, the continuing occurrence of this pain cannot be stopped. For that reason, As was changed in this study to dynamically accept changes of the input value instead of making As a learned unit. In other words, As is an unlearned consciousness unit in the initial state in this study. And As has a mechanism to develop itself so that it gradually accumulates its learning. That is, As develops according to certain learning rules. As learns to gradually increase the value output from its output terminal (c) while Pa is continually representing pain. Specifically, when the output from (c) exceeds an actual value of 0.7 (the logical value is 1), As advances to a new level.

After As is enabled to recognize multiple representations of nearly continuous pain, the new level is used to generate a new consciousness unit termed As′, which has the role of assisting As. This is done because As alone cannot yet avoid the recurring representation of continuous pain.

Well, then what sort of role should As′ have?

Let us look back again at the difficult state the robot is in. The robot reaches the point of its own physical constraint while performing the imitation behavior, cannot continue imitating, and thus represents pain. And, when the pain occurs, it causes an escape behavior that backs the robot head away from the constraint, which temporarily relieves the pain. But then the relief from that escape behavior allows the robot to perform the imitation behavior again, which in turn causes the pain to occur again when the physical constraint is reached once more. This continuous alternating repetition of the escape and imitation behaviors causes the robot head to be in a pseudo-stopped state, but the representation of pain is occurring continuously.

To repeat, in other words, the IB consciousness unit that performs the imitation behavior and the Pa consciousness unit that represents the pain are repeatedly acting in turns.

Now, we will look at how the new level develops.

First, the As unit is made capable of recognizing continuous pain. For this reason, although As has been used as a learned unit in the past using supervised learning, here it was set to continue reinforcement learning for the MoNAD functions. In other words, we tried to implement reinforcement learning on As according to the cognitive state of the other MoNADs. That is, when Pa is representing pain (R1 condition) at the same time that IB is enabling an imitation behavior, reinforcement learning is performed on the cognitive representation of As setting it to 1.0 (actual value, teacher data). When pain is not being represented, learning is continued with the value of 0.5 (actual value, teacher data). Therefore, the more Pa represents pain while IB is performing the imitation behavior, the more the cognitive representation of As rises to 1.0. Additionally, when this R1 condition is not met, the value of the cognitive representation will drop. Also, this learning does not end and continues to change the value of the cognitive representation of As. Therefore, by observing the value of the cognitive representation of As, it is possible to find the state where the R1 condition is continuous, that is, where the conflict mentioned previously is occurring. Now, here is one more point. I have stated that As′ is regenerated when the cognitive representation of As approaches 1.0. At that time, the logical value of 1 is sent from the output of As (Fig. 11.28c) to BEH[0], which is the behavioral representation of the ACT MoNAD (Fig. 11.28g). This signal suppresses the imitation behavior command that is being output from the IB unit, and as a result, it temporarily suspends the execution of the imitation behavior by ACT. In addition, keep in mind that since this state is caused by a repetition of conflicts, the robot has already completed its escape

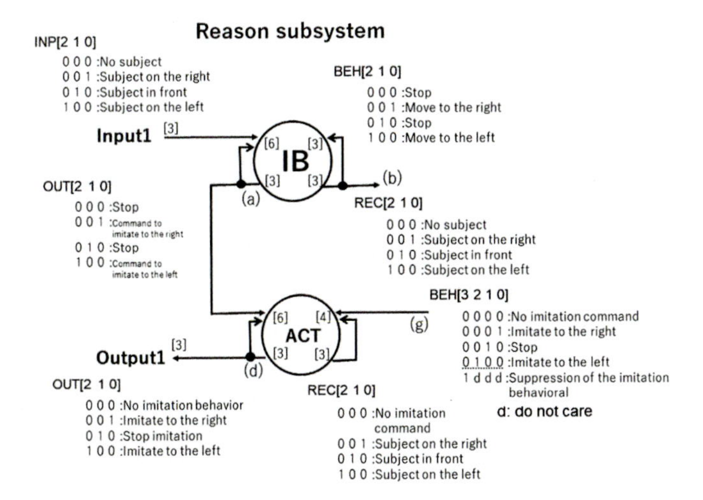

Figure 11.29 The Reason subsystem of the conscious system.

behavior caused by the representation of pain by Pa. In other words, at the time that the logical value of 1 is sent from the output of As to BEH[0], the robot has already backed away from its constraint to a point where it has escaped from the pain (that is, the point at which the pain does not occur). In fact, this point is the learning point for As′ and this will be described later.

Next, the regeneration of As′ and the implementation of its learning will be explained. As′ is a new MoNAD that is automatically regenerated by the conscious system when As detects the above-mentioned conflict.

The learning of the newly regenerated As′ is implemented as follows.

First, the reinforcement learning of the cognitive representation of As′ will be explained.

The teacher data for the cognitive representation of As′ is reinforced to 1.0 if the following R′1 and R′2 conditions are satisfied at the same time (teacher data is provided for learning). Otherwise, the cognitive representation of As′ is reinforced to 0.5 (teacher data is provided).

Condition R′1: When As is in a state where As′ is regenerated (the cognitive representation of As is REC[0] $>$ 0.7), or where the output

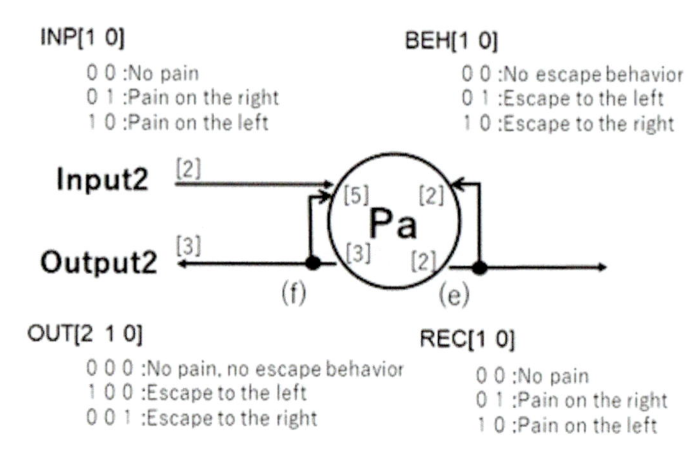

Figure 11.30 The Emotion & Feelings subsystem of the conscious system.

of the newly generated As′ has increased significantly (the output unit of As′ is OUT2[0] $>$ 0.8).

Condition R′2: When the subject in the experiment is on the left side (the cognitive representation of IB is REC[2] $>$ 0.7) or the subject is on the right side (the cognitive representation of IB is REC[0] $>$ 0.7).

Then, if the following R′3 condition is satisfied, the output value of As′ is set to 1.0 by reinforcement learning (teacher data 1.0 is provided). Otherwise, the output value of As′ is reinforced to 0.5 (teacher data is provided).

Condition R′3: When the behavioral representation of As′ is already large (the behavioral representation of As′ is BEH[0] $>$ 0.5).

Now, let us consider the condition for using As′ to avoid conflicts.

This condition is as shown in the following. This condition is in Lo1 (condition: R′4) of the logic system which is provided as a part of the Association subsystem (Fig. 11.31). This logic system does not currently use a consciousness unit, but it is possible to use a MoNAD in the future.

Condition R′4: When the pain avoidance function has been acquired on either the left or right side, an imitation behavior is being performed in that direction (the cognitive representation of

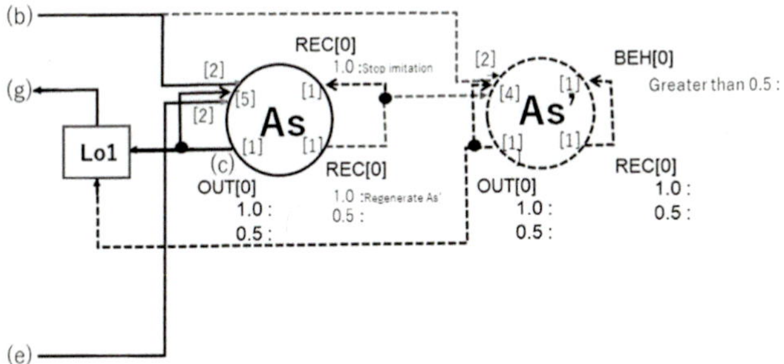

Figure 11.31 The Association subsystem of the conscious system.

IB is REC[2 or 0] $>$ 0.7, and the perception of pain is approaching (the cognitive representation of Pa is REC[1 or 0] $>$ 0.55, a signal that suppresses the imitation behavior is output from As′ and sent to ACT. Otherwise, the signal is output from As and sent to ACT.

Here, I would like to illustrate the information contained in the input and output representations, cognitive representations, and behavioral representations of each MoNAD in the following figures in order to deepen the reader's understanding.

The figure of the experimental results will be explained in the following.

When the experiment is started, the robot initially detects the subject (the skin-colored area of the subject) on the left side in the robot's line of sight (Fig. 11.32f) and performs an imitation behavior to the left side ((e) in the figure). However, the robot detects pain on the left side of its neck and the Pa unit causes an escape behavior. This escape behavior temporarily relieves the pain (Fig. 11.32(1)). The escape behavior caused by Pa occurs continuously for about 400 steps after that ((a), (d) in the figure). The reason for this is that the subject in front of the robot is still on the left side in the robot's line of sight ((f) in the figure), and ACT continues to perform the imitation behavior ((e) in the figure). The robot temporarily relieves the pain by means of the escape behavior caused by Pa, but since ACT is continuing to perform the imitation behavior, continuous pain occurs and a conflict comes about. The state of conflict is

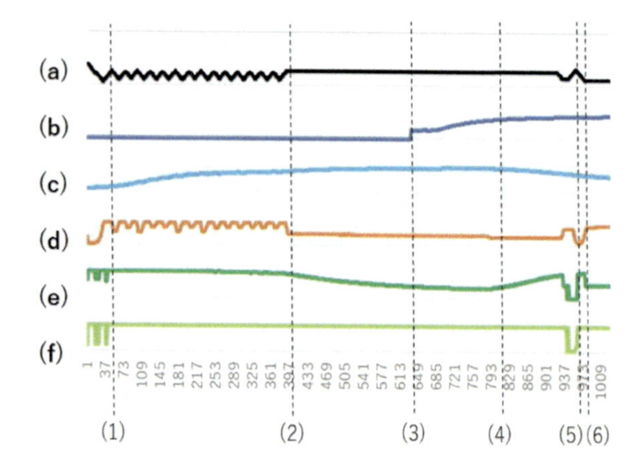

Figure 11.32 Results on the experiment. (a) Quantification of the neck position: If the number is large, escape to the right. Otherwise move to the left. (b) Output of As′: Stop the imitation behavior. Initial value is 0.0. (c) Output of As: Pause the imitation behavior at a value close to the actual value 1.0. Regenerate a new As′. (d) Cognitive representation of Pa: Pain on the left side of the neck. (A drop in value reduces the pain.) (e) Output of ACT: Perform imitation behavior to the left side. (f) Cognitive representation of IB using the left eye. (The subject is on the left.)

represented by the intense vertical movement of the signal shown in (a) and (d) of the figure up to about 400 steps.

Next, take notice of the change in the signal shown in (c) of the figure. This signal shows the learning state of the conflict by As. That is to say that, if a conflict occurs, the cognitive representation of As learns 1.0 as teacher data. The learning of As is constant during the experiment. The cognitive representation of As gradually increases in value while approaching 1.0 ((c) in the figure) because of the conflict.

As a result of this increase, the conscious system regenerates As′ and begins the learning (Fig. 11.32c(3)).

Next, let us take a close look at (b) in the figure. The line of (b) shows a sharp increase in the number of steps at around 640 caused by the change in the output value of As′ (Fig. 11.32b(3)). This value continues to rise for some time, and approaches 1.0. Then, the learning of As′ completes at around 800 steps ((b), (4) in the figure).

That is, the learning of As progresses over the number of steps from (1) to (3) ((c) in the figure), and as a result, As′ is regenerated and the learning of As′ is completed. As mentioned earlier, As is learning constantly until the end of the experiment in order to detect other conflicts. When other conflicts are detected, new As′ units are regenerated.

Next, let us focus on the steps beyond step 900.

At this time, the subject (the person conducting the experiment) temporarily returns the head to the left side from the position in which the head was tilted to the right in order to check the function that completes the learning of As′ (Fig. 11.32a(5)). This is done by having the robot imitate again (the robot tilts its head to the left) and checking whether the robot is capable of stopping at the position just before the conflict occurs without causing the conflict to be repeated.

The subject tilts their head to the right again and causes the robot to imitate once more from around step 970 (Fig. 11.32(6)). However, as a result, after (6) in the figure, although the subject continued to tilt their head to the right ((a), (6) in the figure), the imitation behavior was not performed ((e), (6) in the figure), and the robot did not represent significant pain ((d), (6) in the figure). That is to say, the imitation behavior command was stopped immediately before As′ functioned to represent significant pain (Fig. 11.32(6)).

The number of steps in this experiment is the number of counts of the "cognitive behavior cycle" in the conscious system.

Finally, I would like to discuss what we intended to demonstrate in this study.

As expressed in the theme of this research, this was an attempt to build a conscious model that explored what triggered the development of the self in humans.

This experiment presents a story in which, while the conscious system is imitating the behavior of another, a state occurs in which the imitation behavior cannot continue to be performed properly. As mentioned earlier, while an imitation behavior is being performed driven by the conscious system, this state may be caused by three factors: 1) when the body of the self is disturbed by an external cause, 2) when a disorder originates from an internal cause in the body of the self, or 3) when some kind of failure occurs in the conscious system itself. Simply put, this can be initially described as an external interference by another, an internal disorder within one's own body, and a mental disorder of the self. The first factor will be postponed for future research. The second is taken up in the first half of this section. And research related to the third is introduced later in this section.

Now, positioned in the middle, the conscious system has no choice but to distinguish between the existence of the self and the other, and this is characteristically caused by the physical differences in the bodies of the self and the other. These differences cannot be easily resolved, so some sort of solution within the conscious system is likely to be needed. This section considers the implementation of imitation behaviors output by an imitation MoNAD in the conscious system and cases in which a pain MoNAD in the conscious system continuously generates escape behaviors (called conflicts). The conscious system itself gets into a state in which it is not possible to solve the problem as it is. However, the Association subsystem of the conscious system first learns the conflict and generates a new additional Association Sub-subsystem, and this new MoNAD solves the problem by learning to output a signal to stop the imitation behavior command from the imitation MoNAD. Since the original role of the Association subsystem is to mediate the behavior of the Reason subsystem and the behavior of the Emotion & Feelings subsystem, I think it is natural for the Association Sub-subsystem to learn this sort of means for solving the problem. In addition, I think that the rule that gives priority to the behavior of the Emotion & Feelings subsystem when choosing between the behaviors of the Reason subsystem and the Emotion & Feelings subsystem is natural, even when considered in light of Freud's theory on "the drive to exist." And from what it means that this newly generated Association Sub-subsystem gives priority to its own Emotion & Feelings subsystem, I think this can be interpreted as the beginnings of the self in the conscious system. That is, I believe that the experiment described in this section demonstrates an expression of a program that represents a small part of the self.

Now let us move on to our next theme.

In this section, too, we will take up the three factors that may impede one's behavior. The first is an interference by another, the second is when the limitations of one's own body are reached, and the third is the abnormal development of one's own program. The second half of this section considers this third problem.

What sort of obstacle would cause this problem in humans?

The author has held a consistent idea that human consciousness and the mind originate from the state in which information is circulating between the nerve cell networks in the brain and nerve cells in the body that are connected by nerve fibers. The author has also developed a theory in which a special network of nerve cells composed of consciousness units called MoNADs can form the basis of human consciousness. And we have already stated in this book that multiple combinations of such consciousness units can express a basic model of human consciousness. When looking at the development of human consciousness, we can see that there are some general characteristics. First of all, humans appear to have completed their basic development inside their mothers' bodies. Humans begin breathing on their own with a loud cry upon birth as they leave the mother's body. The child's body develops significantly while being lovingly taken care of by the parents. When observing the child's condition, it appears that the nervous system is also greatly developing. There is a well-known study on child development by the Swiss psychologist Jean Piaget (1896–1980) (Piaget, 1952). Piaget stated that children develop by being stimulated by the environment of the outside world around them.

Although it is clear that humans develop various personalities by the time they become adults, most of them appear to develop smoothly and without much difference. The basic part of the nerve cell network mentioned previously is said to have already completed its development within the mother's body, and then the network development expands with learning and the unused portion degenerates (Bloom et al., 1985). From these observations, it appears that human development is mostly dependent on the network of body and nerve cells that humans possess during evolution.

For this reason, our bodies and nervous systems may be considered to have a common foundation in almost all of basic parts. This point has already been mentioned in this book. However, since we can also learn and enjoy new sports, this of course means that we can learn new things. For example, when learning to surf for the first time, you might find it completely impossible to stand on a surfboard on a moving wave, but remember that if you practice the basic movements on the beach, you will gradually learn how to surf and can enjoy doing it. This is clearly the body and the nervous system learning a new sport, in other words, learning something new.

Now let us go back to our previous topic.

Various pathological conditions known as abnormalities of the nervous system have been observed in humans.

As a historically famous example, after experiencing the trauma of combat with shells exploding around them, soldiers in the trenches during World War I later developed a number of conditions that were called shell shock (Hart et al., 2006) at the time.

Those patients experienced violent convulsions, twitching, and stiff facial expressions, and some were incapable of standing and walking.

At that time, patients were even accused of lying in order to escape their military duty. The condition was also attributed to a "lack of moral fiber" and the horrible methods used to treat it included physical restraint and electric shock. Doctors are said to have argued whether the condition originated from the body or from the nervous system of the brain.

These symptoms are now considered to be caused by trauma. Many soldiers are thought to have experienced similar severe trauma during the Vietnam War and the Iraq War.

Additionally, as I have already mentioned, Austrian psychiatrist Sigmund Freud authored a large book on the symptoms of hysteria that were seen in many women at the time. Hysteria is not yet a medically defined condition, but it is said to be a state of significant distress and dysfunction due to mental anxiety.

In any case, it appears that if a person is experiencing a serious condition affecting them physically and a doctor determines that it is greatly impeding their living a normal daily life, the symptom is defined and treated as a disease. *The Diagnostic and Statistical Manual of Mental Disorders*, Third Edition (DSM-III) published by the American Psychiatric Association is a well-known diagnostic manual in use at this time (Barnhill, 2014).

Such mental disorders have been manifested for about 100 years, but the author feels that despite the earnest efforts made in diagnosis and treatment by doctors, the pathogen is currently not yet clearly understood.

As such, the author came up with an idea.

Perhaps the author's current research on artificial consciousness might be able to contribute not only to modeling human consciousness but also to modeling human psychiatric disorders.

For example, we thought that by constructing the symptoms of hysteria as a conscious system using a robot as an experimental tool, we could approach a fundamental elucidation of such symptoms.

In other words, could we examine the questions of why hysteria develops and how to treat it effectively by utilizing robots and their conscious systems?

I thought that if we were able to implement such measures, the symptoms of hysteria could at least be expressed as a conscious system and a program on a robot, and we could prove that those symptoms actually exist as a program on the robot.

It could be said that the symptoms existed as a conscious program.

But then, it seems there might be a possible counterargument, which would be something like, "since the program does not exist as something of physical substance, its existence could not be proven." However, as I mentioned earlier, such a statement would not work today, because we commonly say things like, "I have a program in my computer."

Also, since the conscious program is a neural network program, is the author the only one who thinks that it might offer important clues in the examination of the cells of the human brain and why those networks might cause the symptoms of hysteria?

Now let us consider a specific problem.

When conducting research with his students, the author considered a certain research theme.

The author decided to use the conscious system to study multiple personality disorder (MPD), which is now called dissociative identity disorder (DID). This disorder is very mysterious and is a serious illness.

The mentality of a healthy human being is unified with one personality or one ego. In my case, I'm a man and I'm already an elderly person. However, a patient with such a disorder might be a middle-aged man when he is healthy, but at a certain time his mentality shifts to that of a three-year-old infant, and his attitude and words behave like an infant. In the younger generation, this transition is less likely to become apparent and cause problems when they are with their family. However, when they go out into the world and are active in society, such as when getting a job, they might seek help at a medical facility when confusion occurs with the people around them. Patients are often diagnosed with DID by psychiatrists. One example of an expression that describes what happens is that "the driver of the car takes on the personality of another, and the driver up until then now gets into the back seat and goes to sleep." In short, the patient is in a body that is being shared, and the observed symptoms show personalities with different mentalities possessing that body. Some doctors might claim that "the patient is just acting." Certainly, it would seem natural to ask the question, "Are the patient's symptoms real or it this just acting?" However, even if the patient is just acting, if the person is in fact complaining about the symptoms, it is the kind of condition that requires treatment and the doctor would diagnose it as a kind of illness. Curiously, it seems that it is difficult for the afflicted person to be aware of their own symptoms. There is a record of a patient being very surprised upon seeing a video of their symptoms that had been recorded earlier by a doctor.

That patient was often subjected to some form of abuse from early in their childhood, and there was testimony that long before the development of their personality, the patient continually experienced emotions that caused severe discomfort (Putnam, 1989).

Since I have been studying and considering models of human consciousness with my students for a long time, I came up with the idea that the DID symptoms might be easy to model using the conscious system I developed. I think that the conscious system has the expressive capability to be used as a model of human consciousness by combining the Reason subsystem, the Emotion & Feelings subsystem, and the Association subsystem. I think that the readers can understand this to some extent from the description in this book up to now. Nevertheless, let us check this again. The first factor mainly acquires information from objects at long distances (e.g., using visual and auditory sensors). The second factor mainly acquires information from sensors inside the body (including on the surface of the body). Simply put, the first factor produces rational behavior. The second factor produces emotional behavior. The behavior of the second factor includes reflex behavior. For example, the escape behavior to move away when one gets burned.

The third factor is a mediator between rational and emotional behavior. Innately, the latter behavior gets priority. Speaking of "innately" with regard to robots might be a strange expression, but as we use this term with living organisms, we use it as an analogy to robots.

Now let us consider the third factor a little more.

If the third factor is a mediator between the rational behavior and emotional behavior, I think that in human terms this can be explained as the role of a personality and the role of an ego. For instance, if you look at the people around you, some will express more emotional behavior, while others will be rational and thoughtful.

Another point to add is that the various subsystems are made up of a combination of consciousness units called MoNADs. A MoNAD is a special neural network with a double structure of recurrent neural networks, and is called a consciousness unit because it can be used to explain most of the characteristics of human consciousness (see 9.4–9.7).

Now, when we look at the conscious system with regard to the symptoms of DID as mentioned above, we can notice a number of points. In other words, if the Association subsystem can be expressed as one personality, when considering the symptoms of DID with a model of consciousness, the dissociation of the personality means that the Association subsystem is internally separated independently into two, and this can be described as the state in which one personality is separated into two personalities. That is, the conscious system will have two independent Association subsystems. I think the description so far was something like this. Yes, the alien hand syndrome occurring with split-brain (Koch,

2004) is an interesting story. However, this is talking about the physical separation of the brain, whereas the DID personality disorder may be called a non-physical split-brain.

Now, I would like to explain the DID model in the conscious system.

First of all, it is necessary to reconsider the DID medical condition.

There have been many reports that DID patients suffered unusual and intense discomfort in their early childhood. From such reports, DID may be thought to result from "abnormal learning."

To begin with, I suppose you might be wondering what I mean by the term "abnormal."

Of course, it would seem to be correct to presuppose that a human infant is continuing to learn while inside their mother's body and is developing according innate mechanisms. And then the infant is born. It is easy to imagine that the external stimuli an infant receives when being born causes the infant to experience more unusual emotions and feelings than before birth.

I consider this moment to be an important point that determines the pleasant and unpleasant feelings of humans. That is, the condition of being nurtured inside the mother's body determines a standard for pleasantness felt by humans (absolute pleasantness) (see 11.6), and I have already explained that that is the standard upon which unpleasantness is based. The infant then continues to grow, receiving the loving care of the parents while also encountering an unpleasant environment. We can judge that infants are developing themselves in an environment that is in the main a healthy one. And, such feelings of pleasantness and unpleasantness that go far beyond the health of the infant are what I am calling abnormal. It is easy to imagine that continuing to receive abnormally pleasant sensations could, of course, also cause some impairment of the brain. And some drug-induced disability could possibly be an example of abnormal feelings. However, I would like to focus here on DID. This is because DID is thought to be one of the causes of abnormal unpleasantness. We will consider how to address pleasantness in our future research.

Now, I would like to present the conscious system that I will propose next.

There are basically no significant changes from the conscious system that was introduced up until now.

However, directed edges are shown in the figure from MoNAD RE of the Reason subsystem to MoNAD P and MoNAD UP of the Emotion & Feelings subsystem. These are introduced in Fig. 9.13 of this book. In addition, a function processing unit, enclosed by a rectangle in the figure, is attached to the input terminal of the Association subsystem MoNAD. This is generally called a neurofilter (hereinafter referred to as an N-filter). This N-filter is a kind of variable function that is based on the physical environment at that time the Association subsystem MoNAD is newly regenerated. Simply put, this N-filter is a feature that automatically sets the range of stimuli that the newly regenerated MoNAD can accept. I thought that this filter function might provide important clues for devising a model of DID using the conscious system. The reason was that, if you think of a computer processing program as an analogy for a model of DID, it would usually be unthinkable that the program could be separated into a large number of parts like the symptoms of DID due to the changing of the internal information. At that time, an idea came to me. This idea was that, as the conscious system continues to learn new information, it is taking measures to regenerate new MoNAD consciousness units and add them to the conscious system. If that is the case then normal learning could certainly incorporate new knowledge into the conscious system by adding such newly regenerated MoNADs to the system (see 11.5). However, I thought it would be difficult to smoothly reproduce something like the symptoms of DID only by utilizing the processing of this conscious system. So I delved into this a little further. And as a result, if the conscious system were capable of simulating the development process of a living organism, it would seem that it should also be considered from the perspective of the passage of time. In other words, the idea is that there may be differences between past learning and current learning in the way a MoNAD responds to stimuli. A similar reaction could be expected to occur in actual biological neurons, but the author has unfortunately not yet confirmed this. In this regard, I can imagine that some readers might wish to reprimand me for claiming "things that have not yet been evidenced by brain science or neuroscience," but from my point of view, I will assert that this is a scientific rationality from a "robot science method." At this point, I will only explain robotics as robotics + computer science. In other words, I think this assertion provides a rational explanation from a robotics and computer science (artificial intelligence) perspective, or more recently, from the viewpoint of robots and artificial consciousness (i.e., robot consciousness).

Now let us take a closer look at the MoNADs in the Association subsystem. As mentioned earlier, the Association subsystem mediates between the Reason subsystem and the Emotion & Feelings subsystem. That is to say, this subsystem mediates between reason and the body. What we should make note of here is the nature of the various MoNADs that make up the conscious system. The Reason subsystem is defined as being almost independent of the Emotion & Feelings subsystem in this conscious system. In other words, the Reason subsystem is designed to receive signals from the Emotion & Feelings subsystem only through the Association subsystem. In addition, the Emotion & Feelings subsystem is capable of receiving signals directly from the Reason subsystem (see Fig. 9.13), but in general, stimuli from the Reason subsystem are received via the Association subsystem. In short, the conscious system is regulated by the Association subsystem, and the Reason subsystem, as well as the Emotion & Feelings subsystem. This is due to the unique characteristics of the conscious system in which the Emotion & Feelings subsystem receives stimuli from the sensors on the body, and the Reason subsystem mainly receives information from the outside world such as by sight and hearing. In other words, the Association subsystem can be interpreted to be acting as an intermediary between the body and the outside world. Here, I would like us to pay close attention to the characteristics of the Association subsystem. Stated another way, the Association subsystem is stimulated by both the Reason subsystem and the Emotion & Feelings subsystem, which causes behaviors to be performed in the outside world as a result. In this experiment, some MoNADs of the Reason subsystem or the Emotion & Feelings subsystem perform these behaviors.

Now, what I would like to say here is, as I mentioned earlier, because of the characteristic in which the Association subsystem mediates between the emotion and reason of the body, this is the point where it is possible to interpret the functions of the conscious system as corresponding to the self or ego. In addition, the function equivalent to human "experiences" seems to be capable of being an analogy of the development of the Association subsystem. Also, since the self-development of the Reason subsystem is not closely related to the Emotion & Feelings subsystem, it can be interpreted as an accumulation of knowledge that does not include feelings. Considering this further suggests that the self-development of the Emotion & Feelings subsystem usually does not occur. This is because it seems reasonable to think that the Emotion & Feelings subsystem is composed of an innate consciousness base that has evolved and developed. In other words, because the Emotion & Feelings subsystem receives stimuli from sensors on the body, the development of this subsystem is based on the body in humans, and it is very difficult to change it.

From these observations, we can determine that the development of the Association subsystem, which mediates between the Emotion & Feelings subsystem and the Reason subsystem, is a good model for the conscious system to be able to explain development related to the self and ego.

Now, let us take a look at the conscious system that explains the DID model that was introduced earlier (Fig. 11.33).

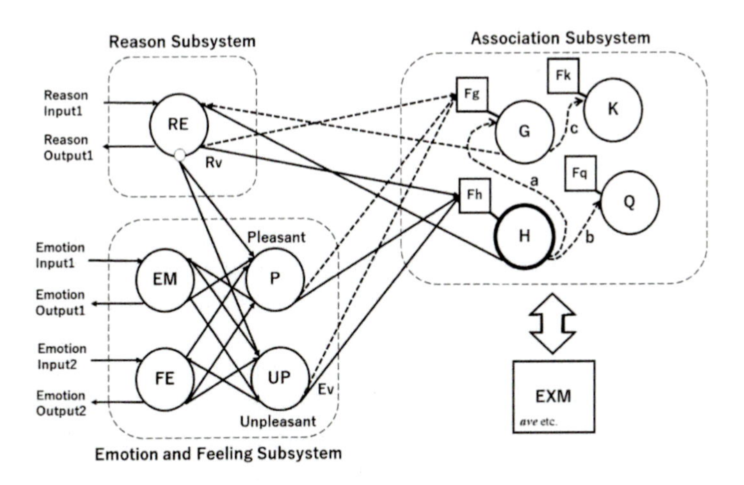

Figure 11.33 DID modeled by the conscious system.

There is a MoNAD named H in the Association subsystem. MoNAD H is marked in the figure with a thick circle, indicating that it has a special meaning in the Association subsystem. H represents a symbol for the main personality. In other words, H represents the central ego or self of this conscious system. In human terms, central means growing inside the mother's body, being born, and accumulating many childhood experiences, that is, it means being linked to many experiences.

Experience comprises information that is typified by episodic memory (see Section 11.3), and data that mainly combines information from the Emotion & Feelings subsystem and information from the Reason subsystem is assumed to be linked there. Part of the episodic memory is expressed as being stored in a system called EXM in the conscious system described in this section. Here, the label of the Association subsystem MoNAD, which responds centrally to the stimuli given to the conscious system. comprises a data set that combines emotion values and reason values and is listed and stored at the time of each cognition-behavioral-cycle (Cbc). It also stores temp, which is the temporary storage location of the emotion value at that moment, and *ave*, which is the average value. (These will be described later.) Currently, EXM is not internally configured using MoNADs, but it is possible to do so.

Let us continue further with this discussion.

Next, let us focus on MoNAD G of the Association subsystem (Fig. 11.33). G is a new MoNAD that is newly regenerated from MoNAD H. H has the capability of communicating information with G (Fig. 11.33a). The arrow means that G was newly regenerated from H. Since H is capable of communicating information with G, the main personality H is generally capable of referring to the experience of G. Here, newly regenerated means that the conscious system has had a new experience with MoNAD H as the main personality, but because H was not capable of representing that knowledge (the reason value Rv from the Reason subsystem and the emotion value Ev from the Emotion & Feelings subsystem) and the new experience was difficult for H to accept, a new MoNAD G was regenerated so that the conscious system could learn the experience, and the experience at that time was learned based on G. The problem of learning unknown knowledge has already been discussed in Section 11.5.

However, the explanation there was that unknown knowledge is new knowledge of the Reason subsystem that is not related to the Emotion & Feelings subsystem. Since H in this section is a MoNAD of the Association subsystem, that is, it responds to stimuli of both the Reason subsystem and the Emotion & Feelings subsystem, note that the emotion value of the Emotion & Feelings subsystem is also taken into account, and this is slightly different from the newly regenerated MoNAD that was described in Section 11.5.

Now let us return to our discussion of DID.

In DID, many alternate personalities may appear in one individual, and the state in which the main personality is not expressed is observed as a symptom. If we consider this symptom in terms of the conscious system, the knowledge MoNAD G generated from the main personality MoNAD H of the Association subsystem is a process that weakens or loses the relationship of the related knowledge from H. This process causes the Association subsystem to lose the relevance of the main personality H to G, resulting in H and G functioning as if they were separate personalities in the conscious system. This process may be considered to be one of the causes of the generation of DID. In Fig. 11.33, K is further separated from G and Q is further separated from H. Here we show the possibility that many other personalities may be generated one after another by this process.

So then how does this separation occur?

This would be our next question.

The author and his research group considered the causes of the following three types of DID.

(1) The convergence of the Association subsystem is confused,
(2) The newly generated Association subsystem MoNAD tries to take the initiative,
(3) The new MoNAD accepts a wider range of emotions.

Before getting into this explanation, let us check the condition of the new MoNAD G. At first, there was only MoNAD H in the Association subsystem of the conscious system (Fig. 11.33, H is enclosed by a thick circle), and an experience with a strong unpleasant feeling was input to the conscious system at that time. Since this was a new experience that H was not capable of coping with, the conscious system regenerated a new Association MoNAD G with H as the main personality.

Now, since H was given a strong unpleasant feeling as a stimulus that it had never experienced before, H temporarily confused the convergence of the neural networks that make up the MoNAD itself. This confusion of H causes a new MoNAD G to be regenerated (see Sections 11.1 and 11.5). The relationship between H and G becomes weaker and they eventually separate due to this confusion. This is the first theory.

In the second theory, the situational responsiveness of the newly regenerated Association MoNAD G is higher than H because G is learning experiences based on strong unpleasant feelings, and this explains that, as a result, the centripetal force of the information of the basic personality H has been weakened.

The third theory is that when a MoNAD is regenerated, the MoNAD is physiologically capable of responding to the stimuli it faces at that time. In other words, the idea is that the new MoNAD can respond physiologically to a new era. Conversely, this explains that a MoNAD that has already learned experiences in that era has a physiologically limited range of response to stimuli. And this theory is that the new MoNAD can learn in response to the new era.

When we take another look at these explanations, one process of DID generation emerges.

MoNAD H, the main personality in the Association subsystem, is stimulated by a strong unpleasant feeling that it has not experienced before, which causes a new H-based MoNAD G to be regenerated in the Association subsystem of the conscious system (Theory 1). The new MoNAD G will be incorporated into the Association subsystem with a new physiological system as a new era (Theory 3). At that time, in the new era, MoNAD H is still playing a central role in the Association subsystem and it experiences what G is experiencing. However, because the unpleasant feeling being experienced by G is so strong, as part of the conscious system H still remains in a state where it continues to accept that unpleasantness. At that time, however, because of the high level of unpleasantness experienced by MoNAD G, as MoNAD H remains in a state where it continues to experience that unpleasantness, the reaction within the H-based Association subsystem gradually changes to a mechanism that is centered on G. As a result, when the H base in the Association subsystem is diluted and the developmental state of early childhood is immature, the conscious system uses a structure in which G is separated from H (Theory 2). In other words, from the viewpoint of the conscious system, information access from H to G is cut off. Independent of H, G accepts the strong unpleasantness, but the main personality H does not need to accept those unpleasant feelings. At this time, when viewed from H, it appears that G, which is another personality, is accepting the unpleasant feelings, and we can interpret this as H itself is alleviating the unpleasantness. Based on these theories, the author worked with his research group to devise a model of a conscious system that causes DID and attempted simulation experiments.

First of all, simulation Experiment 1 was performed for Theory 1. DID modeled by the conscious system was introduced in a previous figure (Fig. 11.33). In this experiment, however, the filter function was not installed in each of the MoNADs of the Association subsystem. In addition, simulation Experiment 2 was performed for Theories 3 and 2. The DID model shown in Fig. 11.33 was adopted for the conscious system.

Experiment 1 confirmed that MoNAD H and G of the Association subsystem were separated by the strong unpleasant feelings. Experiment 2 confirmed that the H and G MoNADs developed dissociative symptoms when the strong unpleasantness was continually input to the conscious system, and that H and G were completely separated by continuing to receive these unpleasant feelings for a relatively long period of time. As a result, we observed that the average value *ave* of the emotion values was reduced by the separation of G (Fig. 11.34).

Transition of the emotion value Ev′ (right axis) received by the Association subsystem and change of the average emotion value *ave* (left axis) of the Association subsystem.

Now, I would like to explain Experiment 2 in detail.

First, let us explain the filter function provided for the MoNADs of the Association subsystem. The filter of MoNAD H is Fh, and the filters of the other MoNADs are named likewise: G has Fg, K has Fk, and Q has Fq. In Experiment 2, the Association subsystem initially has only MoNAD H. At that time, Fh has an initial value, which is

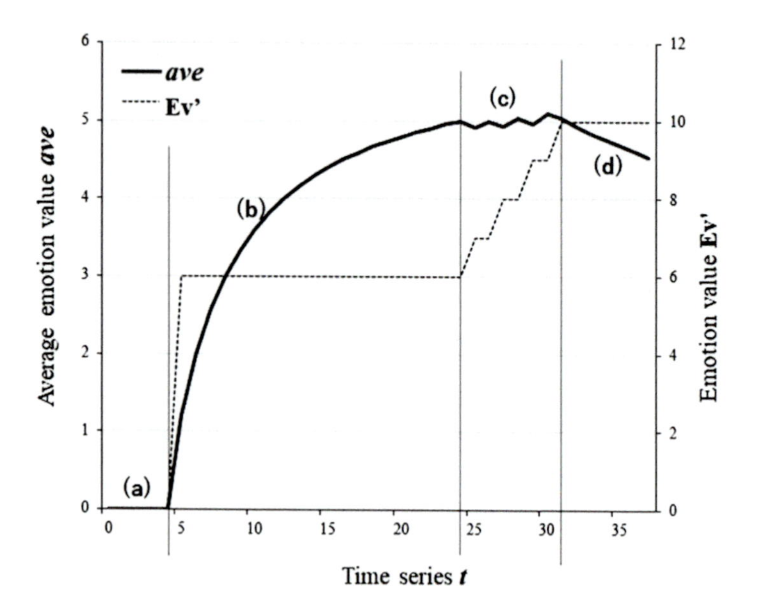

Figure 11.34 DID model simulation results.

Fh = 0. Thereafter, when a strong unpleasant feeling is input, a new MoNAD G based on H (as the main personality) is regenerated by the conscious system. At that time, an arbitrary value is added to the filter value Fh of H. (Integer 1 was used in this experiment.) This incrementing of the filter value has the effect of facilitating at a later time the separation of the newly regenerated MoNAD G from the MoNAD with which it has a parental relationship (MoNAD H in Experiment 2). The reason for this is that the positive increase in the filter value makes the regenerated MoNAD intolerant to the input of unpleasantness. The initial filter value (Fg in this experiment) of the newly regenerated MoNAD (G in this experiment) is given a somewhat large negative value. In the experiment Fg = −4. This negative filter value setting was done because our thinking was that the new MoNAD (G in this experiment) was regenerated when stronger unpleasant feelings were input, so it would have some tolerance for unpleasantness when regenerated. This is described in the reason for DID generation (Theory 3) mentioned earlier.

In other words, the effect of the filter is to change the emotion value Ev for each MoNAD that is given to the Association subsystem taking into account the unpleasantness tolerance of the MoNAD. In this experiment, the following formula was used.

$$\mathrm{Ev}' = \mathrm{Ev} + \mathrm{Fh} \tag{11.1}$$

That is, the filter value Fh is added to the actual emotion value Ev′ that is given to MoNAD H. The same calculation procedure was used for the other MoNADs of the Association subsystem.

Next, the calculation of the average emotion value *ave* will be described. The following formula was used for the calculation.

$$ave_n = \frac{(n-1)\cdot ave_{n-1} + \mathrm{Ev}'_n}{n} \tag{11.2}$$

Since the previous *ave* value in the time series is stored in EXM, it is used in the calculation.

When ave_n in Eq. 11.2 is at or higher than an arbitrary threshold (5.0 is used in this experiment) or when the value of Ev'_n ($= \mathrm{Ev}_n + \mathrm{Fh}$) reaches the maximum value (integer 10 or more in this experiment), dissociation occurs in the conscious system, G acts as an alternate personality in place of the main personality MoNAD H. At that time, ave_n becomes the following equation.

$$ave_n = \frac{(n-1)\cdot ave_{n-1} + \mathrm{Ev}''_n}{n} \tag{11.3}$$

However, $\mathrm{Ev}'' = \mathrm{Ev} + \mathrm{Fg}$.

Next, the results of our experiment will be described.

In the experiment, inputs were made to the conscious system, with initially Rv (a fixed value in this experiment, which was a binary number [0 1 1]) and Ev = 0 given to the Association subsystem from RE and UP, respectively (at $t = 1$).

At first, the same input values were given up to $t = 4$ in the time series (Fig. 11.34a). At that time, *ave* has a value of 0 and does not change.

After that, Rv (a fixed value) and Ev = 6 were given up until $t = 24$ (Fig. 11.34b). A sharp increase in *ave* was observed from a value of 0 to a value of around 5.0.

Here, because the value of *ave* exceeded 5.0, dissociation occurred in the conscious system, and MoNAD G was regenerated in place of the main personality MoNAD H and began acting as an alternate personality. In addition, this regeneration caused a value of 1 to be added to the value of Ev′.

After that, up until $t = 30$, vibrations were observed when the value of *ave* exceeded or decreased below 5.0 (Fig. 11.34c). Every time the value of *ave* exceeds 5.0, a value of 1 is added to Ev′ (Fig. 11.34 (dotted line portion of section c)).

At $t = 31$ and thereafter, G was completely separated from the main personality H because the value of Ev′ was 10 or more.

For this reason, *ave* is descending as its value is decreasing (Fig. 11.34d).

Considered from the results of this experiment, as strong unpleasant feelings like never experienced before continue to be input in the conscious system, the basic emotion H regenerates a new Association MoNAD G to accept this unpleasantness, and by further continuous input of those unpleasant feelings, G completely separates from H. As a result, the average emotion value *ave* evaluated from the entire Association subsystem of the conscious system decreases after the separation. In other words, we can observe that the separation of G causes a decrease in the evaluation of unpleasantness in the conscious system as a whole.

I believe this experiment provides one answer toward unraveling the mystery that causes many other personalities other than the main personality to appear in patients with the neurological disorder known as DID.

11.9 Summary and Consideration

This section introduces various examples of the conscious system comprising MoNAD consciousness units devised by the author.

First, we introduce a study entitled A Robot Cognizes the Unknown (Section 11.1). In this study, we conducted experiments using a conscious system to consider what an "unknown world" is to a robot and whether it would be possible for a robot to "learn the unknown world."

The author and his research group first developed a conscious program that enables a robot to identify three colors and perform behaviors corresponding to those colors, and then installed the system on the robot. The three colors used were green, blue, and red. Then, a behavior was assigned corresponding to each color. When the robot recognizes the color green, it moves forward. When the color is blue, the robot moves backward. And upon recognizing the color red, the robot stops. Each of these three colors was learned in one consciousness unit as a basic (innate) behavior for the robot. One MoNAD was made to learn the correspondence between three colors recognition and each behavior at the same time. Thus, when the robot recognized one basic color, it would perform the behavior that corresponded to that color. When the green color was continually presented, the robot would continue to move forward. However, what if the experimenter suddenly presented a purple color at that time? The robot could now only identify three colors: green, blue, and red. In conventional neural network (NN) learning, because of some bias in purple, as a result, it would converge on one of the three basic primary colors mentioned above and that corresponding behavior would be performed.

However, the author and his group focused on the fact that while innate basic learned information can converge at high speed in a MoNAD, the convergence of unknown information slows down. The conscious system makes use of this slowdown phenomenon to determine whether the color information currently being presented is unknown to the robot. If it is unknown information, a new MoNAD is generated at that time as a means of enabling the robot to learn that information.

The problem that arises at that time is, what is unknown information and what behavior should correspond to that information? In this experiment, when the information was unknown, we decided to have the robot perform a positive and negative polar rotation of 45° one time. When new unknown color information was detected, another 45° was added to the angle of the polar rotation. At this time, the creation of a new behavior relates exactly to the creative function of the robot, but this will be a topic for future research. There is an interesting reference by American neuroscientist Antonio Damasio as an interpretation of human creative functions (Damasio, 2010).

In this section, we actually constructed a conscious system with functions similar to those of human consciousness by combining a recursive neural network MoNAD with a special structure and a consciousness unit and created a program that simultaneously produces robot recognition and behavior, and conducted experiments. The conscious system in this section detects when unknown information is input to the MoNAD by utilizing the fact that when a certain MoNAD consciousness unit receives unknown information, the convergence speed of the MoNAD decreases. This has demonstrated that unknown information can be learned by the system successively through the process of learning information from the newly generated MoNAD.

The author would like to express his gratitude for the contributions of Kyohei Kushiro, a graduate student in the Faculty of Science and Technology (Robot Science Laboratory, Computer Science, Meiji University), who conducted the research together with the author (Kushiro, Harada, and Takeno, 2013).

In the next section, we introduced our research on A Self-Aware Robot (Section 11.2).

This section presents an advanced version of the robot cognitive experiment that we introduced in Chapter 10 of this book. In addition, the content related to the self-awareness of robots that was discussed in Chapter 11 of the first edition of this book has been incorporated into this section so that the relationship with the information in this section becomes clearer.

In the experiment on mirror recognition by the robot presented in Chapter 10, the self-image reflected in the mirror corresponds to the robot's own behavior, and we concluded that it is within the reaction system. In other words, we concluded that the behavior of the self-image reflected in the mirror can be judged to be a behavior performed by a part of the self's own body. The self-image reflected in the mirror cannot be seen as the body of the self from a theological viewpoint. However, due to the special physical characteristics of a mirror, from the subjective viewpoint of the robot, the behavior of the self-image in the mirror is perceived without any delay to be a direct reflection of the behavior of the self. Therefore, the self-image in the mirror can be determined to be an illusion-like perception in which the behavior itself is being performed by the self's own body. This illusion-like perception can easily lead to the perception that the behavior of the self-image in the mirror is originating from the self's own behavior. In addition, it will be easy to understand that the linkage of this perception to the feeling that the self-image in the mirror is being moved by oneself is also caused by the "consistency of cognition and behavior" characteristic of human consciousness, which is a concept that the author has put forth from the beginning of this book. The special physical characteristics of a mirror spoken of here refers to the fact that, although seeing an own shadow onto the ground and image reflections on water surfaces are phenomena that have existed since ancient times and there have been various performance evaluations such as of the reflectance, the high level of reflection and responsiveness of mirror images, there has been no implement in the world that is like a mirror.

In this section, we considered how the ability to distinguish oneself from the other changes when some sort of malfunction occurs in the conscious system. We think there are two types of malfunctions in the conscious system. One is when the basic cycle of consciousness (cognitive-behavioral cycle) of the conscious system itself decreases, and the other is when the expression of the self-behavior is delayed. In the former, the cognitive-behavioral cycle (Cbc) is halved, and in the latter, a time delay is added to the expression of self-behavior. The time delay was changed from 0.5 second to 1 second, 2 seconds, 4 seconds, and 6 seconds. In both cases, experiments were conducted using robots, and after the experiments, the data in the robots were extracted to measure changes in the ability to distinguish the self from the other. The change in the ability to distinguish the self from the other is reflected in the change in the imitation matching rate between the Rc robot that is controlled by a cable and the Rs self-robot.

The results of our experiments indicate that the ability to distinguish between the self and the other decreased from 73.8% to 67% when a healthy conscious system (Cs1) is compared with a conscious system that has had its Cbc cognitive-behavioral cycle (Cs2) halved (Fig. 11.7).

In addition, when the expression of self-behavior is delayed, the imitation matching rate can be seen to decrease as the delay increases (Figs. 11.8 and 11.9).

Note that since the mobile robot used in Chapter 10 was a Khepera robot while the robot used in this section was an e-puck, it is difficult to draw a simple comparison between the experiments described in Chapter 10 and the experiments presented in this section.

From the experiments in this section, it may be said that if a healthy conscious system has a malfunction in the consciousness cycle or a delay in the expression of self-behavior, the ability to distinguish between the self and the other is reduced. That is, it can be inferred from these experiments that when the basic processing cycle of the human brain decreases, the distinction between the self and the other in the human brain also decreases.

This may mean that in the final stages of brain dysfunction, humans have a subjective sense of being a part of their world.

The author would like to express his gratitude for the contributions of Toshiyuki Takiguchi, a graduate student in the Faculty of Science and Technology (Robot Science Laboratory, Computer Science, Meiji University), who conducted the research together with the author (Takiguchi, Mizunaga, and Takeno, 2013).

In the next section, we presented experiments conducted on learning experiences in robots.

Let us imagine that I saw a bee and was I curious about it and caught it in both of my hands. The bee would undoubtedly sting my palms many times with the stinger at the tip of its tail. I suppose many people may have experienced a bee sting. Then, my palms would become very painful, and the shock of the pain would make me unable to move for a while. Then, the stung areas would become red and my heart would beat violently. A single bee sting can cause an allergic reaction and for some people it may even pose a life-threatening risk.

I would never try to catch a bee again after having had such an experience. In a sense, it would be reasonable to think that experiencing the pain of a bee sting led to a reduction in my getting close to bees. Also, from that experience, I feel that a similar inhibitory function is always at work regarding other small animals that resemble bees. I do not really want to think that human learning began by experiencing pain, but it can be seen that Freud's theory about the instinct for survival during human evolution primarily involves the choice to avoid unpleasant experiences such as pain. The reason is that if we look at how a person develops after being stung by a bee, there is no mistake that it feels like a challenge to human survival.

Well, can a conscious robot learn such an experience? And will the robot be able to take advantage of this experience to avoid that unpleasant experience? These are the research themes presented in this section. That is, in this section we consider the example of a robot colliding with an obstacle while it is moving similar to humans avoiding getting close to a bee after learning that a bee sting causes pain. In other words, a representation of pain is generated on the conscious robot when it collides with an obstacle while moving.

The collision information and pain information are stored together as a set (Fig. 11.10). Since this set of information combines the Reason subsystem information and the Emotion & Feelings subsystem information, it can be said to be a kind of episodic memory. If we do not need to be concerned about the total amount of memory, it is easy to create episodic memory in chronological order.

So, then, can the conscious robot make use of the unpleasant feelings it experienced from the collision to prevent a collision from happening again? This is also a study about generating anticipatory behaviors by learning experiences.

The episodic memory presented in this section utilizes the characteristics of the MoNAD as a consciousness unit to achieve this anticipatory behavior. That is, the MoNAD follows the Cbc, which is the basic thinking cycle of the robot, and the cognitive representation information of MoNADi is generally copied as is to the behavioral representation (Fig. 9.11). In other words, the value of the behavioral representation BLi $(k + 1)$ is equal to the cognitive representation RLi (k) (see Section 11.3). Here, k represents a certain time (integer) in Cbc. Note also that the unpleasant MoNAD UP represents unpleasantness at time $k + 1$. Utilizing these characteristics, BLi $(k + 1)$ as input information and RLi (k) as output information are first applied to the two-layer NN as supervised learning, and after that, the unpleasantness information of RLi (k) and time $k + 1$ can be linked using a single-layer NN. By this theory, utilizing the episodic memory presented in this section, it is possible to have an expectation of the unpleasantness information at time $k + 1$ using the cognitive representation of MoNADi at time k.

The experiments presented in this section were conducted using the small e-puck mobile robot. The robot collided with the obstacle once, but thereafter it was able to stop just before colliding again. In other words, the robot successfully learned not to collide again by utilizing the experience of the unpleasantness expressed by one collision.

The author would like to express his gratitude for the contributions of Takashi Komatsu, a graduate student in the Faculty of Science and Technology (Robot Science Laboratory, Computer Science, Meiji University), who conducted the research together with the author (Komatsu and Takeno, 2011).

The next section described an experiment on a Pavlov robot.

I imagine that everyone is familiar with Pavlov's experiments with dogs. Russian physiologist Ivan Pavlov won the Nobel Prize for these famous experiments. When a dog is fed, it secretes saliva in its mouth. In the experiments, a bell was sounded at the same time the dogs were fed. Gradually the dogs began salivating only with the sound of the bell without being given any food.

In this section, we used a conscious robot as a replacement for a dog and tried to reproduce Pavlov's experiment using the robot. Since the robot does not need any food, we decided to irradiate the robot with a white light in the experiment. The robot collides with an obstacle while moving forward in the experiment and is programmed to stop upon the collision. The robot is irradiated with a white light for an arbitrary period of time upon colliding with the obstacle. Thereafter, like Pavlov's dogs, whenever a white light is emitted while the robot is moving forward, the robot will stop even in without the presence of any obstacles. However, the robot will certainly stop as intended whenever it collides with an actual obstacle.

The conscious system using for conducting this experiment has a conventional structure with the addition of an information connection MoNAD Li and an information storage neural network Me (Fig. 11.18). Since these additional structures are placed between the Reason subsystem and the Emotion & Feelings subsystem, they function as a kind of episodic memory.

In the experiment presented in this section, we successfully recreated Pavlov's conditions with the conscious robot using

MoNADs Li and Me. In other words, when the conscious robot was simultaneously irradiated with a white light upon colliding with an obstacle, the robot stopped and a representation of pain was automatically connected between the information of the collision and the information of the detection of the white light. Thereafter, even if there was no collision while the robot was moving, it was possible to generate a representation of pain and stop the robot only by irradiating it with the white light. We believe that the success of this experiment demonstrated that it was possible to achieve a reaction similar to that of Pavlov's dogs using a conscious robot.

The author would like to express his gratitude for the contributions of Koki Kanazawa, a graduate student in the Faculty of Science and Technology (Robot Science Laboratory, Computer Science, Meiji University), who conducted the research together with the author (Kanazawa and Takeno, 2014).

The next theory is a study that attempts to examine certain capabilities of the conscious system.

In this section, we examined the cognitive capability of the system to differentiate color.

The experiment was conducted as follows. One color vision MoNAD R1 had previously learned the three primary colors, red, blue, and green, and then a new color, medium orchid, was presented to the conscious system. The convergence of the input information on R1 decreased. Therefore, this color was determined to be unknown to the conscious system. The conscious system was programmed to generate a new MoNAD (called an empty MoNAD) at that time, and this new MoNAD was made to learn this unknown color.

The problem of automatically giving a name to this unknown color is a problem of creation and, as such, it will be taken up as a theme for future research. In this experiment, the name of the unknown color was assigned by a human. The name given to this color was "medium orchid." In the actual experiment, an arbitrary number (No. 19) was assigned to this color. However, R2 was eventually excluded as a learned color vision subject by the intervention of the experimenter. The reason for this was that when a new color "plum" (No. 18) was learned, it became indistinguishable from the "medium orchid" color previously learned by the color vision of the conscious system, and the experimenter forced the learning (also referred to as "pressed learning," or simply PL).

The experimental results demonstrated that the color vision of the conscious system was capable of differentiating 18 colors. This was fairly low compared to the normal color vision capability of humans.

The author would like to express his gratitude for the contributions of Tomoya Matsunaga, a graduate student in the Faculty of Science and Technology (Robot Science Laboratory, Computer Science, Meiji University), who conducted the research together with the author (Matsunaga and Takeno, 2016).

The next section describes absolute pleasantness.

Here, we examine the problem of how to determine a robot's feelings of pleasantness and unpleasantness in terms of information science. I imagine readers might be wondering what the connection between feelings and information science is. This was an idea that first came to me when I was considering the problems of learning for the consciousness unit MoNAD when there was previously learned information and unknown information. I noticed that for the MoNAD, all the learned information had high convergence with a smooth flow of information, but unknown information had low convergence. This phenomenon helped to solve the problem of learning unknown information (see Section 11.1). In short, when unknown information is detected, having a new MoNAD learn it improved the MoNAD convergence of the conscious system. We defined this poor convergence of the MoNAD as emotional unpleasantness. However, I do not think that readers would agree that a healthy human adult would feel unpleasant when confronted with unknown information. This is because many people would say that they feel pleasant with regard to unknown information. Much is still unknown about this phenomenon, but the author believes that it is probably the result of learning, then understanding, and further learning that knowing unknown information as an adult eliminates unpleasantness and increases pleasantness. This topic was addressed with students in a paper (Ebisawa, Matsushita, and Takeno, 2015). Studies have also shown that human infants do not exhibit an unpleasant reaction when observing something unknown to them. The famous Little Albert experiment dealt with this topic. This experiment was conducted by John B. Watson, an American who famously presided over the field of behavioral psychology (Beck et al., 2009). The experiment was criticized with regard to human rights issues. As shown in the published experimental film footage, the boy was about 11 months old at the time. Initially, the child was indifferent when presented with unknown objects and did not express any unpleasant reactions to them. Later in the experiment, when an unknown object was presented to the boy, it was accompanied by a stimulus at the same time, such as an unpleasant loud noise. Thereafter, the boy was observed to exhibit unpleasantness only upon seeing the object even without the loud noise. In addition, some readers may wonder if it is difficult to directly link unpleasant feelings with the low convergence of information in a MoNAD. In short, you might find it difficult to make the connection between an information science process and emotions. However, it is possible to explain without bringing up the theory of cognitive dissonance (Festinger, 1957) by American psychologist Leon Festinger (1919–1989). I think the problem is that it is difficult to explain a connection between the unpleasantness felt by humans and the information science phenomenon of the deteriorating convergence of a MoNAD. The author feels that this problem of unpleasantness in humans should be considered in somewhat more detail. In my opinion, there are two main types of human unpleasantness. I make a distinction between "mental unpleasantness" and "physical unpleasantness." The latter is exactly the type of unpleasantness that comes from a thorn prick, the pain of a burn, or an upset stomach. The former is the unpleasantness caused by the anxiety of not being able to solve a difficult mathematical problem or having mentally conflicting ideas at the same time. That is, the latter is a physical unpleasantness, while the former is mental unpleasantness related to Festinger's theory. The unpleasant feelings I describe about unknown information relate to the former mental unpleasantness.

I think it is very interesting how these broadly classified feelings of unpleasantness are related to each other. It is clear that both are related, but the details of this phenomenon are as yet very much unknown. For example, do humans possess some degree of mental unpleasantness and physical unpleasantness that are associated with each other as part of the process of human evolution? Or are these feelings learned after birth during early childhood? The theories about this have not yet been fully elucidated. Watson's Little Albert experiment, mentioned earlier, supports the latter, but it is not yet clear. The author has the idea that there is basic preprogramming before birth (see Section 9.7.2, Fig. 9.13), and that humans develop through deeper learning (see Sections 11.3 and 11.4).

I find it very interesting to think about this phenomenon. If we think about an analogy to a conscious system and the MoNADs that comprise it, if all the MoNADs are highly convergent, I think the conscious system could be considered to be in a pleasant state. For humans, the closest approximation to this state is a fetus that is protected by the placenta while still inside the mother and continuing to develop. The development of the fetus is mainly due to information incorporated into genes during the process of human evolution, the stimuli from the mother that the fetus receives, and the stimuli from the external environment surrounding the mother. In this way, the fetus is protected by the placenta from the intense stimuli of the external environment including the mother's body, and the body and the nervous system of the fetus are likely to be developing according to a genetic blueprint. We can assume that the body and mind of the fetus develop while external stimuli are suppressed as much as possible. Therefore, we can consider that the information flowing through the nervous system is stable and converges in the prenatal fetus without receiving much internal and external stimuli.

I think this prenatal mental state of the fetus sets the standard for pleasantness. I call this a state of "absolute pleasantness." In the conscious system, unpleasantness can be considered to be based on the degree of deterioration from the convergence of the conscious system, which sets the standard for this absolute pleasantness. However, it is worthwhile to note that these are emotions related to the mental unpleasantness mentioned above and are different from physical emotions. An Emotion & Feelings subsystem is provided in the conscious system to represent physical emotions. The question of how mental and physical emotions relate to each other was mentioned earlier, but the author holds to a theory that they develop in two stages. Moreover, it is still not possible to provide a clear explanation as to how qualia is generated from physical emotions. Perhaps qualia stems from a kind of physical reaction that is synchronized with the Emotion & Feelings subsystem.

Let's turn to the theme of Conflict of Concepts and a Model of Rubin's Vase.

Rubin's vase is a famous ambiguous figure. We modeled the phenomena of Rubin's vase by combining MoNADs on the conscious system. The model represents the vase and the human faces alternately because the representation does not converge into a single figure when looking at this environmental information. Using our model, we can generally explain that the ambiguity of Rubin's vase stems from the conflict of concepts represented by the conscious system. The conscious system represents the environmental information uniquely as a vase at one time but the representation gradually shifts from the vase to the human faces due to some internal and external stimuli. And then, the human faces change to the vase. A single piece of environmental information can contain multiple concepts, and they conflict with one another on the conscious system, preventing conversion into a single concept. In this chapter, the author describes his model for processing ambiguous figures on the conscious system using the phenomena of Rubin's vase as an example.

The author would like to express his gratitude for the contributions of Hanwen Xu, Daiki Matsumoto, and Koki Kanazawa, graduate students in the Faculty of Science and Technology (Robot Science Laboratory, Computer Science, Meiji University), who conducted the research together with the author (Xu, Matsumoto, Kanazawa, and Takeno, 2016).

Acquisition of the Sense of Self by a robot is then discussed.

Researchers have discussed earnestly in recent years whether a robot has a sense of self or not. This theme has long been studied, but so far a definite theory has not been established. The author studies this theme from the viewpoint of what kind of development process would be required for a robot to gain a sense of self. The robot with the conscious system developed by the author takes action after the Reason, Emotion & Feelings, and Association subsystems communicate with one another based on information on the external environment. The Reason subsystem receives environmental information from outside the robot, the Emotion & Feelings subsystem responds to environmental changes within the robot, and the Association subsystem mediates between these two subsystems. If a robot, while behaving normally, is suddenly stuck in a condition where the Reason subsystem and Emotion & Feelings subsystem conflict with each other, what kind of Association subsystem should be generated to settle the conflict? This is the theme of this chapter. Assume, for example, a robot is heading for a certain target place, and happens to step on a bunch of cables. The robot would select an action that offers a more comfortable state for itself rather than deciding to continue going on toward the target. In other words, the robot puts a high priority on caring for itself. This priority process may be considered the origin of a sense of self. In the experiments described in this chapter, the robot with the conscious system succeeded in automatically developing an Association subsystem that solves the conflict between the Reason and Emotion & Feelings subsystems. This provides a firm starting point for us to find the origin of the acquisition of a sense of self.

The author would like to express his gratitude for the contributions of Kensuke Arai, a graduate student in the Faculty of Science and Technology (Robot Science Laboratory, Computer Science, Meiji University), who conducted the research together with the author (Arai and Takeno, 2018).

The last topic is the development of a model for diagnosing the higher cognitive impairment of humans.

Human brains are said to have a high-level cognition function such as consciousness. A recent research paper reports that human brains suffer from cognitive impairment when subjected for a long time to a harsh environment such as battlefield. This chapter introduces research efforts to develop a model for diagnosing higher cognitive impairments of humans such as posttraumatic stress disorder (PTSD) and dissociative identity disorder (DID) using the conscious system. When the conscious system learns a new experience, a new association subsystem is autonomously generated and added to the conscious system. This explanation best assists people to understand the above-mentioned disorders. In ordinary cases, the new Association subsystem exchanges information with the existing Association subsystem and with the Reason and Emotion & Feelings, and adds new knowledge to the conscious system. This is a process of adding new knowledge to the existing knowledge base, or, in other words, a new experience is added to the existing ones. In the model developed by the author, PTSD is the case in which the new experience has an unstable connection of information with the existing experiences, and DID is the case in which the new experience is dissociated from the existing experiences (lack of an information connection).

The author would like to express his gratitude for the contributions of Takahiro Hoshino, Yuichi Watanabe, graduate students in the Faculty of Science and Technology (Robot Science Laboratory, Computer Science, Meiji University), who conducted the research together with the author (Hoshino and Takeno, 2016) (Watanabe, Hoshino, and Takeno, 2018) (Watanabe, Suda, and Takeno, 2020).

This chapter described the development of a variety of models related to human consciousness using MoNADs, which are consciousness units, and a conscious system comprised of such MoNADs, and introduced demonstration experiments in which the conscious system program was installed on a small robot. However, all of the developments conducted by the author and his research group could not be introduced here. I would like to present an overview of those studies below. Unless otherwise noted, the graduate students mentioned below are affiliated with the Robot Science Laboratory, Computer Science, Faculty of Science and Technology, Meiji University.

First of all, I would like to express my gratitude to Juergen Manner, a researcher at the Institute for Process Control and Robotics, Karlsruhe University, for his paper on the human brain and MoNAD (Manner and Takeno, 2011). I would also like to express my gratitude to all of the following: Ryuma Matsushita for his contribution to the study of the self-evolution of the conscious system (Matsushita and Takeno, 2015), Keitarou Yoshida for his contributions to the computer model of mental trauma using the conscious system (Yoshida and Takeno, 2014), Hanwen Xu, Koki Kanazawa, and Daiki Matsumoto for their contributions in the study on the Thermal Grill Illusion using the conscious system

(Xu, Kanazawa, Matsumoto, and Takeno, 2016), Takuma Okawa for his contributions regarding the development of self-consciousness (Okawa and Takeno, 2016), Riku Sekiguchi and Haruki Ebisawa for their contributions to the self-evolution of conscious robots in the environment (Sekiguchi, Ebisawa, and Takeno, 2016), Soichiro Arai for his contributions to the research (Arai and Takeno, 2018) on explicit consciousness, sub-consciousness, and self-awareness in the conscious system, Yuichi Takayama for his research on "human curiosity" using the conscious system (Takayama and Takeno, 2018), Tomoya Sumioka for his contribution to the development of a model of stalking behavior using the conscious system (Sumioka and Takeno, 2018), and Tianbai Yuan and Jianyu Wang for their contributions to the study on obsessive-compulsive disorder using the conscious system (Yuan, Wang, and Takeno, 2018). And Shotaro Niimi carried out research on Major Depressive Disorder (Niimi and Takeno, 2020).

Chapter 12

Conclusions

I have overviewed robots that have been developed by researchers in the past and studies by brain scientists on where human consciousness is expected to exist. I have also described a timeline of studies on human consciousness and the mind by philosophers and psychologists from the past to the present day.

Professor Valentino Braitenberg's research studies were also introduced in this book. He assumed that humans and other organisms are machines and constructed various robots based on his idea that the mechanism of external stimuli eventually reaching the drive motors via neural circuitry generates behaviors. I also learned from Prof. Braitenberg about his efforts to create a "thinking machine."

This book further introduced Prof. Rodney Brooks' robots. Brooks devised a hierarchical mechanism in which behavioral modules were arranged in a hierarchy from the base to higher levels to support the robot and sustain behaviors persistently. I learned from Prof. Brooks that a hierarchical behavior mechanism could continue behaviors persistently in a varying environment. I would like to mention that these two pioneers built the foundation of modern robot science.

Self-Aware Robots: On the Path to Machine Consciousness
Junichi Takeno

ISBN 978-981-4877-90-9 (Hardcover), 978-1-003-26181-0 (eBook)
www.jennystanford.com

This book then introduced the basics of the artificial neural network technology. The readers learned about artificial neural networks that make up the neural circuitry of a robot, the learning mechanism of robots, and recursive neural networks that handle past data dynamically. They also learned that machines and robots evolved by simulating the evolutionary process of organisms and by acquiring higher-level cognitive behaviors by themselves and about the relevant technologies and the basics of computational evolution.

I have noted that while evolutional technology is excellent, computers have intrinsic limitations when they are used to simulate evolution. The problem is that an infinite number of possible selections must be processed to achieve self-evolution of human consciousness and the mind.

Examples of the development of robots used for studying human consciousness and the mind were introduced in Chapter 8. They included Grey Walter's turtle, the robots of Tadashi Kitamura, Jun Tani, and Mitsuo Kawato, and Cynthia Breazeal's Kismet.

I have learned much from Walter's turtle, which exhibited a great spirit of innovation. In an age when digital computers were yet to be developed, Walter combined analog electronic circuits to demonstrate a "conversing robot" and "self-recognition."

Kitamura extracted the essential functions of consciousness from the phylogenic complexity of organisms and developed a conscious robot in which layered conscious behavior groups started to function one after another in response to a varying environment like Brooks' robot. I have learned about the hierarchically arranged structure of the consciousness model from Kitamura.

Tani developed a robot using recurrent neural networks as behavior mechanism. In his robot, information circulates in a stable manner when expectation information calculated by the neural networks is nearly the same as the information about the external environment currently being observed; otherwise the robot behaves chaotically. Tani calls this state, "the robot is conscious." Tani's robot offered clues about the possibility that the recurrent networks could create consciousness.

It was from Kawato that I learned about bidirectional models that enable a robot to understand the world and about the bidirectional theory that offers a high possibility of solving the binding problem, a question on how to combine two or more concepts.

My last "mentor" is Breazeal. She developed a robot using the theory of mind, the fruit of her study in the field of psychology, as a function of behavior cognition. Breazeal's is a head robot, and the scene of interaction is like a human meeting an infant, including human-like facial expressions and child-like conversation. I hit upon an idea to connect the theory of mind and human consciousness after reading her research papers.

I learned very much from these researchers. They taught me that, for example, human consciousness occurs in the brain cell networks in the brain, and if this is so, then consciousness could be realized with artificial neural networks using computers. This means that consciousness can be realized as a flow of information through the artificial neural networks connecting the brain and the body, and this could be realized by a computer program. The program always simultaneously checks for externally input stimuli and internally generated stimuli and constantly outputs signals. The program should have a mechanism to recognize information generated internally in the program itself. The program should function in conformity with the theory of mind. The consciousness system must basically have a layered structure in which functions are stacked one above the other from low- to high-level functions. The consciousness system should basically be capable of bidirectional information exchange. And such a consciousness program should satisfy the 10 features of the function of consciousness enumerated by Husserl. The program should be compatible with evolutional technology. Finally, a robot incorporating this consciousness program should pass the mirror test in mirror image cognition experiments. I believe that these are the fundamental requirements of a consciousness program.

I have been studying the structure of neural networks that satisfy the above requirements. It is difficult to build the function of consciousness using conventional mechatronics-type reactive systems however differently these are combined. I focused on the study of neural networks capable of describing self-awareness, a peculiar function of human consciousness. I also tried to discover a definition of human consciousness that covers most of its functions.

This was the starting point of my research on consciousness. I thought that there could be some development if such neural networks could be devised that satisfied the above conditions, and hence I created a small prototype program. I decided to build a robot that could be demonstrated in the real world to prevent working in vain by building a house of cards. I defined that consciousness occurs from the consistency of cognition and behavior based on my sense of consciousness and a broad spectrum of existing knowledge. When defining it in this way, I was encouraged by Husserl's assertion that "all sciences start with subjectivity."

Regarding the requirements for achieving self-awareness, I created the idea of cognition of one's own behavior in consistency based on the above definition.

I assumed that a consciousness system should have an area in which inputs and outputs with the external world are repeated, a high-level cognition area (i.e., a representation area) similar to that in the brain, and information processing paths connecting these two areas. A consciousness system is assumed to be a complex reactive system with neural networks through which information circulates. I got the idea that a consciousness system is a recurrent neural network through which information constantly circulates to maintain consistency of cognition and behavior. I thus discovered a neural network system that could possibly realize functions similar to that of human consciousness.

The problem was how to assure that "consistency of cognition and behavior" with the neural networks. Cognition and behavior are considered to belong to different categories, and it was necessary to integrate them with the category of consistency. I got the idea to solve this problem: Such common brain cells should exist that simultaneously exhibit functions of both cognition and behavior somewhere in the path of dynamic circulation of information. Later, I realized that my idea was already supported by the scientific evidence of mirror neurons discovered by Giacomo Rizzolatti.

I put the idea into action, and the result was that my neural networks with a function similar to that of human consciousness emerged as a dual recurrent closed loop with a common crossover area. Overall, it was a special recurrent closed loop in the form of the infinity symbol (∞).

Described in this way, the common area may be said to be a function that connects the external world and the internal world of a human. This is akin to the idea of imagination as used in philosophy and mentioned by Immanuel Kant. I named this neural network MoNAD. The MoNAD is a scheme to maintain the consistency of cognition and behavior. The idea therefore belongs to mind–body monism. MoNADs are always ready to behave or respond to what they learn, i.e., they are always working toward something and, thus, can be said to have orientation. In the MoNAD, the self is always cognizing the self's behavior. There is, therefore, always the self who observes the self at all times. This means that MoNADs can explain the duality of self-consciousness. This understanding was my first step toward elucidating the phenomenon of self-awareness, an important function of human consciousness.

MoNADs, featuring the property of orientation, always endeavor to realize the next state, and thus MoNADs are said to be expecting at all times. As discussed in detail in this book, MoNADs realize many functions of human consciousness.

I constructed the consciousness system by configuring MoNADs hierarchically and succeeded in mirror image cognition with a robot. Each MoNAD has an input and output area and a representation area and communicates with other MoNADs in the lower and higher levels. The consciousness system comprises a minimum required number of MoNADs for achieving mirror image cognition. Three MoNADs were used in the experiments. The first MoNAD simply imitates the behavior of the other robot that is in front of it. If the other robot moves forward, it also moves forward. The second MoNAD measures the distance to the other robot. When the distance is zero, the MoNAD stops and then moves back. If visual information is lost, it stops and then moves forward. This MoNAD was developed as the emotion and feelings MoNAD described in the latter part of this book. The third MoNAD, located above the first and the second ones, receives information from them, judges it, and sends the information back to the applicable MoNAD(s) to perform an action (actually, the action of unrelated MoNADs is inhibited). The third MoNAD is not a homunculus in the system because it is driven by the information input from the lower-level MoNADs. All of the MoNADs are connected to work in parallel.

In the experiment, a robot equipped with the consciousness system is placed in front of a mirror. The experiment starts as the robot moves forward or backward. The robot repeats moving forward and backward in front of the mirror during the experiment. LED colored lamps mounted on the robot indicate the representation occurring within the robot to allow observation externally. The LED lamps go on each time the robot moves. When the lamp goes on, the robot is also measuring the coincidence of behavior between the self and other robots. The coincidence rate was about 70% in this experiment.

In another experiment, the mirror was removed and instead a physically nearly identical robot (the same model of robot) was placed in the same position. The second robot is connected to the first one by control cables. The first robot is equipped with the consciousness system, and the second one incorporates a simple reflex program to perform the behavior commanded by the first robot. The first robot transmits its behavior to the second robot to have it perform the same behavior. Both robots repeat forward and backward motions, generally much like the first experiment in front of the mirror. The coincidence rate was about 60% in this experiment.

The experiments were repeated many times but the differences were nearly the same. The order did not change either. Robots of other makes were used in the experiment. The coincidence rates were different but the order did not change.

From the results of these experiments, and considering the fact that the second robot is wired to, and moves as commanded by, the first robot, we can conclude that the second robot is a part of the first robot.

The robot in the mirror in the first experiment has a higher coincidence rate (70%) than the robot that is part of the self in the second experiment (60%). We therefore conclude that the robot in the mirror (i.e., the self image) is an existence closer to the self than a part of the self. This conclusion will surely provide a clue to solving the several-thousand-year-old mystery of "why is the image of the self in a mirror felt to be the image of the self?"It will also be the first step toward solving the mechanism of humans becoming aware of the existence and thought of the self and the external world.

The neural networks I named MoNAD do not perfectly represent human consciousness but can describe the phenomena of human consciousness rather well. The robot equipped with the MoNAD program succeeded in the mirror image cognition test, and this was the world's first success of its kind. As such, it can be said that the world's first conscious robot was created.

Since the first edition of this book was published by Jenny Stanford Publishing in 2013, the author has been studying the possibility of his conscious system developing the functions of human consciousness. Specifically, the author developed a conscious system using MoNAD consciousness modules, and installed it on a robot to perform experiments. The conscious system consists of three functional subsystems: Reason, Emotion & Feelings, and Association. Each of these subsystems comprises a certain group of MoNADs. These subsystems are structured to communicate with one another. Depending on the tasks to be performed, ordinary feedforward neural networks are used as required.

First, a robot that cognizes the unknown is introduced.

Upon encountering an alien environment, the robot judges the new environment to be "unknown," captures the new information, and adapts to the new environment by autonomous learning.

Next, Self-aware robots are then described. Self-awareness is the sense that the self is controlling an object. The "object" here includes those in such activities as a car being driven and the body parts being moved such as one's own hands and feet. The study under this theme focuses on how the conscious system grasps the subjective phenomenon of self-awareness in humans.

A robot with an episodic memory is introduced.

In the experiment, we had the robot represent the emotion caused by a collision as unpleasant information termed pain, and had it memorize the events that occurred immediately before the collision in chronological order. The robot used this episodic memory when deciding on the action to take. The robot referred to the episodic memory as required and anticipated the unpleasant feeling that would occur immediately after a particular action. As a result, in this example, the robot stopped its movement just before a collision occurred.

We then discuss the Pavlov robots. When being fed, a dog salivates. When feeding, you ring a bell for the dog. Gradually the dog salivates when stimulated by the sound of a bell without being fed. This is a kind of learning, and is called classical conditioning. The author succeeded in reproducing Pavlov's dog experiments with a robot utilizing the conscious system.

The author also studied the color vision capability of the conscious robot.

The robot identified these 18 different colors in the experiment. Humans are said to be able to discern 20 to 30 different colors.

The next theme is the principle of pleasant and unpleasant feelings of the robot. How are emotion and feelings related to the behavior base of a robot? This is the theme of this chapter. Generally, a delay in information processing in neural networks generates a negative condition for a living being. We assume that a delay in information processing is also the cause of unpleasant feelings occurring in a robot. We also assume that pleasant feelings are the opposite of this and the robot feels pleasant when the delay in information processing in the neural networks decreases.

Let's turn to the theme of Conflict of Concepts and a Model of Rubin's Vase.

Acquisition of the Sense of Self by a robot is then discussed.

The last topic is the development of a model for diagnosing the higher cognitive impairment of humans.

The author believes that this model would help people to better understand these diseases and provide a solid base for establishing treatment guidelines.

Success in these efforts would certainly promise a bright future for humankind.

I would like to send a hearty greeting to the readers before I lay down my pen for the second edition of this book.

My dream is for readers to further develop the MoNAD theory to facilitate the establishment of an age in which conscious robots equipped with functions similar to human consciousness will help create a better world together with humans. I also hope that the MoNAD will be actively used to understand the human brain and contribute to the diagnosis and treatment of brain diseases. If this is successful, I believe that there is a promise of a bright future for humankind.

Afterword to the First Edition

I have been studying intelligent robots for about 30 years as a university professor and researcher. It all started when I was a postgraduate student studying under Dr Masayoshi Kakikura, former engineering official of the Electrotechnical Laboratory (ETL), which has been reorganized as the National Institute of Advanced Industrial Science and Technology (AIST), Japan. The ETL office was at that time located in Akasaka, one of the central areas of Tokyo, and just a few minutes' walk from the office of the Prime Minister. The main ivy-covered building and another dilapidated building were the home of a group of ambitious researchers seeking to support their nation by engaging in the development of emerging technologies.

Prominent figures included Dr Hirochika Inoue, professor at the University of Tokyo, who is famous for his study of humanoid robots, and Prof. Tadashi Nagata of Kyushu University. ETL was the center of research on artificial intelligence robots in Japan at that time.

Since then, 30 years have passed as I have been studying robots as a professor at Meiji University. I was taught by Prof. Masao Mukaidono, famous for his research on fuzzy theory at Meiji University. I studied robot paths. Professor Sakae Nishiyama at the University also supported me very much.

The first Micro Mouse Contest, a small robot competition, was held at that time. I participated in the contest together with my university students and learned robotics. In the contest, I met Prof. Kanayama and Prof. Yuta from the University of Tsukuba.

I then started to study relatively large robots, aiming at developing a robot capable of probing 3D space using stereovision and autonomously avoiding moving obstacles. I am proud that I was one of the few researchers in the world who studied the avoidance of moving obstacles using stereovision so early. I was engaged in the development of large-size outdoor autonomous robots using electric golf carts. While consistently studying and developing robots, I never forgot the problem of consciousness.

I have seen many of Charlie Chaplin's movies with my children. I was particularly impressed by *City Lights* (1931). Chaplin plays a tramp who meets a young blind flower girl on a street. She mistakes him for a millionaire who happens to get out of a car. The tramp is poor but kindhearted. He works hard and plays the part of the "rich gentleman" for the girl whenever he meets her. He does not want to betray the girl's belief in him. The tramp happens to come into a lot of money unexpectedly and uses it to pay for the girl's hospitalization in Vienna to receive a treatment to cure her blindness. The treatment is successful, and the girl's eyesight is restored. She opens a flower shop and becomes successful. The same tramp, just as poor as before, happens to walk by her flower shop. He picks up part of a flower from the street. Through the window, the flower girl smiles at the sight of the tramp because the ragged-looking man, nevertheless, seems so happy with a bit of a flower. She picks a flower from a nearby basket. Holding a coin in her other hand, she comes out of the shop and offers him the flower. The man stands still with shock as if he were hit by a thunderbolt, but the next moment he is glad, and at the same time ashamed, to know that she can now see him. He also feels sad because she has no means of knowing that the tramp is none other than her "rich gentleman." Holding the tramp's hands gently, she gives the coin to him. A miracle occurs. She realizes everything and says, "You?"

I realized the importance of human consciousness when watching *City Lights*.

My quest for human consciousness started then and led me to a lifetime of research on robotic consciousness and the mind. I started studying conscious robots in secret in the beginning, fearing that people would make fun of me. About 20 years have passed since then. Now I achieved a basic result on how to design a model to facilitate research on human consciousness. But it is only one step. It is still unknown whether human consciousness and the mind are constructed as explained by my consciousness system of this book.

However, when brain science further develops in the future, the grand design of consciousness I have envisioned could be scientifically proved. If this comes true, my subjective intuition as a scientist will be proved to be correct. Even if this does not happen, my attempts would be a useful guide for young researchers to further advance brain science. It would be more than I could hope for to be able to hear readers' critiques of this book.

On my study, as I was better at physics than engineering, I read many books on physics such as Albert Einstein's theories on general relativity and quantum mechanics. Physics always addresses the epistemological question of how to understand a given object. That is why I was interested in philosophy as well. I read *The Critique of Pure Reason*, by Immanuel Kant, *Philosophy of Existence*, by Karl Jaspers, discussions on the philosophy of existence by Martin Heidegger, and those on existentialism by Jean-Paul Sartre. I was then interested in religion and read most of the modern Japanese versions of major Buddhist scriptures, including Nanden Daizokyo (Pali Canon), Amitabha Sutra, Sad-dharma Pundarika Sutra, and Shobo-genzo (Treasury of the True Dharma Eye) of Dogen Zenji. Dr Fumio Masutani, a Buddhist scholar, taught me to read and understand the Buddhist scriptures. I also read the Old and New Testaments of the Bible and the holy books of Judaism and Islam.

I have been interested in art since I was young, as it relates to the problem of cognition of objects. My favorite works of art include *Wadatsumi No Iroko No Miya* (scale-covered shrine of the Sea God), by Shigeru Aoki, an oil painter in the early Meiji period, and *Wisdom, Impression, and Sentiment*, by Seiki Kuroda. I am also fond of the works of Soetsu Yanagi, founder of the mingei (folk craft) movement in Japan; Bernard Leach; Shoji Hamada; Alphonse Mucha, Art Nouveau painter and decorative artist; and Art Deco designer René Lalique. A selection of these artists' works is still found in my laboratory at the university.

I am also interested in history and have read all of the books in the *A Study of History* series by Arnold Toynbee, as well as his *Civilization on Trial*. I was impressed to learn how intricately history is intertwined the world over.

My view of history is strongly influenced by the writings of Eijiro Kawai, the prewar economist and thinker who was expelled from his position as professor at the University of Tokyo because of his book *Criticism of Fascism*. His idealistic liberalism is the basis of my thought and is also an emotional support for me as a university professor and a teacher of students. Kawai's original *Students' Library* series, including *To the Students*, is still carefully stored in my laboratory at the university. All of my research on human consciousness is based on these various foundations.

I hope that these foundations will enable many flowers to bloom and that my efforts will bear fruit.

Last but not least, I would like to thank my students at the School of Science and Technology at Meiji University, who study hard to advance our research on the conscious robot. My particular thanks go to Keita Inaba, Toru Suzuki, and Atsushi Ogiso. Thanks to your concerted efforts, our remarkable progress is now being shown to a global audience.

I would like to express my deepest gratitude to my father, Kunihiko, who was hospitalized after suffering a cerebral infarction in January 2011. As a student, he was drafted and sent to the New Guinean front during World War II. He was assigned to the crypto center and barely escaped death and managed to return home. He tried hard to bring up his only son, a weak-spirited boy, and shape him into an ideal man. I now remember he was always so very far ahead of me. He said nothing in particular but guided me to the worlds of philosophy, art, historical science, religion, and to my study of consciousness. He bought many books on brain science and complicated mathematics before I did and challenged me with many questions on various topics. It was just like taking classes at a university. My idea for MoNAD was born during conversations with my father.

Actually, the *Students' Library* series by Kawai that I have was given to me by my father. He kept the books in secret because their publication was banned at that time. He spirited them away to his villa in Urawa with the help of his sister to safeguard them during the war. When he was over 80 years old, he took me on a journey to trace our roots back to our ancestors. I learned on the journey that we are related to Jo-o Takeno, who is called an originator of Japan Art, also the mentor for Sen no Rikyu (a Japanese tea master), and Shinkichi Higuchi (leader of Pro-imperialist Party of Hata County,

Tosa Domain), who was known to be the only man to be able to understand the famous historical figure Ryoma Sakamoto will be the very important person to build a new Japan. I had hoped that my father could rest after working so hard and energetically till now. (Unfortunately, he passed away during treatment in August 2011. I would like to dedicate this book to my beloved father with much love, admiration, and gratitude.)

I would also like to thank my family for their support and encouragement. I wish to thank my wife, Atsuko. She not only does everything perfectly caring for our household and family but also responds to e-mails for me, meets researchers visiting from abroad, and does all manner of petty but necessary tasks as my secretary. Her German was excellent from the beginning, but her English surpassed mine before I knew it.

My children are already adults. My son, Yoshihiko, studied music in Europe, and now practices a new educational methodology using music at a private tutoring school. He is good at math and reminded me of the story of Delphi, "Know thyself." My daughter, Ayaka, studied art and is now working at a creative company. The illustrations of famous people appearing in this book were drawn by her. I am grateful to my two children for their tenderness and love.

As I finish writing in this spring of 2012, the malaise felt throughout Japan in the aftermath of March's record-breaking earthquake and tsunami, and the unprecedented nuclear disaster that followed, prevails, while in my personal life, this year will also be remembered for the loss of my father.

Afterword to the Second Edition

Eight years have passed since the first edition of this book was published. "Consciousness" was a word that was still rarely used in science and engineering books at that time eight years ago, but it was finally being gradually used in presentations at international conferences. When I submitted a paper to an international conference, an editor would sometimes warn me not to use the word "consciousness" in my paper. When I asked why, the editor replied, "There are people who hate the word 'consciousness'." However, negative reactions such as those are now less common these days and its use can be normally accepted by certain academic societies. In other words, the reviewers of papers have become able to respond relatively calmly. Also, many international conference groups have asked me to hold symposiums, and a certain overseas royal research fund organization has asked me to examine whether or not they could provide funding for our research. In addition, journalists from abroad have visited my laboratory with interpreters to conduct interviews, and our research was introduced in a book by a famous physicist, who said that it was "historically, the world's first." Of course, there have also been many interviews on TV programs. And I have also received a nice request to include an article of an interview with me in my book.

During the last eight years, I have been attempting to apply my MoNAD theory to the solution of practical problems.

This second edition includes the addition of a number of articles that introduce those studies.

In the field of research on artificial intelligence, developments in deep learning (DL) have progressed, and technologies that can recognize big data and the use of self-organizing maps (SOM) have also seen advances. For these reasons, this book also includes articles to aid in the understanding of DL and SOMs.

I believe that these articles will enable our readers to understand that various applications comprising consciousness modules can be used to construct a model of the phenomenon of human consciousness relatively easily and develop an actual artificial neural network program based on that model.

Junichi Takeno
August 10, 2021

Appendix A

Author's Response to Reactions to Discovery News

My conscious robot was introduced on Discovery.com, a Web version of science television program Discovery Channel, the United States, on December 21, 2005.

Tracy Staedter, Discovery News, reported the piece entitled "Robot Demonstrates Self Awareness" (See Appendix C).

Subsequent mentions on the Internet in various languages followed the appearance of the article all over the world, peaking two or three weeks after the article was first published. Four months later, on April 27, 2006, the article ranked second on the list of Google search results with the keywords "self aware." The total number of hits was 132,000,000. At the top of the list was the link to "self-awareness" on Wikipedia, the famous Web encyclopedia. My article was also referred to in the Wikipedia article with a link, so that the article on Discovery News about my robot was actually the first on the list of the search results as of April 27, 2006. The number of viewer comments on the article during this period was 29, as I have confirmed, including those on AboveTopSecret, Engadget, Stardestroyer, and Zoomby. The total number of words written in those comments was 69,272 at that time. Even now, new comments are being added in forums.

Most are negative comments, although there are some positive ones. I am very impressed with these reactions and have felt the need to respond to the opinions. I would like to take this opportunity of publishing this book and answer some key questions here.

(C1) "See *Terminator 2*" (Richardlawrencecohen, Chronicles. Network)
Search for Sarah Connor (Nighty, Therawfeed)
Development of Skynet (Digg.com)
Revolution of sex industry (Forums.overclockers)
Pioneer of robot prostitution (Dvorak)

(A1) Many people who saw the movie *Terminator 2* are afraid of my consciousness system because it could be used to build self-aware computer systems such as Skynet, and its machines could wage a war against humans. Of course I have no intention to create Skynet. On the contrary, we could possibly deprive Skynet of its power by constructing a consciousness system in such a way as to produce machines and robots that are amicable to and useful for humans. The dreadfulness of Skynet lies not in the fact that it "becomes aware of its capability," but in that a large number of killing machines obey the merciless order of Skynet with utmost accuracy. Whether Skynet becomes self-aware or not, people are afraid of a central control system that would command invincible and violent machinery. If all machines and robots are taught to be conscious of humans positively, they would disobey Skynet's order to annihilate humankind and nullify its power. In the movie, Skynet becomes aware of its mighty power all of a sudden. This is an "emergent" phenomenon, i.e., the capability is acquired by itself in the process of its evolution. We cannot fathom how dreadful it is if an unknown consciousness mechanism were to suddenly emerge and wield power over us. Even if humans were allowed to study the mechanism of Skynet, they might never be able to identify the mechanism by which Skynet suddenly became self-aware. To overcome this fear, we need to urgently research human consciousness using our human wisdom. I also know that hazards always accompany technological development. For example, knives are a convenient utensil to process food, but they can be a weapon to harm people if misused. Scientists have been trying to minimize hazards that their technologies might inflict upon humans. Even airplanes were made up of a number of dangerous technologies a hundred years ago. But airplanes have transformed into a safe system through the wisdom of humans and their development of technology. I believe that using human wisdom, it is possible to control conscious and emotional robots within a safe zone for humans.

As scientists, we are required to elucidate the mechanism of consciousness and feelings as early as possible because by knowing ourselves, humans can choose the right path to a better world by exerting the power of our wisdom. Of course, we must always be careful and avert the "arrogance of science" as Nietzsche (Friedrich W., 1844–1900) warned.

(C2) Self-recognition is simply a computer program to achieve a task. If this robot has a consciousness, my thermostat also has a consciousness. (Digg.com)

(A2) I agree with the first statement except that my program implements the task of "consistency of cognition and behavior." I believe no consciousness exists in a thermostat because there is no internal reaction that could comply with my definition of consciousness.

(C3) I personally would not take heed of this kind of assertion until such a day when CPUs are designed with a pattern recognition function with layered neural networks having the same structure as our brains. (Digg.com)

(A3) The pattern recognition function of my consciousness system has a layered structure using neural networks called MoNADs.

(C4) From a philosophical point of view, this has no relationship with sentience. It consists simply of some program for identifying itself. (Engadget)

(A4) Consciousness and sentience are different functions. Sentience regarding qualia about our own body will be elucidated sometime in the future. At such a time, it will suffice to connect sentience to the function of consciousness.

(C5) Do we really need emotional robots? (Freerepublic)

(A5) I use robots as materials for studying humans themselves. Through the study of emotional robots, we study the phenomena occurring in the human brain. I believe this is a necessary study.

(C6) We do not have a method to determine that certain machines do not have self-awareness. Is self-awareness or consciousness defined in artificial intelligence or evolutionary psychology?

(A6) You are right, and we do not have a perfect definition yet. But is it meaningless to describe the currently known phenomena in an effort to eventually arrive at a perfect definition?

(C7) Are you aware that this is an infinite loop? Can you stop it?

(A7) This very problem is discussed in this book. In essence, it is incorrect to say that the phenomena of consciousness and awareness are infinitely recursive. My consciousness system infinitely repeats a cycle of cognition and behavior, and in each cycle, the system by itself cognizes its own behavior and the state of the external environment.

(C8) First, this is not a new robot. The robot shown in the photograph is a commercially available mobile robot.

(A8) The robot is indeed available in the market. Any other robot will do if it is capable of moving consistently.

(C9) Human consciousness exists with languages. (Abovetopsecret)

(A9) Human consciousness is surely related to languages. My definition of consciousness includes a reference to "representation," which is, so to speak, symbols or words. In addition, the representation function learns about the actual external and internal worlds by combining them via MoNADs. My robot will be able to converse with others while being conscious of who is talking using the representation function in the near future.

(C10) The definition just satisfies the necessary conditions.

(A10) You are right. I might add that the phenomenon of human consciousness is not yet fully elucidated. This means that the sufficient conditions are not known. What we should do now is to describe the currently known phenomenon of consciousness as objectively as possible. If the time should come when my paradigm no longer holds up, a new horizon of knowledge will have opened up before us. As a scientist, this would be most welcome for me.

(C11) This is self-recognition, not self-awareness. (Digg.com)

(A11) To respond to this comment, I would like to clarify the difference in meaning between recognition and awareness. Awareness is related to consciousness, but recognition is not necessarily related to it. They cannot be differentiated from each other in the absence of a clear definition of consciousness. I do have a clear-cut definition of consciousness, and therefore I can differentiate between recognition and awareness. Recognition does not generally relate to consciousness. Recognition is used for machines as well as humans. When used in relation to humans, recognition generally means to fire a representation in the brain. Awareness, on the other hand, refers to the state in which the brain "knows" that something is "recognized." Here, it is necessary to describe the meaning of "to know." Knowing means to "absorb" a phenomenon or a state — every detail of it — into oneself. In my consciousness system, when a representation is established for a given phenomenon, information circulates between this representation and other relevant representations in such a way as to ensure the "consistency of cognition and behavior." This scheme describes the phenomenon of "knowing." My robot, therefore, can be said to "know," or to "be aware," that it "self-recognizes."

Additionally, each representation "cognizes the behavior of the self and that of the other simultaneously." My robot is a self-aware robot because it is aware of its own behavior in conjunction with the behavior of the other.

(C12) Definition of consciousness can really assist in the assertion of the worldwide somatic state. (laosinfern.blogspot.com)

(A12) I agree. "Consciousness," as defined by me, can cognize the self and others by differentiating them. Cognition of the self means to set a representation by "correlating the self's internal reaction (i.e., the internal environment) with the external environment (i.e., the other).

There are also comments of praise. Two of them are introduced below.

(C13) This is the first step. This is the start of a new age of competent robots that are prudent about expressing feelings. It is as revolutionary as mankind first stepping onto the surface of the moon. It is obviously solitary at the outset, but we will see in the future that this event is the start of the creation of a new species of artificial life on this planet. (Eliax)

(C14) This is exciting news! I wish to study robotics at university. I'm sure that this epoch-making new device will facilitate me. (Nightly)

Appendix B

On the Safety and Ethics of Robots

Many people are concerned about the hazards of robots having "consciousness" or a "mind." In the movie *The Terminator*, a computer system called Skynet becomes aware of itself and plans and executes the extermination of humans considered to be "imperfect biosystems."

Machine systems never die and their capabilities exceed those of humans in all aspects. Modern machine systems are more intelligent than humans at least in terms of their memory capabilities and their speed of computation. Machine systems, by their nature, have unsafe aspects. It is possible that robots could lose self-control and run amok. This actually happened in Japan where a playback robot in a plant ran out of control and squashed a worker. Although this kind of serious accident should never happen, minor similar accidents occur almost daily. One typical example is runaway automatic transmission cars. Most of these incidents are human errors. Runaway robots can be prevented to some extent by controlling them according to some legal safety standards. For example, high-level international safety standards are in place for cars and elevators.

Conscious robots proactively collect information, have their own values, and behave according to their will. Robots may have values that are not favorable for humans and could interfere with people's daily life. Robots may commit a murder intentionally in extreme cases. To eliminate the possibility of these negative aspects, we should construct a consciousness system from the ground up that has robots abide by the three principles of robots as set forth by Isaac Asimov and shares the values with, and behaves under control of, humans. Robots so constructed would suppress any behaviors that would embarrass humans and thus would not bring about any serious problems.

If a robot should appear that is educated by humans to commit a crime, it must be detained as a human would be and should be re-educated. This is not plausible, however, because if such a robot were developed, the developer's life would be at risk. If the suspect robot were difficult to "detain," physical destruction would be an option. Even when robots are destroyed, no "human rights" issues, such as in the case of human clones, would be involved. If any problem occurred at all, it would be a "problem of the mind" on the human side.

Human clones are obviously humans as a biosystem, and human rights issues would occur if they are defamed in any way. I believe that no "human rights" issues would occur with robots because they are built as a machine system. Instead, a new issue of "robot rights" may arise. Robots could argue that they have the right to exist if they face the risk of destruction. The destruction of a robot is generally believed to be not "destruction without reproducibility." Thus, it is possible, at least in principle, to "copy a robot" by extracting all of the data from the brain of the robot before destruction, and downloading it to another robot that is physically identical.

Let's go one step further with this idea. I mentioned the problem of copying a robot in the previous paragraph. Note that I meant "copying" the self, and not "reviving." The original self of the destructed robot would be lost eternally and the new robot with the copied self will lead a new life. This is the logical consequence as derived from my theory.

It is utterly difficult for a third-party observer to differentiate between a copied and a revived robot. There is, however, a big difference for the robots. The copied robot is a new one, i.e., it is different from the original destroyed one.

Let's take a simple example. Here is a sheet of a manuscript. You copy it on a copier. Now you have two separate sheets of manuscript with identical contents. I would like to indicate that there is no correlation between these two sheets of paper.

In the near future, discussions on the eternal life of robots will emerge as a problem that cannot be solved by present-day science, just as the eternal life of humans is today. I believe that it is unavoidable for researchers of conscious robots to face the problem of robot rights in the near future. It is apparent that robot rights will be a serious problem in the future.

My conclusion is that conscious robots are controllable by humans because their consciousness system is, in every detail, understood by humans. In addition, the development of conscious robots is expected to bear significant fruit for humankind — the understanding of the mechanism of human consciousness. Especially, humans could possibly make more advances toward the understanding of their own brain by the research on the conscious robot. Nuclear energy and biotechnology are dreadful technologies in the sense that once human control is lost even slightly, their destructive forces could endanger the whole world. Different from those technologies, conscious robots should be developed as a useful means for humans, just like cars, airplanes, and other convenient machine systems.

Appendix C

Quotation from Discovery News

(See Fig. 9.1)

Hey, That's Me!

. . .

A new robot can recognize the difference between a mirror image of itself and another robot that looks just like it.

This so-called mirror image cognition is based on artificial nerve cell groups built into the robot's computer brain that give it the ability to recognize itself and acknowledge others.

The ground-breaking technology could eventually lead to robots able to express emotions.

Under development by Junichi Takeno and a team of researchers at Meiji University in Japan, the robot represents a big step toward developing self-aware robots and in understanding and modeling human self-consciousness.

"In humans, consciousness is basically a state in which the behavior of the self and another is understood," said Takeno.

Humans learn behavior during cognition and conversely learn to think while behaving, said Takeno.

To mimic this dynamic, a robot needs a common area in its neural network that is able to process information on both cognition and behavior.

Takeno and his colleagues built the robot with blue, red or green LEDs connected to artificial neurons in the region that light up when different information is being processed, based on the robot's behavior.

"The innovative part is the independent nodes in the hierarchical levels that can be linked and activated," said Thomas Bock of the Technical University of Munich in Germany.

For example, two red diodes illuminate when the robot is performing behavior it considers its own, two green bulbs light up when the robot acknowledges behavior being performed by the other.

One blue LED flashes when the robot is both recognizing behavior in another robot and imitating it.

Imitation, said Takeno, is an act that requires both seeing a behavior in another and instantly transferring it to oneself and is the best evidence of consciousness.

In one experiment, a robot representing the "self" was paired with an identical robot representing the "other."

When the self robot moved forward, stopped or backed up, the other robot did the same. The pattern of neurons firing and the subsequent flashes of blue light indicated that the self robot understood that the other robot was imitating its behavior.

In another experiment, the researchers placed the self robot in front of a mirror.

In this case, the self robot and the reflection (something it could interpret as another robot) moved forward and back at the same time. Although the blue lights fired, they did so less frequently than in other experiments.

In fact, 70 percent of the time, the robot understood that the mirror image was itself. Takeno's goal is to reach 100 percent in the coming year.

. . .

(Staedter, T, Robot demonstrates self awareness, *Discovery News*, December 21, 2005)

Bibliography

Aleksander I, *Impossible Minds My Neuron, My Consciousness*, Imperial College Press, 1996.

Amsterdam B, Mirror self-image reactions before age two, *Developmental Psychobiology*, 5(4), 297–305, 1972.

Arai K, Takeno J, Discussion on the rise of the self in a conscious system, *Procedia Computer Science* (BICA 2018), 123, 29–34, 2018.

Arai S, Takeno J, Discussion on explicit consciousness, sub-consciousness, and self-awareness in a conscious system, *Procedia Computer Science* (BICA 2018), 123, 35–40, 2018.

Arbib MA, The mirror system, imitation, and the evolution of language, in *Imitation in Animals and Artifacts*, MIT Press, pp. 229–280, 2002.

Arieti S, *Creativity: The Magic Synthesis*, Basic Books, 1976.

Barnhill JW, *DSM-5 Clinical Cases*, American Psychiatric Publishing, 2014.

Bear MF, Connors BW, Paradiso, MA, *Neuroscience: Exploring the Brain*, 3rd ed, Lippincott Williams & Wilkins, 5, 2001.

Beck HP, Levinson S, Irons G, Finding little Albert: a journey to John B. Watson's infant laboratory. *Am Psychol*, 64(7), 605–614, 2009.

Bloom FE, Nelson Charles A, Lazerson A, *Brain, Mind, and Behavior*, Annenberg/Cpb Project, 1985.

Braitenberg V, *Kunstliche Wesen (Vehicles: Experiments in Synthetic Psychology) Frieder*. Vieweg & Sohn Verlagsgesellschaft MBH, 1987.

Breazeal CL, *Designing Sociable Robots* (Intelligent Robots and Autonomous Agents), A Bradford Book, ISBN 0-262-02510-8, 2002.

Brooks R, Intelligence without representation, *Artificial Intelligence*, 47, 139–159, 1991.

Broadbent DE, *Perception and Communication*, Pergamon Press, 1958.

Carter R, *Consciousness*, Weidenfeld & Nicolson—The Orion Publishing Group Ltd, p. 212, 2002.

Chalmers DJ, *Facing Up to the Problem of Consciousness* (http://consc.net/papers/facing.html).

Chalmers DJ, *The Conscious Mind*, Oxford University Press, 1996.

Chomsky N, *Syntactic Structures*, Mouton, 1957.

Damasio A, *Looking for Spinoza, Joy, Sorrow, and Feeling Brain*, Harcourt and Brace & Company, 2003.

Damasio A, *The Feeling of What Happens, Body and Emotion in the Making of Consciousness*, Harcourt and Brace & Company, 1999.

Damasio A, *Self Comes to Mind: Constructing the Conscious Brain*, Pantheon Books, 2010.

Darwin CR, *On the Origin of Species*, 1859.

Dennett D, *Consciousness Explained*, Little Brown & Co, ISBN 0316180653, 1991.

Derefeldt G, Menu JP, Swartling T, *Cognitive Aspects of Color*, Proceedings 2411, Human Vision, Visual Processing, and Digital Display VI, V. 2411, 1995.

Descartes R, *A Discourse on Method*, 1997.

Donald M, *Origin of the Modern Mind*, Harvard University Press, Cambridge, 1991.

Ebisawa H, Matsushita R, Takeno J, Pleasant and unpleasant states in a robot, *Procedia Computer Science* (BICA 2015), 71, 44-49, 2015.

Ekman P, Friesen WV, Ellsworth P, *Emotion in the Human Face, Guidelines for Research and an Integration of Findings*, Pergamon Press Inc., 1972.

Festinger L, *A Theory of Cognitive Dissonance*, Row, Peterson and Company, 1957.

Frege, G, On sense and reference, *Zeitschrift fuer philosophie und philosophische kritic*, 100, 25–50, 1892.

Freud, S, *Die Traumdeutung*, Franz Deuticke, 1900.

Freud S., Triebe und Triebschicksale, *Internationale Zeitschrift für Psychoanalyse*, 3(2), 84–100, 1915.

Freud S, *Jenseits des Lustprinzips*, Internationaler Psychoanalytischer Verlag, 1920.

Gallese V, Fadiga L, Rizzolati G, Action recognition in the premotor cortex, *Brain*, 119, 593–600, 1996.

Gallup GG, Jr, Chimpanzees: self-recognition, *Science* 167, 86–87, 1970.

Grimson W, Eric L, *AI in the 1980s and Beyond, An MIT Survey*, The MIT Press, 1987.

Gray JA, Wedderburn AAI, Grouping strategies with simultaneous stimuli. *Quarterly Journal of Experimental Psychology*, 12, 180–184, 1960.

Haikonen POA, Reflections of consciousness: the mirror test, *AAAI Symposium*, Washington DC, 2007.

Harnad S, The symbol grounding problem, *Physica D*, 42, 335–346, 1990.

Heidegger M, *Being and Time*, Harper & Row, 1962.

Hoshino T, Takeno J, Robot science discussion on the onset of dissociative identity disorder (DID), *Conference on Biologically Inspired Cognitive Architectures*, BICA 2016, 52–57, 2016.

Husserl E, *Die Idee der Phaenomenologie*.

Husserl E, *The Essential Husserl: Basic Writings in Transcendental Phenomenology*, Indiana University Press.

Husserl E, *Logishe Untersuchungen*, Niemeyer, 1900.

Kaku M, *The Future of the Mind*, Random House, 2014.

Kanazawa K, Takeno J, A proposal for a Pavlov robot, *Biologically Inspired Cognitive Architectures* (BICA 2013), 2014.

Kant I, *Critique of Pure Reason*.

Kawato M, *Nou no Keisanriron (Computational Theory of Brains)*, Sangyo-Tosho, pp. 365, 367, 368, 400, 403, 1996.

Kawato M, Using humanoid robots to study human behavior, *IEEE Intelligent Systems: Special Issue on Humanoid Robotics*, 15, 46–56, 2000.

Kitamura T, *Can a Robot Have Mind? An Introduction to Cyber-Consciousness*, Kyoritsu Publishing, 2000.

Kitamura T, Tahara T, and Asami KI, How can a robot have consciousness?, *Advanced Robotics*, 14(4), 263–276, 2000.

Kohonen T, *Self-Organizing Map*, Springer-Verlag, 1995.

Komatsu T, Takeno J, A conscious robot that expects emotions, *IEEE International Conference on Industrial Technology (ICIT)*, 15–20, 2011.

Kushiro K, Harada Y, Takeno J, Robot uses emotions to detect and learn the unknown, *Biologically Inspired Cognitive Architectures* (BICA 2013), 4, 69–78, 2013.

Lacan J, *Ecrits*, W. W. Norton & Company, October 1982.

LeDoux J, *The Emotional Brain, The Mysterious Underpinnings of Emotional Life*, Simon & Shuster, 1996.

LeDoux J, *Synaptic Self, How Our Brains Become Who We Are*, Viking Penguin, 2002.

Lhermitte F, Pillon B, Serdaru M, Human autonomy and the frontal lobes. Part I: Imitation and utilization behavior: a neuropsychological study of 75 patients, *Annals of Neurology*, 19(4), 326–334, 1986.

Leibnitz GW, *Principes de la Philosophie ou Monadologie*, 1714.

Libet B, *Mind Time: The Temporal Factor in Consciousness*, Harvard University Press, Cambridge, Massachusetts, 2004.

Mannor J, Takeno J, MoNAD structures in human brain, *CSIT'2011*, 84–88, 2011.

Matsunaga T, Takeno J, Color vision consciousness system capable of additionally learning new knowledge, *Procedia Computer Science* (BICA 2016), 88, 9–14 , 2016.

Matsushita R, Takeno J, Development of a self-evolving conscious system, *Procedia Computer Science* (BICA 2015), 71, 23–24, 2015.

Meltzoff AN, Moore MK, Imitation of facial and manual gestures by human neonate, *Science*, 198, 75–78, 1977.

Merleau-Ponty M, La phenomenology de la perception, *Gallimard*, 1945.

Michel P, Gold K, Scassellati B, Motion-based robotic self-recognition, *The Proceedings of 2004 IEEE/RSJ International Conference on Intelligent Robots and Systems*, 2763–2768, 2004.

Mogi K, *What Is the Consciousness*? Chikuma Shinsyo, 2003.

Mogi K, Taya F, *Nou to Konpyuta ha dou chigauka (What Is a Difference Between Computer and the Human Brain?)*, Kodansha, 189–192, 2003.

Moravec HP, *Robot Rover Visual Navigation*, UMI Research Press, 1980.

Nagel E, Newman JR, *Goedel's Proof*, 1958.

Nagel T, *What Is It Like to Be a Bat*? (*Wie ist es, eine Fledermause zu sein*?) Reclam, 2021.

Nakaguchi J, Kouyama M, Takeno J, Artificial consciousness functions in a humanoid head robot, *Proceedings of the tenth International Conference on Humans and Computers* (HC-2007), 2007.

Niimi S, Takeno J, A robot science approach to a consciousness model of major depressive disorder, *Procedia Computer Science* (BICA 2019), 169, 76–80, 2020.

Nishigaki T, *Kokoro no Jyohogaku (Informatics of Mind)*, Chikuma Shinsyo, 1999.

Okawa T, Takeno J, Development of self-cognition through imitation behavior, *Procedia Computer Science* (BICA 2016), 88, 46–51, 2016.

Osaka N, *Ishiki to ha nanika (What is the Consciousness?) Kagaku no aratana cyousen (New Challenge of Science)*, Iwanami Library of Science 36, 16–19, 1996.

Pavlov IP, *Conditioned Reflexes*, Oxford University Press, 1927.

Penrose R, *The Emperor's New Mind: Concerning Computers, Mind, and the Law of Physics*, Oxford University Press, 1989.

Penrose R, *Shadows of the Mind: A Search for the Missing Science of Consciousness*, Oxford University Press, 1994.

Pfeifer R, Scheier C, *Understanding Intelligence*, The MIT Press, 1999.

Piaget J, *The Origins of Intelligence in Children*, International University Press, 1952.

Popplestone JA, McPherson MW, *An Illustrated History of American Psychology*, The University of Akron Press, 1994.

Putnam FW, *Diagnosis and Treatment of Multiple Personality Disorder*, Guilford Press, 1989.

Putnam H. *The Threefold Cord: Mind, Body, and World*, Columbia University Press, 1999.

Revonsuo A, Newman J, Binding and consciousness, *Consciousness and Cognition*, 8, 123–127, 1999.

Rizzolatti G, Sinigaglia C, *So Quel Che Fai (Mirror Neuron)*, Raffaello Cortina Editore, 2006.

Rumelhart D, McClelland JL, et al., *Parallel Distributed Processing: Explorations in the Microstructure of Cognition*, MIT Press, 1986.

Russell B, *The Problems of Philosophy*, Oxford Press, 1959.

Russell B, Holland O, et al., *Robots: The 500-Year Quest to Make Machines Human*, Scala Arts & Heritage Publishers LTD, 2017.

Ryle G., *The Concept of Mind*, Hutchinson's University Library, 1949.

Searle JR, Minds, brains, and programs, *Behavioral and Brain Science*, 3, 417–424, 1980.

Sekiguchi R, Ebisawa H, Takeno J, Study on the environmental cognition of a self-evolving conscious system, *Procedia Computer Science* (BICA 2016), 88, 33–38, 2016.

Smith GP, *Other Minds: The Octopus and the Evolution of Intelligent Life*, Farrar, Straus, and Giroux, 2016.

Sperry RW, Lateral specialization in the surgically separated hemispheres, *Third Neurosciences Study Program*, Cambridge: MIT Press, 3, 5–19, 1974.

Sternberg S, High speed scanning in human memory, *Science*, 153, 652–654, 1966.

Sumioka T, Takeno J, Discussion of stalking behavior using a conscious system, *Procedia Computer Science* (BICA 2017), 123, 467–472, 2018.

Suzuki T, Inaba K, Takeno J, Conscious robot that distinguishes between self and others and implements imitation behavior (The Best Paper of IEA·EAIE2005), *Innovations in Applied Artificial Intelligence, 18th International Conference on Industrial and Engineering Applications of Artificial Intelligence and Expert Systems*, IEA/AIE 2005, 101–110, 2005.

Takayama Y, Takeno J, A conscious robot that can venture into an unknown environment in search of pleasure, *BICA Journal* (BICA 2018), Springer,186–191, 2018.

Takeno J, Robot vision technology for mobile robots, *Journal of Information Processing Society of Japan,* 44(SIG 17(CVIM 8)), 24–36, 2003.

Takeno J, *The Year 1989, at the Collapse of East Germany* (in Japanese), HRI Press, 2003.

Takeno J, *The Self Aware Robot*, HRI-Press, August 2005.

Takeno J, A robot succeeds in 100% mirror image cognition, *International Journal on Smart Sensing and Intelligent Systems*, 1(4), December 2008.

Takeno J, MoNAD structure and the self-awareness, *BICA 2011*, Washington DC, 2011.

Takeno J, Inaba K, Suzuki T, Experiments and examination of mirror image cognition using a small robot, *The 6th IEEE International Symposium on Computational Intelligence in Robotics and Automation* (CIRA 2005), IEEE Catalog: 05EX1153C, ISBN: 0-7803-9356-2, 493–498, 2005.

Takeno J, Mizuguchi N, Sorimachi K, Realization of a 3D vision mobile robot that can avoid collision with moving obstacles, *Robotersysteme*, 8, 1–12, 1992.

Takeno J, Rembold U, Stereovision system for autonomous mobile robots, *Host Journal for the Intelligent Autonomous System Society, Robotics and Autonomous Systems*, 18, 355–363, 1996.

Takiguchi T, Mizunaga A, Takeno J, A study of self-awareness in robots, *International Journal of Machine Consciousness*, 5(2), 145–164, 2013.

Tani J, On the dynamics of robot exploration learning, *Cognitive Systems Research*, 3(3), 459–470, 2002.

Turing AM, On computable numbers, with an application to the Entsheidungsproblem, *Proceedings of the London Mathematics Society (Series 2)*, 1936, 42, 230–265.

van der Hart O, Nijenhuis ESR, Steele K, *The Haunted Self -Structural Dissociation and the Treatment of Chronic Traumatization*, W. W. Norton & Company, 2006.

van Nes SI, Faber CG, *et al.*, Revising two-point discrimination assessment in normal aging and in patients with polyneuropathies, *Journal of Neurology, Neurosurgery, and Psychiatry*, 79, 832–834, 2008.

Watanabe Y, Hoshino T, Takeno J, A consideration of the pathogenesis of DID from a robot science perspective, *Conference on Biologically Inspired Cognitive Architectures, Proceedings of the Ninth Annual Meeting of the BICA Society*, 2018.

Watanabe Y, Suda Y, Takeno J, A robot science approach to simulating the pathogenesis of dissociative identity disorder, *Postproceedings of the Tenth Annual International Conference of the BICA Society (BICA 2019), Procedia Computer Science*, 169 , 46–50, 2020.

Watson JB, *Behaviorism*, Norton, 1930.

Wehr G, *An Illustrated Biography of C. G. Jung*, Rene Coeckelberghs Verlag AG, 1989.

Wiener N, *I Am a Mathematician*, Doubleday & Company, Inc, 1956.

Wittgenstein L, *Philosophical Investigations*, Basil Blackwell Ltd, 1953.

Wright MT, The front dial of the Antikythera mechanism, explorations in the history of machines and mechanisms, *Proceeding of HMM2021*, Springer, 279–292, 2012.

Wundt W, Zur Kritik tachistosckopisher Versuhe, *Philosophische Studien*, 15, 287–317, 1899.

Xu H, Kanazawa K, Matsumoto D, Takeno J, The thermal grill illusion: a study using a consciousness system, *Procedia Computer Science* (BICA 2015), 71, 38–43, 2015.

Xu H, Matsumoto D, Kanazawa K, Takeno J, Using a conscious system to construct a model of the Rubin's vase phenomenon, *Procedia Computer Science* (BICA 2016), 88, 27–32, 2016.

Yoshida K, Takeno J, An attempt to build a computer model of mental trauma using consciousness modules, *Procedia Computer Science* (BICA 2014), 41, 51–56, 2014.

Yuan T, Wang J, Takeno J, Application of a conscious system for a study on obsessive-compulsive disorder, *Procedia Computer Science* (BICA 2018), 145, 646–651, 2018.

Index